DENDRO-CHRONOLOGY IN EUROPE

edited by

John Fletcher

National Maritime Museum, Greenwich,
Archaeological Series No. 4

Research Laboratory for Archaeology and
History of Art, Oxford University, Publ. No. 2

BAR International Series 51

1978

Dendrochronology in Europe

Principles, interpretations and applications to Archaeology and History

Based on the Symposium held at
the National Maritime Museum, Greenwich,
July 1977

edited by

John Fletcher

National Maritime Museum, Greenwich,
Archaeological Series No. 4

Research Laboratory for Archaeology and
History of Art, Oxford University, Publ. No. 2

B A R International Series 51

1978

B.A.R.

B.A.R., 122 Banbury Road, Oxford OX2 7BP, England

B.A.R. International Series 51, 1978: "Dendrochronology in Europe".

Price £8.50 post free throughout the world. Payments made in currency other than sterling must be calculated at the current rate of exchange and an extra 10% added to cover the cost of bank charges. Cheques should be made payable to 'British Archaeological Reports' and sent to the above address.

ISBN 0 86054 038 3

National Maritime Museum, Greenwich, Archaeological Series

General Editor—Sean McGrail, B.A., Ph.D., Master Mariner

1. Sources and Techniques in Boat Archaeology (BAR-S29)
2. Logboats of England and Wales (BAR 51)
3. The Graveney Boat (BAR 53)
4. Dendrochronology in Europe (BAR-S51)

The Museum publishes shorter papers as Monographs, which may be obtained from the Museum Bookshop.

BAR publishes monographs and collections of papers on all aspects of archaeology and on related subjects such as anthropology, classical and medieval history and numismatics.

BAR publishes BAR International Series which is concerned with world archaeology and BAR British Series which covers the British Isles.

For details of all BAR publications in print please write to the above address. Information on new titles is sent regularly on request, with no obligation to purchase.

Volumes are distributed direct from the publishers. All BAR prices are inclusive of postage by surface mail anywhere in the world.

Printed in Great Britain

FOREWORD

The proposal that a symposium on dendrochronology should be held at Greenwich came from the Research Laboratory for Archaeology and the History of Art, and was readily agreed by the National Maritime Museum. With encouragement from the Royal Society and the Council for British Archaeology, the Museum and the Laboratory jointly planned the 4 day symposium, and the proceedings of that conference, with additional matter, are published in this volume edited by Dr. John Fletcher.

Members of the symposium came from many European countries, from Scandinavia to the Alps, and from the USSR to Ireland, and also from the USA. As well as being international the symposium was also interdisciplinary in character. In addition, there were as 'consumers' of dendrochronology as well as 'producers', and the dialogue between these groups—some of which is summarised in the Discussion sections of this publication—proved highly constructive.

The papers in this volume cover a wide range of topics including the early development of dendrochronology, the problems and the techniques of the discipline, and applications to climatology, ecological studies, archaeology and the history of art. This publication should thus have a wide appeal and we believe should become a valued work of reference.

Basil Greenhill
Director
National Maritime Museum
Greenwich

E. T. Hall
Research Laboratory for
Archaeology and the History of Art
Oxford

CONTENTS

LIST OF PAPERS

Papers marked + were presented at the Symposium

INTRODUCTION

The symposium held at the National Maritime Museum, Greenwich, from 11 to 14 July 1977 provided Europe's first regional conference on dendrochronology. Its purpose was to focus attention on the progress of this science, and bring together European dendrochronologists.

During the last century the growth pattern derived from felled trees has been a subject of scientific study over a wide area from Odessa to Oslo. The difficult step of extending tree-ring chronologies to trees whose life-span ended in bygone centuries and millennia was taken by Huber who began research in this field in 1938. He established dendrochronology in Europe as a powerful new dating technique for archaeologists and historians and provided, with his co-workers, dated European samples for calibrating the radiocarbon scale. His team at Munich also examined the extent to which annual growth could be linked with climatic variations and drew a number of important conclusions.

Lack of familiarity with the German language may explain unawareness in the British Isles of that research. It had made so little impact that its achievements were not represented, alongside U.S.A. work in this subject, at the British Academy-Royal Society meeting, "The Impact of the Natural Sciences on Archaeology", held in London in 1968. That was unfortunate. Within Britain, ecological conditions change over a relatively short distance from the hilly, and even mountainous, areas in the west to the lowlands of the east. Climatic conditions that are dominated by proximity to the Atlantic ocean alternate with ones influenced by continental air masses.

It follows that the principles needed to establish dendrochronology for the different parts of Britain are more likely to be those that have been evolved in Germany than those developed for conifers in arid parts of the U.S.A. One peculiarity, for example, had emerged from research in continental Europe by the end of the 1960s. Similarities existed in the growth pattern of certain individual species from the Vosges to the Carpathians (a range of over 1000 km) but yet there was a great difference over distances of a few tens of kilometres between the pattern for oak growing in a lowland maritime area and the pattern for oak in a hilly inland area.

In recent years, dendrochronology and dendroclimatology have also been studied in many parts of Russia and the subject has long been regarded as of importance in Scandinavian countries. Work of a highly original character has emanated from France and new centres of research created in Switzerland. The theme of dendrochronology in Northern and Central Europe was therefore a matter of common interest, and by mid-1974 it seemed to me that a meeting in Britain would be appropriate.

The National Maritime Museum offered many advantages as the venue. Its ample lecture theatre, the encouragement being given to meetings related to its numerous activities, its prestigious buildings and setting—near to but not in central London—were obvious attractions. Apart from the 'dendro' link with wooden ships and 'chronology' with Greenwich time, the museum had recently set up an archaeological research centre to which the Graveney boat had been a notable addition. I am sure I am expressing the view of those who participated in the symposium in extending gratitude to the Trustees, the Director and the staff of the Museum for contributing so much to its success.

The Greenwich symposium was the third meeting in Western Europe confined to the theme of dendrochronology. The previous two were national rather than international. The first, on 1-2 March 1968, just before Huber's retirement from his Chair, was organised by the Central Institute for Art History and the Forestry-Botany Institute of the University of Munich. The value of dendrochronology in dating objects of medieval art were brought to the attention of German archaeologists and historians and discussed. Extracts from the two contributions by Hollstein (whose work since 1960 to the west of the Rhine had run parallel to that of Huber to the east of the Rhine) are included in the present volume. The second meeting, two years later at Reinbek near Hamburg, enabled progress for the lowlands of north Germany and Schleswig, areas hitherto regarded as unsuited to dendrochronology, to be presented and for further results in dating to be reported. The texts of the papers were published as a separate volume.

Apart from the meetings mentioned, the publication in volume 32 of the Tree-Ring Bulletin of the article by Eckstein on Tree-Ring Research in Europe created an important milestone in 1974. It provided the first summary in English on this subject and listed numerous papers. In 1977, as a preliminary to the Greenwich symposium, abbreviated translations of 24 of the Russian papers presented at the All-Union Conferences of 1968 and 1972 at Vilnius and Kaunas respectively, carried publications in English a stage further. The book, Russian Papers on Dendrochronology and Dendroclimatology, translated and edited by J. M. Fletcher and W. Linnard, formed this Laboratory's publication No. 1 in dendrochronology.

The topics given to those wishing to present papers to the Greenwich symposium were as follows:

1. Characteristics (width, density etc.) of the annual rings of Northern European trees as chronological indicators. Their geographical variation and ecological interpretation. Techniques for their study.

2. The suitability and reliability of dendrochronology for solving problems in maritime and terrestrial archaeology and in architectural and art history. Coordination with other physical and biological chronologies.

These topics did not stress post-glacial climatic swings in Europe revealed through radiocarbon dating and dendrochronology as these had been the subject of an international conference at Mainz in 1974 (Proceedings published, 1977).

The symposium at Greenwich was inaugurated on Monday, 11 July 1977 by an evening lecture given by Prof. Dr. Liese in honour of Prof. Bruno Huber, who had refused to be dispirited by scepticism and incredulity in the early days of his work on dendrochronology. On the following two days, 31 papers were presented and for Thursday 14 July arrangements were made for participants to travel by boat from Greenwich to central London past the waterfront from which archaeological timber had been tree-ring dated. Applications of dendrochronology in art and architectural history were respectively on display at an Exhibition at the National Portrait Gallery and in parts of Westminster Abbey.

The contents of the seven sections which form this book do not follow exactly the papers and sessions at Greenwich. It has seemed appropriate to devote much of the opening section to extracts from previous publications which give the methods, principles and conclusions evolved in Germany. These, hitherto, have been almost entirely unavailable in English. Some short papers on recent research relevant to the topics of the symposium are also included. Thus it is hoped that this volume will prove of value as a comprehensive record, in English, of the development, methods and present status of research in dendrochronology in many parts of Europe.

I am grateful to Philip Adland, Peter Wood and Howard Wright of the Dept. of Forestry, Oxford University who acted at the symposium as rapporteurs for the discussions: to Sean McGrail for much assistance with the Discussions and with decisions concerning the form and content of this volume; also to Ruth Morgan and to Wendy Bell who made available their translations into English of papers published in German journals.

Acknowledgement is made to the publishers of the German journals concerned—Springer-Verlag of Heidelberg, Umschau-Verlag of Frankfurt am Main, VEB Gustav Fischer Verlag of Jena— for their permission to use copyright material for these translations; also to the authors Frau Giertz-Siebenlist, Herr E. Hollstein, Dr. A Delorme, Dr. O. Fürst for their permission to use translations of their papers or extracts from them; to Dr. B. Schmidt for permission to use a published map; and to Mr. F. S. Walker for help with the Index.

The cost of editing and preparing the papers for publication has been helped by generous contributions from the Marc Fitch Fund and from Prof. E. T. Hall which are gratefully acknowledged. The successful conclusion of these aspects of publication owe much to the wisdom and advice of Margaret Tapper, to the typing of Margery Mills and to the skill of Judith Takacs in perfecting numerous figures and their legends.

November 1978

John Fletcher

Polished section of a Spessart oak (SP 60). The 567 Annual Rings span the years 1382-1948 A.D. Length 45 cm.

Prof. E. T. Hall presents the Huber Memorial Medal to Prof. Dr. Walter Liese (left) at the National Maritime Museum, Greenwich on 11 July 1977.

1.1. BRUNO HUBER: THE PIONEER OF EUROPEAN DENDROCHRONOLOGY

Prof. Dr. W. Liese

(Huber Memorial Lecture, delivered 11 July 1977)

Forty years ago, in 1937, Professor Bruno Huber became interested in evaluating the changing pattern of the annual rings of trees for the dating of timber stimulated by the work of Douglass in the USA. He soon realized the great importance of this method and its impact for many fields of natural and historical sciences. Therefore, Huber has to be regarded as the pioneer of European dendrochronology. In the meantime about 400 papers on dendrochronology have been published in Europe alone, improving the techniques and extending its application. This first International Symposium on Dendrochronology in Northern Europe is certainly the most significant expression of the importance dendrochronology has gained within these few decades. If Professor Huber could have been with us, at the age of 78, he certainly would have considered it as the highlight of his life, a life which was full of scientific research, fascinating discoveries and recognition.

The organizers proposed to honour the merits of this great man by introducing the Conference by the Huber Memorial Lecture. Mrs. L. Huber felt deeply moved about this most honourable acknowledgement and wants to express her gratitude. Being asked to present this lecture as one of his former coworkers, I feel equally touched.

Since only a few of the distinguished participants will have known the personality of Bruno Huber closely, it seems appropriate to illuminate first of all his biodata briefly, and then to elaborate his foundation of dendrochronology in Europe.

Born in Hall/Austria in 1899, Huber studied Natural Science in Innsbruck and Vienna, obtained his Doctorate in 1921, and worked at the Botanical Institutes in Vienna, Greifswald and Freiburg. In 1932, at the age of only 33, he became Professor for Botany in Darmstadt and two years later he moved to Tharandt as Professor for Forest Botany. After the Second World War, in 1946, he went from East Germany to München, where he retired in 1964, a few years after he had recovered from a severe heart-attack. Only 5 years later, in 1969 he died shortly after his 70th birthday, which had brought together once more his many colleagues, friends and pupils.

During a scientific life of nearly 50 years Huber wrote several standard books and about 230 papers. His range of interest was remarkably wide, but concentrated in particular on the interrelationships between the structure and function of trees and the influences of environmental factors. Characteristic of his manifold activities was his imagination, his sense for new lines of research and for the application of modern methods. He was an initiator in many ways, paving the way for others. Only some areas of his many contributions can be mentioned here.

Being the father of "tree physiology" he investigated the influence of water and temperature on trees, introducing for example the thermoelectric measurement of water transport still in use. He explored the biomass-production of trees, nowadays intensively studied everywhere, introducing modern physical methods for measuring the gas exchange of plants, e.g. the first phytotron.

He was also the first who used radioactive substances as indicators for metabolic path-ways and he was among those who first applied the electron-microscope for investigating the fine-structure of plants. Furthermore, he contributed significantly to the structure of phloem and its function.

Bruno Huber was, however, above all an impressive personality, a professor in its best traditional sense, a gifted teacher and a fascinating leader of excursions with students, whom he impressed with his extensive knowledge of plants, especially in his beloved homeland, the Alps. He was a dynamic and inspiring professor, and many scientists, now well known, were guided by him in their earlier days. His scientific achievements were honoured in many ways, only to mention the Doctorates, honoris causa, by the University of Vienna and the University of Istanbul. However, it is in particular his contribution to dendrochronology that merits our attention.

Bruno Huber was gifted with a distinct inclination for history, probably stemming from his grandfather, Alfons Huber, professor of history at the University of Vienna and later Secretary-General of the Austrian Academy of Science. His interest in dendrochronology, aroused exactly 40 years ago, lead in 1939 to his first paper in the subject. It dealt with the dating of a medieval settlement and in it he mentioned already the possibility of dating timber constructions in Europe by applying the method developed by Douglass in 1919.

Subsequently Huber continued this line of work with increasing intensity. Only a few years later, at the Annual Meeting of the German Academy of Forestry in 1941, he presented in an invited lecture his ideas and intentions to establish dendrochronology for Europe; he reported three successful investigations which had already led to a master curve of 250 years and to the synchronisation of palisade posts of a neolithic settlement. A rather sceptical discussion followed this presentation, but Huber's firm answer was: "The dating of wood samples according to their growth rings is also possible in Europe!"

In 1957 he visited the Laboratory of Tree-Ring Research in Tucson, Arizona, where he met A.E. Douglass and E. Schulman, and saw at the White Mountains the oldest trees on earth. This personal contact with the founders of dendrochronology was most impressive for him. Especially during his last years, dendrochronology became Huber's favourite subject and he expected to continue for a long time after his retirement, stimulating the work of others with his experience and the philosophic wisdom. Altogether, Bruno Huber and his immediate co-workers have published around 50 papers on dendrochronology and its application. Among his co-workers special mention should be made of B. Becker, W. Holdheide, W.v. Jazewitsch, and V. Giertz-Siebenlist.

Let us consider now the magnitude of Huber's dendrochronological work and the impact he had on the methods applied. His contributions can be grouped into several areas, which still merit our attention, as the programme of this Conference shows.

To mention the methodology first; Huber soon realized that the variations in annual ring-widths are much less pronounced in the balanced climate of Germany than in the semi-arid climate of Arizona, where most of the pioneering work was done. Consequently, the skeleton plot, developed by Glock in 1937, could only seldom be applied. Instead, the whole length of the curves had to be compared. For the evaluation of the similarity of tree-ring series he introduced the 'percentage of agreement, ' or that of disagreement, counting how often a ring width decreases when the comparison curve shows an increasing ring or vice-versa. He also applied 'fractional disagreement percentages' by counting the separate disagreements for different strengths of year-to-year ring width variation. In this way he used statistical evaluations for dendrochronology. For plotting ring widths he proposed a semilogarithmic scale, whereby the characteristic minima are intensified by the maxima reduced. This also made the curves independent of the magnitude of the absolute growth and helped the visual comparison of two tree-ring sequences. The method is still generally applied in Europe. In addition, he emphasized the importance of "signatures" - unusual sequences of rings for years.

Huber and his group even developed an automatic instrument for the measurement of ring width after the Eklund model, and introduced an apparatus for counting automatically the number of disagreements, thus facilitating the synchronization of curves of unknown age. He had already thought of using computer technology for more efficient storage and comparison of curves. After synchronising recent material, he soon successfully used the 'bridge method' to obtain chronologies for more than ten species.

One early question in his research concerned how far growth curves might be similar over great geographical distances - the problem of teleconnection. Being a Forest Botanist and a Tree Physiologist he opposed strongly the hypothesis by Ebba H. de Geer that the variation in width of tree-rings depends on cosmic factors and, therefore, agree over the whole earth. His conclusion was that each country must set up its own chronologies. It is also worthwhile mentioning Huber's response to the introduction of radiocarbon dating by Libby (1952) which, in the first enthusiasm, was regarded as likely to outstrip the dendrochronological method. He immediately realized that both methods have to work together, complementing each other. And he expressed clearly the potentials of dendrochronology for refining the carbon-14 dating, especially with regard to changes in the C-14 content of the atmosphere over past centuries.

Even before Huber became interested in dendrochronology, he had investigated the growth pattern of trees (thereby leading to an interest in annual rings). Therefore it was only natural that aspects of the biology of wood formation were another field to which he contributed. Only a few topics may be mentioned - the growth pattern of oaks and of alder; the differentiation of the two main European species of oak; the occurence of false and double rings (of importance to accuracy in dating); the impact of ecological conditions, or, the difference and similarity between species in their environmental responses.

Huber's favourite species was oak, the most important building timber in Central Europe, where no rings are missing; unlike alder, often used in earlier settlements, where up to 40% of the rings may be absent (Elling 1966). Even though the oldest oaks, from the famous Spessart Forest, reach an age of only 400 to 500 years, already by 1963 Huber had established a chronology of more than 1000 years. Oak also gives good crossdating - from France to Poland - similar to fir.

As a tree physiologist, Huber was naturally very interested in relating ring-widths to temperature and to precipitation. He realized the difficulties for Central Europe compared with North America, where, in the semi-arid climate of Arizona, ring widths largely follow the rainfall. This led to Schulman's tree-ring hydrology over many centuries for the catchment area of the Colorado River. Based on the work of Wellenhofer (1948), Artmann (1949), and Brehme (1951), three Ph.D. students, and of v. Jazewitsch (1961), he established for Central Europe the dependence of the growth of oak, beech and larch on the main environmental factors, using, at an early stage, multiple regression analysis. One result may be mentioned: at Berchtesgaden, in the Alps, the growth of larch before 1600 A.D. was twice as high as thereafter underlining the drastic climatic changes, also known from glacier research. Another student, Fürst (1963), demonstrated that along an East-West section the variation of tree-ring widths decreases with continentality. Consequently, the more pronounced variation of neolithic samples indicate that the climate was then more continental. Special emphasis was also given to the question of periodic cycles in the growth rhythm of various species, by applying suitable mathematical treatments. Thus Huber has broadened considerably the biological basis especially by taking into account climatological influences.

From the beginning of his interest in dendrochronology, Huber applied it as a method for dating and emphasized its supporting role for other fields of science. A few successful results may be mentioned. The dating of historic and prehistoric buildings, settlements, and villages was a favourite subject of his research. The village of Ziegenhain, the moated fortification of Buchau (Bronze Age) and the neolithic settlements of Zug-Sumpf, Thayngen-Weier and Burgäschisee in Switzerland are well known. Also he was the first to date objects of art.

When Bruno Huber died on December 14, 1969, dendrochronology had already spread from Munich to several other places in Germany and to other European countries, as was impressively underlined when he was invited to present the lecture "Dendrochronology in Middle Europe" at the 12th Nobel Symposium in Uppsala 1969. Therefore, I should like to mention finally the centres in Western Germany that are applying his method and their main lines of work.

The centre for dendrochronology at Munich no longer exists, thereby underlining how much scientific development depends on persons and not on institutions. Huber's one time co-worker, V. Giertz-Siebenlist, however, is still engaged in applying the standard chronology for the dating of prehistoric objects, mainly in co-operation with scientists from Switzerland and Austria. The most valuable material of Huber's work, the tree-ring sequences, wood samples, and reprints were brought to Stuttgart, where his pupil, B. Becker is a dedicated dendrochronolgist. His work concentrates on the establishment of an oak-chronology for the whole of the post-glacial period with application to geomorphological processes, especially to the history of the South-German river systems. Furthermore, his samples will enable radiocarbon to be calibrated by European dendrochronological dated material.

In the west, one has to mention E. Hollstein in Trier; from his work which started in the late fifties, an oak-chronology back to about 700 B.C. has been created. Working at the Rheinisches Landesmuseum in Trier, his approach is characterized by a careful technological evaluation of the timber used and the methods of handicraft applied to it: this has led to substantial contributions about the utilization of wood in prehistorc and historic times. He also widened our knowledge about the architectural history of the famous Cathedral of Trier, covering the whole period from Roman times to the present day. Further North, in Göttingen, A. Delorme has elaborated a regional chronology for the hilly region of the rivers Weser and Leine and also co-operates with Becker in building up the post-glacial oak-chronology.

Lastly, the group in Hamburg should be mentioned, but briefly because four papers from this group are in the programme. In 1963 J. Bauch and myself came from Munich, the dendrochronological centre of that time, to Hamburg and investigated the possibility and the limitations of dendrochronology in this more maritime-influenced area, which had been a blank spot on the - dendrochronological map so far. Soon D. Eckstein joined us, and during the last 10 years more than 40 papers have been published on dendrochronological subjects. Some areas covered are: (a) the development of an improved dating technique using a computer programme; (b) the establishment of dendrochronology in Northern Germany with regional oak-master chronologies from Denmark up to the Netherlands; (c) absolute dating of the famous Haithabu settlement, and several other prehistoric excavations; (d) contributions to the history of buildings in several rural and urban districts; (e) dating of panels from paintings belonging to the Dutch, Flemish and primitive German Schools, and other objects of art; (f) dendro-climatological investigations for the reconstruction of the regional climatic conditions of the past in co-operation with the Laboratory of Tree-Ring Research in Tucson/Arizona; (g) dendro-ecological analysis of trees under various anthropogenous influences; and finally (h) biological investigations on the formation of the annual rings as the basis for all our work. As an offspring of this group B. Schmidt in Cologne is collaborating in our dendro-climatological studies and also is concerned with the establishment of a post-glacial tree-ring series for North-West Germany. Later, he will have to concentrate on dating activities, being a staff-member of an institute for prehistory.

Tree-ring research is certainly not only done in Western Germany, but also extensively in other European countries. This was demonstrated in a symposium during the XIIth International Botanical Congress in Leningrad, 1975, and finds its expression especially at this International Symposium on Dendrochronology in Northern Europe.

Bruno Huber, the pioneer for all these activities, deserves our great respect for his farseeing imagination, his inspiration and his impressive energy in leading the way.

Dendrochronological Publications by Bruno Huber and his Group

ARTMANN, A.: Jahrringchronologische und -klimatologische Untersuchungen an der Zirbe und anderen Bäumen des Hochgebirges.
Diss.Univ.München 1949.

BECKER, B.: Die Baugeschichte der Landshuter Kirchen, dargestellt anhand einer über 1000-jährigen Tannenchronologie.
Kunstchronik 21(1968), 183-187.

BECKER, B. u. V.GIERTZ-SIEBENLIST: Eine über 1000-jährige mitteleuropäische Tannenchronologie.
Flora B 159(1970), 310-346.

BREHME, K.: Jahrringchronologische und -klimatische Untersuchungen an Hochgebirgslärchen des Berchtesgadener Landes.
Z.Weltforstwirtsch. 14(1951), 65-80.

ELLING, W.: Untersuchungen über das Jahrringverhalten der Schwarzerle.
Flora B 156(1966), 155-201.

FÜRST, O.: Über räumliche und zeitliche Unterschiede interannualler Jahrringbreitenschwankungen und ihre klimatologische Auswertung. Ein Beitrag zur Klimageschichte der Bronzezeit und Jungsteinzeit.
Flora 153(1963), 469-508.

HÖFFLER, K. u. W.ELLING: Der Jahrringbau von Linden des Doernhoefferschen Parks in Bayreuth.
Ber.Dt.Bot.Ges. 79(1966), 3-6.

HOLDHEIDE, W.: Über vor- und frühgeschichtliche Werkhölzer.
Holz Roh- u.Werkst. 4(1941), 169-174.

HORNSTEIN, Frh.v., J.: Die Tannengebälke des Konstanzer und Freiburger Münsters und ihre geschichtliche Auswertung.
Diss.Univ.München 1963.
=Allemann.Jb. 1964/65, 239-289.

HUBER, B. u. E.SCHMIDT: Die mittelalterlichen Holzfunde der Wasserburg Obergöltzsch bei Rodewisch.
Tharandter Forstl.Jb. 90(1939), 146-157.

- - : Aus dem Arbeitsgebiet der Holzanatomie.
Holz als Roh- u.Werkst. 3(1940), 137-142.

- - : Aufbau einer mitteleuropäischen Jahrring-Chronologie.
Mitt. H.G.-Akad.d.Dt.Forstwiss. 1(1941), 110-125.

- - u. W.HOLDHEIDE: Jahrringchronologische Untersuchungen an Hölzern der bronzezeitlichen Wasserburg Buchau im Federsee.
Ber.Dt.Bot.Ges. 60(1942), 261-283.

HUBER, B.: Über die Sicherheit jahrringchronologischer Datierung. Holz als Roh- u.Werkst. 6(1943), 263-268.

- - : Die Jahresringe der Bäume als Hilfsmittel der Klimatologie und Chronologie. Die Naturwiss. 35(1948), 151-154.

- - : Ein deutscher Pionier der Jahrring-Chronologie in Amerika. Allg.Forstz. 4(1949), S.93.

- - , W.v.JAZEWITSCH, A.JOHN u. W.WELLENHOFER: Jahrringchronologie der Spessarteichen. Forstwiss.Cbl. 68(1949), 706-715.

- - u. W.v.JAZEWITSCH: Aus der Praxis der Jahrring-Analyse. I. Gerichtsgutachten. Allg.Forstz. 5(1950), 443-444.

- - u. W.v.JAZEWITSCH: ... II. Datierung geschichtlicher und vorgeschichtlicher Holzfunde. Allg.Forstz. 5(1950), 527-529.

- - u. W.v.JAZEWITSCH: ... III. Die klimatologische Auswertung von Jahrring-Kurven. Allg. Forstz. 7(1952), 1-2.

- - : Was versprechen wir uns von der Jahrringchronologie? Die Umschau 51(1951), 331-333.

- - : Beiträge zur Methodik der Jahrringchronologie. I. Gegenläufigkeitsprozent und Gegenläufigkeitsstruktur als Maßstäbe bei der Sicherung jahrringchronologischer Datierungen. Holzforschung 6(1952), 233-235.

- - : Die Jahresringe der Bäume und die Messung der Zeit - Methoden und Ergebnisse der modernen Jahrringchronologie. Universitas 9(1954), 1105-1108.

- - und W.v.JAZEWITSCH: Tree-ring studies of the Forestry Botany-Institutes of Tharandt and Munich. Tree-Ring Bull. 21(1956), 28-30.

- - u. W.v.JAZEWITSCH: Jahrringuntersuchungen an Pfahlbauhölzern. Flora 146(1958), 445-471.

- - : Dendrochronologie. Geol.Rundschau 49(1960), 120-131.

- - , W.MERZ u. O.FÜRST: Dendrochronologie in Europa. VIth International Congress on Quaternary, Warschau 1961 S.677-685.

HUBER, B. u. W.MERZ: Jahrringchronologische Untersuchungen zur Baugeschichte der urnenfelderzeitlichen Siedlung Zug-Sumpf. Germania 40(1962), 44-56.

- - u. W.MERZ: Jahrringchronologische Synchronisierung der jungsteinzeitlichen Siedlungen Thayngen-Weier und Burgäschisee-Süd und -Südwest. Germania 41(1963), 1-9.

- - : New Discoveries in Tree-Ring Research. International Association of Wood Anatomists. News Bull. 2(1963), 2 - 4.

- - u. V.SIEBENLIST: Das Watterbacher Haus im Odenwald, ein wichtiges Brückenstück unserer tausendjährigen Eichenchronologie. Mitt.Flor.-soz.Arbeitsgemeinschaft N.F. (1963), 10, 256-260.

- - : Radiocarbon- und Jahrringforschung im Dienste der Geochronologie. Mitt.a.d.Staatsforstverw.Bayerns (1964), H.34, 162-178.

- - , H.COURTOIS u. W.ELLING: Die Periodenlänge von Jahrringbreitenkurven. Meteorologische Rundschau 17(1964), 122-125, 169-171.

- - , V.SIEBENLIST u. W.NIESS: Jahrringchronologie Hessischer Eichen. Büdinger Geschichtsblätter (1962/63), Band V, 29-82.

- - : Durchschnittliche Schwankung und Periodenlänge von Jahresringbreitenkurven als Klima-Indikatoren. Geol.Rundschau 54(1964), 441-448.

--- , C.FERGUSON u. H.E.SUESS: Determination of the Age of Swiss Lake Dwellings as an Example of Dendrochronologically Celibrated Radiocarbon Dating. Z.Naturforschung 21a(1966), 1173-1177.

- - : Beiträge zur prähistorischen, klassischen und jüngeren Archäologie - Bd.II, T.4. Seeberg, Burgäschisee-Süd - Dendrochronologie. Acta Bernensia 1967, 145-156.

- - : Neues von Radiocarbon- und Jahrringdatierung. Mitt.Staatsforstvers.Bayerns, München (1967),H.36, 97-104.

- - : Dendrochronologie. Kunstchronik 21(1968), 142-143.

- - u. V.GIERTZ-SIEBENLIST: Unsere tausendjährige Eichen-Jahrringchronologie durchschnittlich 57(10-150)fach belegt. Sitzungsbericht Österr.Akad.Wiss.Math.Nat.Wiss.Klasse Abt.1, 178(1969), 37-42.

- - : Dendrochronologie. Handbuch der Mikroskopie in der Technik. 2.Aufl. Bd.V (1969), 171-211.

- - : Dendrochronology in Middle Europe.
12. Nobel Symposium, Uppsala 1969.

- - : u. V.GIERTZ: Central European Dendrochronology for the Middle Ages.
In: Scientific Methods in Medieval Archeology. Ed. by R.Berger Univ. California Press 1970, 201 - 212.

- - u. V.GIERTZ-SIEBENLIST: Dendrochronologische Datierung von Eichengebälken aus St.Veit.
Forschungen und Berichte der Archäologie d.Mittelalters in Baden-Württemberg, Bd.1, 1972, 266-268.

KRÖNER, H.: Die physiologischen Auswirkungen des Dürresommers 1947 auf Höhen- und Dickenzuwachs unserer Waldbäume.
Diss.München 1952.

JAZEWITSCH, W.v.: Über die Möglichkeiten einer jahrringchronologischen Individualdiagnose von Bäumen. Mit Beiträgen zur Methodik der Jahrringforschung.
Diss. München 1948.

JAZEWITSCH, W.v.: Beiträge zur Methodik der Jahrringchronologie. II. Die fraktionierte Gegenläufigkeitsstatistik.
Holzforschung 6(1953), 82-89.

JAZEWITSCH, W.v.: Jahrringchronologie der Spessartbuchen.
Forstwiss.Cbl. 72(1953), 7/8, 234-247.

JAZEWITSCH, W.v.: Jahrringchronologie von Ziegenhainer Eichengebälken.
Z.d.Ver.Hess.Gesch.-u.Landeskunde 65/66(1954/55), 55-72.

Jazewitsch, W.v., H.SIEBENLIST u. G.BETTAG: Eine Synchronisiermaschine zum Vergleich von Jahrringkurven und eine langjährige Eichenchronologie.
Ber.Dt.Bot.Ges. 69(1956), 128-142.

JAZEWITSCH, W.v., G.BETTAG u. V.SIEBENLIST: Eine neue Jahrringmeßmaschine mit elektrischer Meßregistrierung.
Holz als Roh- u.Werkstoff 15(1957), 241-244,

JAZEWITSCH, W.v. (mit Nachwort v.B.HUBER): Zur klimatologischen Auswertung von Jahrringkurven.
Forstwiss.Cbl. 80(1961), 175-190.

LERCHENFELD, M.L. Frhr.v.: Jahrringchronologische Datierung verbauter Eichenhölzer.
Diss.München 1953.

MÜLLER-STOLL, H.: Vergleichende Untersuchungen über die Abhängigkeit der Jahrringfolge von Holzart, Standort und Klima.
Bibl.Botanica 122(1951). 1-93.

WELLENHOFER, W. Untersuchungen über die Ursache der Eichenjahrringschwankungen und Aufstellung einer langjährigen Spessarteichen-Jahrringchronologie.
Diss.Univ.München 1948.

1.2. ANNUAL TREE-RINGS AS AN AID TO CLIMATOLOGY AND TO CHRONOLOGY

B. Huber, Munich

Extracts from Translation by Ruth Morgan, University of Sheffield, of paper in Die Naturwissenschaften, 1948, 35, 151-4.

In our latitudes, wood is laid down as annual rings. At the beginning of the growth period, thin-walled elements are formed under favourable water supply; later there is increasing production of storage material. This variable formation (earlywood and latewood) corresponds to the needs of the tree; the porous earlywood consists predominantly of water-conducting channels for the supply of the young shoots; the latewood provides the increase of strengthening material and, particularly in deciduous woods, storage cells (parenchyma) for the reception of assimilates.....

Dendroclimatology

The average width of the annual ring diminishes with the age of the tree. This is more than compensated by the increasing circumference, certainly in the first decades of a tree's life. So the area increase and also the annual wood production of each trunk is liable to be a maximum at an age, according to the species and environment, of 40 to 100 years (latest in fir).....

The analyses of tree-ring results from areas where a single climatic factor is dominant in restricting wood growth are the most promising: for example, in arid regions the precipitation. In North America, the ring width of the widespread Douglas Fir is related to the winter precipitation level (October to June) with a very high correlation coefficient (0.85), while the summer precipitation (July to September), because it evaporates quickly, has little influence.....

Steep slopes offer the best guarantee that the water supply of the trees is totally related to the current rainfall. However, different species vary in their sensitivity, that is the average fluctuation in ring-width from year to year. For our mean curves, Müller Stoll (1945) found the sensitivity was 28% for beech, about 16% for fir and spruce and about 18% for oak.

In Anatolia the relationship to climate was found by Gassner (1942) to follow a different pattern. The earlywood width of pine (*Pinus sylvestris* and *Pinus nigra*) is proportional to the spring rainfall, while during the subsequent arid summer the latewood width depends on the quantity of stored water and so reflects the winter precipitation. Yet wide rings in South-West Africa (Namibia) are controlled by rainfall for only a few species (Walter, 1940).

On the other hand in parts of Scandinavia the correlation between annual ring width and precipitation was found by Erlandsson (1936) to be definitely negative, as the limiting factor in northern latitudes is often absence of summer warmth. In Germany, the same applies to trees in the high mountains: My pupil Artmann showed that the ring-width of the cembra pine in the Bavarian Alps was related mainly to the mean summer temperature, to the mean summer midday temperature and to the hours of sunshine at the mountain recording stations.

In parts of Central Europe, tree-ring analysis has met with difficulties because warmth and moisture affect growth in a way that is difficult to disentangle. Some years of unusual growth, e.g. 1904, 1911, 1922/23, 1928/29, 1934/35 and 1947 are due to dry summers; but the severe winters 1928/29 and 1939/40 caused a sharp fall in ring-width in the following year. So it is not easy to deduce the cause from a narrow ring. My colleagues Müller-Stoll (1945) and von Jazewitsch (1948) found that the examination of tree-ring curves of several species was relatively promising for climatic deductions in Central Europe and for defining climatically similar areas. For example, minima in fir and beech are related to frosts following a severe winter; minima in spruce and pine, to dryness.

Dendrochronology

The analysis over a period of 40 years of many thousand samples by Douglass and his pupils unexpectedly brought to light the individuality of a chronological sequence derived from ring width fluctuations and so dendrochronology developed into a refined dating method, accurate to a year. As soon as a reference sequence of ring widths has been fixed by dated material from one area and for one species, the dating of other samples of wood on the basis of their ring sequence becomes possible, just as a safety key fits a lock on the basis of the variations of a few teeth.

Contemporary mature samples, owing to their biological diversity, show not identical but only similar annual ring fluctuations. For Central European trees the degree of similarity between the sequences is weaker than for trees in the Pacific South West. Hence special care must be given to the procedure. We measure the annual ring widths by reflected light, but subfossil bog wood needs to be measured in thin sections with transmitted light. The years form the abscissa, the ring widths the ordinate on a logarithmic scale..... Transparent paper and a light-table aid the comparison of the curves which are drawn in Indian ink.

In two unconnected curves, movement is equally probable, from one year to the next, in one direction as in the other. Their disagreement over a large number of years, is approximately 50%. Synchronous curves however have a small disagreement percentage. Thus parts of the same tree or even different radii of the same cross-section disagree by about 10-20% and different trees from the same location by perhaps 20-25%. With a change of region and of species, the disagreement increases further, as shown by the following table, derived by Müller Stoll (1945):

	Disagreement					
	Within the same species			Between different species		
	Fir	Beech	Spruce	Beech/ fir	Fir/ spruce	Beech/ spruce
Within a region (the Beskides)	24	30	32	36	41	44%
Between different Mountain ranges of Central Europe	35	38	40	42	45	45%

The unforeseen result revealed by this table is that the disagreements for one species over a very considerable distance (in fact about 1500 kms from the Beskides in the east, to the Vosges in the West) are less — 35, 38, 40% — than between different species in the same location — 36, 41, 44%. This implies that the Central European climate is more uniform than the physiological response of different species, such as fir/spruce or beech/spruce. On the basis of these results cross-matching within much of Central Europe can be achieved, as proved by numerous 'blind' dating exercises. This is encouraging for the framework of a Central European tree-ring chronology using samples from different regions.....

Yet a falling off with distance is unmistakable..... Between Central Europe and Scandinavia there is scarcely perceptible agreement of more than 50%. Signature Years: The visual similarity of synchronous curves appears greater than indicated numerically by the overall disagreement percent. This arises because weak fluctuations in the curves count numerically as much to disagreement as pronounced fluctuations, i.e. signatures. Recently we have therefore resorted to obtaining the disagreement percentage for those changes in width which are pronounced. For this, we use deviations in the width by over 50%. This showed for spruce, which had an average of 18% disagreement overall, that there was 28.4% disagreement in the years with under 25% of deviation; 8.8% in the years deviating by 25-50%; and no disagreement when the deviation was over 50%. In this way we arrived at an objective definition of the signature years and identified them for particular years and in different species.....

Construction of Central European Tree-Ring Chronologies

We have confirmed the possibility of a tree-ring chronology from recent material and have developed and refined the necessary methods for longer chronologies. Table 1 gives the ring-widths obtained as an index in percent relative to ten year-means by Müller-Stoll (1945) for spruce and beech. Important minima are in bold type and important maxima in italics. For oak, the chronology from recently felled Spessart oaks has been extended (back from 1470) by timbers from the Church of Our Lady at Munich, while there is an (as yet unpublished) oak chronology to 1000 A.D. by Holdheide with samples from the Wolliner-Vineta excavations.

A relative chronology, consisting of a 120 year sequence derived from piles of bog-pine from the palisades of the Bronze Age site of Wassenburg-Buchau in Federsee, has been established and an archaeological dispute settled: namely it has proved that the inner and outer palisades were of uniform construction built within a short time. The results compare with the success of Ording (1941) at Rakuehaugen. The degree of synchronisation of these Bronze Age samples was considerably in excess of that for recent samples. This is interpreted as being due to a variation in the climate, although one cannot say which particular weather factor is concerned.....

REFERENCES

Erlandsson, St. (1936). Stockholm Högskolas Geokronol. Inst., Uppsala.

Gassner, G. (1942). Nova Acta Leopoldina NF. 12.

von Jazewitsch, W. (1948). Diss. Munich.

Müller-Stoll, H. (1945). Diss. Dresden.

Ording, Å. (1941). Meddelelser fra det Norske Skogsforsöksvenen 7, 101.

Walter, H. (1940). Die Naturwiss. 28, 607.

	0	1	2	3	4	5	6	7	8	9	10	11	12	13	14	15	16	17	18	19
										A. Fichte.										
1740	117	101	112	111	119	119	108	91	80	84	75	93	118	109	108	95	90	103	96	94
1760	95	101	108	95	100	114	102	98	101	98	91	98	113	104	110	108	106	102	93	84
1780	85	92	92	94	96	97	102	103	107	115	102	107	95	90	92	81	78	78	86	116
1800	105	86	90	85	99	84	80	110	114	124	113	98	104	119	108	99	97	98	99	91
1820	84	89	86	92	100	109	104	98	107	105	109	125	114	108	113	93	80	80	86	102
1840	101	110	**95**	**87**	97	110	*120*	100	102	88	98	111	105	113	112	107	97	105	**86**	**81**
1860	**84**	102	*116*	108	99	104	*107*	*110*	**82**	**81**	**89**	100	*111*	*112*	91	99	88	**85**	96	*108*
1880	*111*	*108*	*110*	98	*116*	101	106	**88**	**91**	**89**	103	*105*	103	95	105	110	83	106	110	93
1900	91	108	100	*117*	108	88	**68**	89	102	102	*118*	108	**92**	94	101	107	*124*	99	99	88
1920	88	85	**79**	**73**	98	123	*130*	*130*	107	**87**	**88**	113	94	99	*123*	96	99	104	83	102
1940	**88**	91	98	74																
										B. Buche.										
1700	56	78	67	80	98	108	106	114	156	108	**101**	142	113	91	76	96	85	98	100	100
1720	97	85	89	102	107	99	92	117	88	111	116	94	98	113	102	108	89	106	92	108
1740	92	**61**	86	125	122	110	86	89	103	110	109	106	112	94	112	87	91	120	89	87
1760	79	99	63	91	99	100	121	111	96	126	124	110	92	113	105	111	92	97	98	89
1780	96	100	83	112	103	101	94	107	107	94	101	98	118	101	110	**84**	124	111	110	77
1800	67	118	94	82	100	109	91	106	95	104	100	102	113	119	106	89	96	113	92	90
1820	94	96	96	86	93	98	103	86	120	124	100	132	110	105	122	84	74	109	93	99
1840	73	82	108	68	110	121	101	94	105	98	114	115	129	114	94	114	113	114	**64**	87
1860	**57**	107	111	*152*	103	89	74	105	100	**61**	99	*124*	*119*	*125*	**47**	89	88	79	112	*136*
1880	81	104	95	*132*	*142*	94	100	111	**46**	75	103	106	101	75	96	*126*	100	103	106	117
1900	**69**	107	97	101	*114*	**77**	101	105	109	90	99	108	**80**	**75**	139	98	116	*120*	**68**	90
1920	103	98	96	100	*113*	*117*	*111*	*110*	**87**	**82**	**85**	100	*118*	*115*	92	102	**73**	81	98	*121*
1940	99	115	**64**	108																

Table 1 Ring-width indices, as percent of 10-year means, for spruce (above) and beech (below)

1.3. DENDROCHRONOLOGY

B. Huber, Munich, 1970

Extracts from Translation by R. Morgan, University of Sheffield, of paper in Handbuch der Mikroscopie in der Technik, 5, 171-211.

Editorial Note: In this paper, written at the end of his career, Huber summarises the methods and principles he developed and adopted and gives his main findings and conclusions.

I. INTRODUCTION

Dendrochronology aims at the dating of wood samples on the basis of the fluctuations of their annual ring widths. Even though some of the fluctuations are not pronounced, there are other sections of the sequence, particularly in contrasting climates, in which the ring widths fluctuate so sharply that they form unmistakable time-markers ("signatures").

II. PROVISION AND MEASUREMENT OF SUITABLE SAMPLES

A. Starting Material

In the forested areas of the world, wood has been the cheapest and thus the most important building material. While wood in moist air quickly undergoes fungal and animal decomposition, if sealed it can resist decomposition not only for thousands of years but occasionally for millions. Furthermore, if the chemical transformation, particularly carbonization, is gradual the microscopic structure may be preserved. In dry areas, wood may lie undecomposed for a long time, e.g. *Pinus aristata* in the Wormwood White Mts. of California for over 7000 years. In the Wormwood steppe (Israel), perennial shrubs such as *Artemisia tridentata* and *Zygophyllum dumosum* seem useful for tree-ring analysis.

It is not difficult therefore to obtain, in adequate quantity, samples of recent, historic and prehistoric material for dendrochronological analysis. In much of Europe, the principle species is oak, which, as a ring-porous wood, possesses the advantage of annual growth increments recognizable under reflected light. From the well-maintained and cultivated oaks of the Spessart Forest, a thousand very healthy trunks of the 500-year age group are available. With these we began our study for a Central European oak chronology. At present it extends over the last 1200 years and in many parts there are a hundredfold samples (Huber and Giertz-Siebenlist, 1969, this volume, page 27).

Conifers also have annual rings that are clearly recognizable. Fir from the Bavarian Forest has provided samples for the last 500 years, while the cathedrals of Constance, Freiburg and Maulbronn extend the chronology to a 1000 years (Hornstein, 1964/5). Diffuse-porous deciduous species, such as lime, alder and maple, cause greater difficulties, because their annual increase shows little contrast, only recognizable in transmitted light: moreover they are more subject to missing rings (Elling. 1966).

When possible, one collects complete sections (discs), not from a weathered stump where the annual rings are often distorted and wavy, but from a cylindrical trunk, at two different heights. With building timber, one must often be satisfied with cores. Valuable material can be measured along a radially cut surface or even by beta or soft X-rays. Not less than 5 samples from different individual trees from the same place, stand or building, should be obtained in order to establish not only the individual but the average behaviour.

B. Measurements

The surface is smoothed with a razor or plane. For the analysis of discs one measures along radii, e.g. a pair from a suitable diameter; cores are mounted in a suitable holder. One can emphasise the contrast of the rings with a dye or with chalk powder. Our chief preparer planes up to 10 cm long diameters of diffuse-porous wood and uses Fuchsin as a dye. Some magnification is needed even for fast growth. To obtain direct information on building timber, we use a 15 cm long microscope with a lens, containing a scale divided into 0.1 mm units, that can be placed on the object to be measured.

For purposes such as climatic analysis, it is recommended that the width of the earlywood and of the latewood be recorded separately; also that the density of the early- and latewood should be obtained by special methods, e.g. X-radiography. In ring-porous oak, the width of the latewood fluctuates much more than that of the earlywood width.

III. FROM THE SINGLE CURVE TO THE MEAN CURVE

The measured values are plotted on a logarithmic scale as the ordinate, with the years as the abscissa increasing from left to right at 5 mm intervals. Though this uses much paper, curve comparison is much more difficult with narrower spacing and we do not recommend it. We use transparent paper from Holtzmann/Speyer. Logarithmic representation offers the advantage of giving the same vertical distance for a decrease or increase in the ring-width by a given factor, independent of the absolute growth. In our experience, the deviations that apply to a dominant and a suppressed sample, or to a younger and an older sample, are better compared on a logarithmic than on a linear scale. It is surprising that the logarithmic representation has not been universally accepted; the Americans have carried out many mathematical studies for the elimination of age trend.

A. Average Variations

If we look at such a ring-width curve we find, as mentioned in the introduction, a combination of insignificant (weak) sections alternating with marked signatures, just as average weather conditions are mixed with years of extremes which appear conspicuous.

Ring-widths of different species in the same location form very differently-shaped curves (see Fürst, 1963 and this volume page 65). Fig. 1 (after Artmann, 1949) shows that the fluctuations with larch are much greater than for Swiss stone pine (Pinus cembra); similarly, the fluctuations of beech are much greater than those of oak. As a measure of the movement, the average variation (S) for a sequence of n rings is obtained from the following expression

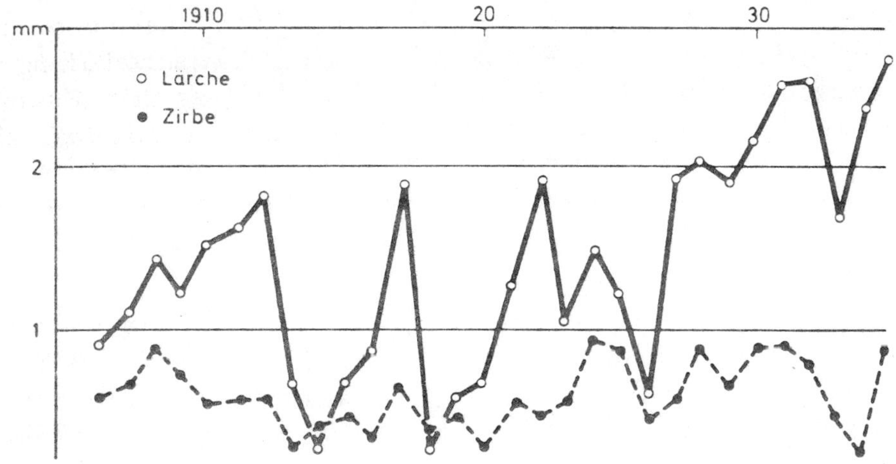

Fig. 1 The contrast in the average variation of the ring-widths of larch and Swiss stone pine growing in the Alps near their habitat limit. Both curves have been correlated with the recorded hours of sunshine.

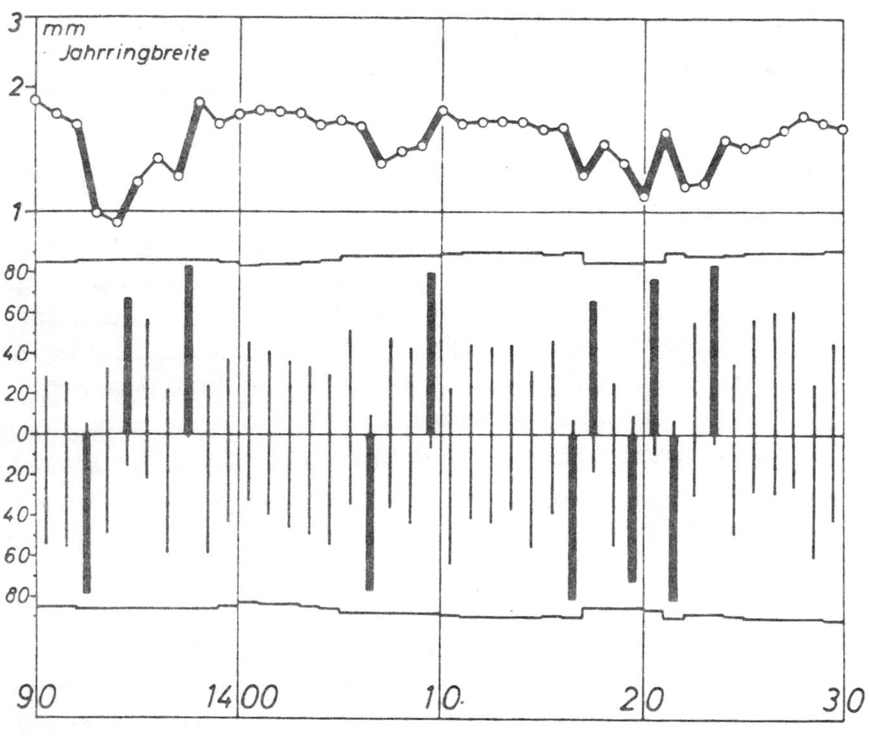

Fig. 2 Above: The oak standard mean curve for 1390-1430.

Below: The number of components and the number showing a rise or fall in ring-width for each year.

Thick Lines: Changes found in > 75% of the component single curves.

(Huber, 1943):

$$\frac{n}{(n-1)} \cdot \frac{\text{The sum of the difference of the successive ring widths}}{\text{The sum of the successive widths}},$$

with the sign of the differences being disregarded when they are summed. The factor n/n-1 is necessary, because the number of differences (n-1) is smaller than the number of ring widths (n).

The Americans call this quantity "sensitivity". They find similar values for different species in the same location. In Europe, the species exhibit geographical and temporal differences in their average variation (Fürst, 1963 and this volume page 65). They were greater in the east (Poland) than in the west (France). Also, they were greater in the Neolithic period than today (Huber and Holdeheide 1942). We retain the mathematically correct term "average variation" and use it with reference to a species or to a climatic difference.

Theoretically, the average variation can move between 0 (no variation) and 200% of the mean value. Values calculated on many samples in Central Europe are:

beech, 31%: oak, 18%; larch, 25%: Swiss stone pine, 18%.

Schulman observed a value of 53% for Douglas Fir in the arid conditions of Arizona.

B. Period length

A second characteristic of each annual ring curve is the "period length". By this we understand the number of years in which the curve continues to fall or to rise. When conditions are such as to make a rise or fall equally likely, half the curves will alter their direction after one year, a quarter after two years, an eighth after three years, etc. However, a return of the direction from an indicator or signature year is frequent and the maxima or minima so created are suitable for the synchronization of curves. We find that a continental climate appears to shorten the period length. Our "saw signature" from 1530 to 1540, a repeated alternation of rising and falling ring widths, has helped considerably in synchronising both oak and fir.

C. Curve synchronisation (Cross dating)

The crucial step is the comparison of many contemporary curves. When a similar deviation is found in many curves, it refers to a common and typical occurrence. With the aid of such conspicuous features one can synchronize contemporary curves (Americans call this cross-dating) and proceed to the making of mean curves.

The mean curves naturally lose some of the mobility of a single curve, but less so the more the single curves agree. In Arizona there are 5 mean curves showing over 90% of the movement of the single curves. In Central Europe, the role of each individual curve is greater as a consequence of the more temperate climate.

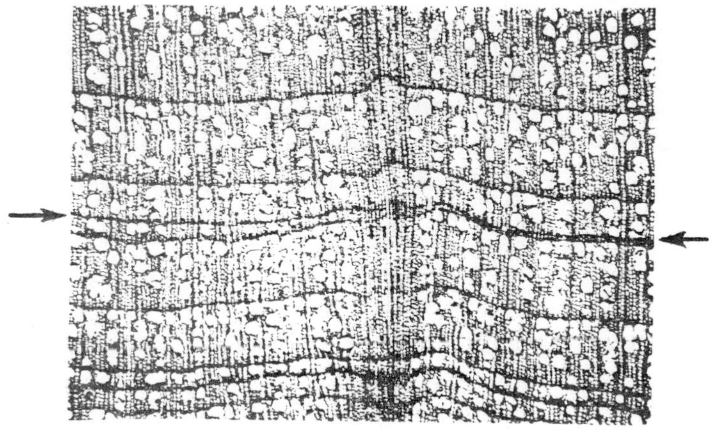

Missing ring (on right only) in alder.
After Elling.

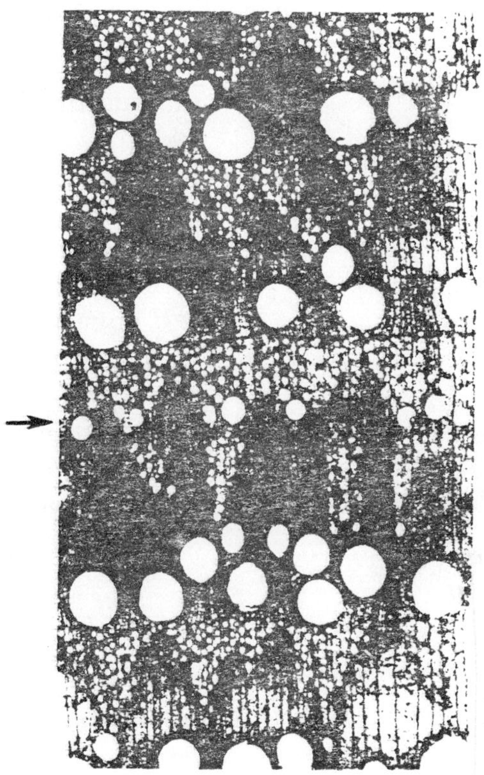

False ring in oak.

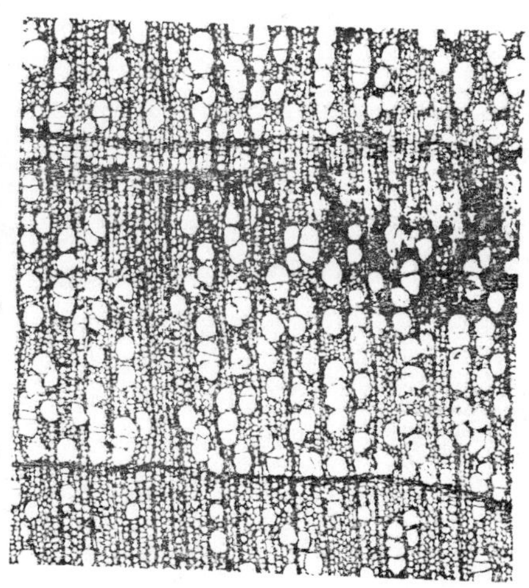

False ring in alder.

Plate 1. Missing and False Rings.

The formation of mean curves contributes greatly both to objectivity and to summarising the otherwise vast and increasing amount of material. But one must remember that each reliable single curve refers to some actual behaviour. It facilitates synchronization to remember that a particular fluctuation in ring width does not occur in exactly the same year in all the single curves, but in some a year or two later. In such cases we go back to the single curves and often find that the resemblance of a curve to be dated is less with a mean curve than with certain individual single curves.

We join the points on the mean curves, as shown in Fig. 2, by thick or thin lines to represent on a common statistical basis, changes which occur in the same direction in the majority of the single curves. Thus for the numerous samples available for the period 1390-1430, represented in Fig. 2, the thick lines apply to years in which 75% or more of the curves move in the same direction. Even for abundant material, 90% or more of the curves move in certain years in the same direction. We call the changes in width that show such uniform behaviour "signatures" and distinguish two types:

(a) When the direction is uniform: trend-signatures;
(b) When a peak or trough on the curve is produced: max. or min. signatures.

A rise or fall in a curve in a particular year usually occurs reproducibly, but the position of minima and maxima may differ in single curves by one or even two years, because a fall in width and a recovery in width can extend over several years.

D. Missing and double rings

It is difficult to make a mean curve when individual curves correspond along most of their length and then suddenly disagree by one ring. A missing ring is suggested. From the beginning of this research we have searched systematically for such missing rings, and have almost always found them by searching around the circumference (Plate 1). The ability to carry out such an examination is the main advantage in having an entire section from a trunk, rather than cores.

In time one gains such experience in the localisation of missing rings that one can interpret them without direct evidence. As a general rule, it is the narrowest ring which may be missing. In larch, Brehme (1951) found and identified up to 25% missing rings, in suppressed alder trunks Elling (1966) found up to 10%. Ring-porous woods have the great advantage, due to peculiarities of their sapstream physiology (Huber, 1935), that missing rings are rare in their trunk-wood. After measuring more than a hundred thousand rings in oak we have found no evidence for a missing one. But defoliation from insects or lammas growth can lead to a false ring in oak; this differs from a true annual ring (Plate 1).

In the latitude with which we are concerned, false rings play a much smaller role than missing rings. However, especially in fir but more rarely in spruce and pine, in a cold summer the latewood appears to indicate the end of growth; then, with a warm August, an increase in growth again occurs and the growth resembles earlywood. As a rule, the blurred boundary prevents confusion between a false ring and a narrow annual ring. This applies also to the boundary, the so-called pith stains, which occur in alder from the channels of burrowing larvae (Plate 1).

How many single curves should be included in a mean curve? Theoretically, the reliability rises with the number included. Practically, the curves for central Europe change only slightly when over 20 single curves are meaned together. For calculating the percentage of fractional disagreement, and hence obtaining accurate cross-dating, we take a batch of 20 curves (Fig. 3).

IV. LONG CHRONOLOGIES. BRIDGING

As trees that are several hundred years old are rare, the possibilities of using dendrochronology were limited until Douglass arrived at the ingenious idea of bridging, that is, extending from the inner rings of younger, to the outer ones of older samples. Experience shows that a minimum of 50 years overlap of younger with older samples is necessary. For Central Europe, because young trees in a stand react more weakly than established ones, it is recommended that not too much weight is placed on the innermost 20 to 50 rings, and that one should aim at not less than a 100 year overlap when trying to bridge two curves. Repeated, independent bridging increases the reliability. Today, our Central European oak chronology covers 1200 years by 50 to 100-fold samples (Huber and Giertz-Siebenlist, this volume, page 27). It is confirmed by the independent oak chronology obtained by Hollstein (1965) at Trier for west of the Rhine (Fig. 3).

A. Relative chronologies

At first, one cannot always establish an absolute chronology from historic and prehistoric samples. Nevertheless one can arrange such samples in a sequence of time, particularly when they show unique signatures. Douglass in his work on timbers of *Pinus ponderosa* from early Pueblo Indian settlements found a conspicuous ring sequence that was only dated absolutely through bridging after ten years further study.

A relative chronology can provide valuable information in archaeological investigations. Our earliest work in this field concerned samples from Switzerland; first from the late Bronze age lakeside settlement of Zug-Sumpf (Huber and Merz, 1963).

The final aim is always absolute dating. For historic buildings one can often obtain this approximately by documentary evidence as was done for the roof beams from the cathedrals of Constance, Freiburg and Maulbronn (Hornstein, 1964/5).

B. Synchronisation

When the possible options for dating lie within a few years it is often easy to recognise the correct dating with certainty. This applied to about a hundred dating applications for which Huber and von Jazewitsch (1950-52) have published results. Thus for a courthouse constructed recently at Karlsruhe it was feasible to decide between seasoned pine wood or 'green' wood (as a last

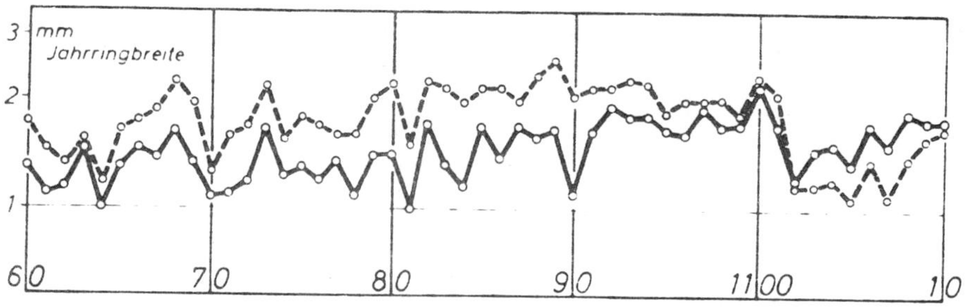

Fig. 3 Oak mean curves for east and west of the Rhine, 1060-1110.
Above: For the former, derived in Forestry-Botany Institute, Munich.
Below: For the latter, derived by Hollstein at Trier.

Especially characteristic are:
The Maximum at 1073; the Minimum at 1081; and the fall from 1100/01 to 1102.

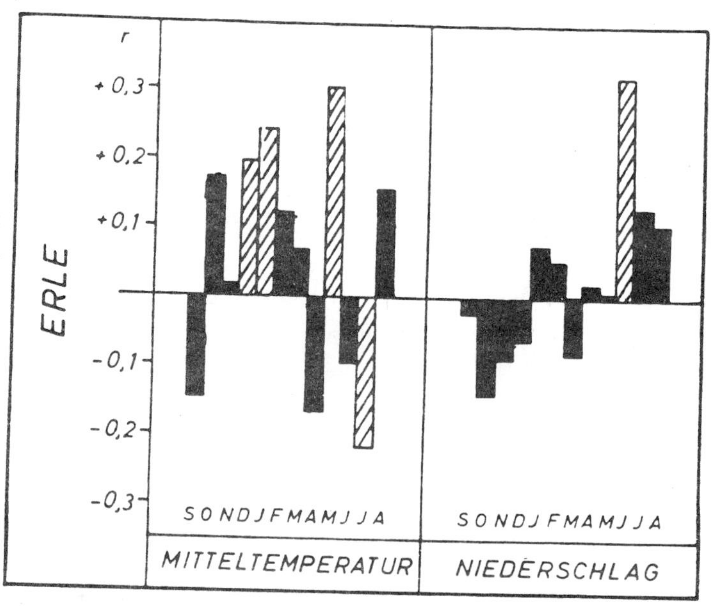

Fig. 4 Correlation (by Elling) of Annual Ring-Widths for Alder with monthly mean temperatures and precipitation.

There is a positive correlation (coefficient, r, of over 0.3) with the May temperature and June precipitation; a negative correlation with the July temperature; and a negative correlation with the September temperature of the previous year.

minute expediency) being used: the evidence from signature years such as 1947 (caused by dryness), 1940/2 and 1929 (after an extremely cold winter), left no doubt as to the exact year of felling. Even one year earlier was excluded, as it produced a maximum of disagreement (Huber, 1943).

However, when we know nothing of the likely date, as with waterlogged oak from river gravels, one is faced with the fact that similar curve development may appear in more than one position. In the earlier stages of this research, we noted possible dates and then waited until, by additional comparisons, the correct position was obtained. Today this is rarely necessary as our experience has taught us that it is only convincingly similar curves that are synchronous; tolerable matches are probably not synchronous. As an aid, an automatic synchronisation device has been developed using punched tape which is progressively shifted year by year.

As access to further material was blocked during the war we made use of the enforced interval to examine year by year the disagreement in all the incorrect positions using material already available. This proved very informative for further dating. For many thousands of such positions, the average disagreement was the expected 50% (i.e. agreement was as probable as disagreement). In a single position, the likely deviation from 50% is dependent on the curve length, A, by the formula:

$$\text{disagreement percent} = 50 \pm \frac{50}{\sqrt{A}}$$

For triple the standard deviation and with curves having rings for 100 years, false positions can give disagreements between 35 and 65%; for shorter, 25 year curves, the disagreement can be as low as 20% and still be false. In the correct position the disagreement usually amounts to 20 to 30% for single-tree curves from the same region; for mean curves based on several trees, to under 25%.

Instead of the disagreement percent, curve similarity is today expressed as correlation coefficients derived by computers. It shows that our 1000 year long oak chronology is statistically reliable with a certainty of well over a million to one. So when Suess (1965) calibrated radiocarbon dating by dendrochronology with 150 deciduous and coniferous samples, some from us in Europe and some from North America, he included no faulty date for our samples.

It is a fact that two tree-ring curves are never identical, but only approximately similar. The search for the correct position is a study for the position of the greatest similarity. It is not—as it may appear to many—subjective by intuition and by "God's glance"; it is more a case of an acquired technique with objective results. Those initially in doubt as to its reliability have repeatedly ended by being convinced.

V. GEOGRAPHICAL RANGE OF DENDROCHRONOLOGY

There remains the question as to the size of the area in which tree-ring behaviour is sufficiently similar to justify a common mean curve. Opinion on this question has occasionally been widely divergent; the truth lies, as often, somewhere in the middle.

Gerhard de Geer, having derived a varve-chronology, was inclined to believe not only that the fluctuations in varve widths, but also those in tree-rings arose from worldwide cosmic factors; his wife claimed to have dated samples from the Old and New Worlds, from the north and south hemispheres, from Swedish church roofs and Viking ships, all with the aid of the North American sequoia chronology. No one really believed in such teleconnection. So I commissioned my colleague, Frau H. Müller-Stoll (1951), to examine the similarity of tree-ring curves from different areas. This showed that not only from Europe to North America, but even from Central Europe to Scandinavia the disagreement exceeds the chance distribution from 50%; so teleconnection, in the sense of de Geer, does not exist.

Within Central Europe, however, one finds a somewhat surprising similarity. In particular, silver fir (Abies alba) gives curves for areas separated by more than 1000 km, e.g. from the Vosges to the Beskides, that can be easily synchronised (Becker and Giertz-Siebenlist, 1970). The similarity over distances of 1000 km exceeds that between fir and spruce in the same stand. Furthermore, stands of oak in the central hilly areas of Germany react so similarly to those west of the Rhine, that Hollstein's results for the latter and our results for trees from Spessart, Franconia and Hesse form almost identical mean curves for a period of over 1000 years (Fig. 3).

By contrast, research for several years at Reinbek/Hamburg has failed to provide a chronology from the extremely abundant oakwood finds on the artificial mounds used for early settlements in the North German lowlands (Weitland 1960). Liese and his colleagues (e.g. Bauch, Liese, Eckstein 1967) have established the causes of this failure. In those lowlands, slight changes of elevation and differences in soil decide whether a warm dry summer checks or advances the trees' growth.

Thus the question of the geographical range cannot be answered in a general way, but must be established in each case. It was fortunate that in constructing our chronology of over 1000 years we worked in an area where the samples react uniformly over a great distance. Success at an early stage gave the research the impetus which allowed a very difficult problem to be mastered.

VI. CLIMATIC ANALYSIS FROM TREE-RING CURVES: DENDROCLIMATOLOGY

The similarity of tree-ring curves in areas lying not too far apart confirms the opinion that tree-ring widths are largely climatically controlled and suggests the idea of a climatic analysis of ring-width fluctuations (dendroclimatology). Leonardo da Vinci wrote that in his area the annual ring width was wide in moist years and narrow in dry. Rather later and with similar good authority, Linne (1745/47) asserted that the contrary held in Sweden, namely that warm summers gave a wide ring, cold ones a narrow ring. Those different assertions make the point that in one region one factor is limiting; in another region, a different factor. In the North and near the Alpine tree-line it is warmth; in dry areas, it is moisture. In Central Europe, an area that is climatically-balanced, areas can be found where the influence of one or the other predominates; thus at the tree-line in the high-rainfall area of the Northern Alps it is warmth; but in xerothermic locations, such as the Würzburg 'wavy' limestone, it is moisture (Jazewitsch 1961).

Recently such correlations with climate have been expressed numerically. With skilful selection of location and with suitable growth periods (not a calendar year, but September to the next August, for example), one can obtain correlation coefficients of over 0.9. Schulman (1945) was able to reconstruct the water behaviour of the Colorado River over many centuries by using results from trees on steep slopes, whose growth was related only to the current rainfall. At the Arctic tree-limit, Hustich (1956) confirmed the strong correlation of the ring widths with summer warmth, and established that Scandinavia on the one hand, Labrador and Alaska on the other hand, rarely react similarly in the same year, but generally alternate. Latewood density, measured by soft X-rays (Polge, 1966, see this volume, page 77) is also related to summer warmth.

Gassner and Christiansen-Weniger (1942) found in Anatolia that the width of earlywood of black spruce was dependent on the rainfall of the current growth period, but that the latewood width depended on the moisture stored during the preceding winter. In light-loving species, such as larch, at the tree-limit in the Alps the entire radial increase can be a measure of the climate over a long period of time. Thus during the "little ice-age" of the seventeenth century, the increase fell to half in the Berchtesgaden area of Bavaria(Brehme, 1951).

For a single climatic factor to limit the annual ring width is unfortunately the exception, not the rule. In the balanced climate of Central Europe, warmth and moisture in individual months act partly with, partly against each other in a way that is difficult to disentangle. But the computer allowed Elling (1966) to account for 90% of the fluctuations of the ring widths in alder over a period of a century five monthly climatic factors. The correlation with two are shown in Fig. 4. Temperature, precipitation and other climatic factors partly promote growth, partly create stresses which hinder growth, the sum controlling the ring width.

So obvious is this result, and so compelling is it to the painful understanding that a ring width is determined by the sum of many factors, that neither man nor calculator can decide which, of the many single factors, have changed in actual cases. Pokorny (1867) considered the annual rings of a tree to be a true meteorological yearbook. We must confess that they are written in a code, which, apart from the special cases with a limiting factor, we have not learnt to decode.

Nevertheless I am convinced that dendrochronology is not at the end of its possibilities. As was stated in our report (Huber, Merz and Fürst, 1961) to the VIth International Quaternary Geology Congress in Warsaw, progress is likely to come first from "comparative" dendrochronology. By this we understand the systematic comparison of the mean curves of different species or of different areas. For example, fir responds to a severe winter by a marked fall in ring-width; while in spruce, the ring-width may fall after a year's dryness. For Colorado, Fritts (1965) ventured to calculate and map 10-year means of the atmospheric distribution from the relationships of the mean ring widths of different small areas. In Central Europe, comparison of the average fluctuations, as studied by Fürst (1963 and this volume page 65), and of the period length of tree-ring curves promises information on secular changes.

References

Those not listed at the end of the Huber Memorial Lecture (this volume, page 7) are as follows:

Bauch, J., Liese, W. and Eckstein, D. (1967). Holz als Roh-u. Werkstoff, 25, 285-91.

Fritts, H. C. (1965). Monthly Weather Review, 93, 421-43.

Gassner, G. and Christiansen-Weniger, F. (1942). Nova Acty Leopoldina N.F., 12, Nr. 80.

Hollstein, E. (1965). Bonner Jahrbücher, 165, 12-27.

Huber, B. (1935), Die physiologische Bedeutung der Ring- und Zerstreutporigkeit. Ber. Dtsch. Bot. Ges., 53, 711-19.

Hustich, L. (1956). Acta Geographica, 15, No. 2.

Linné, C. Oländska och Gothländska Resa. Stockholm und Upsala 1745. Westgöta Resa. Stockholm 1747.

Schulmann, E. (1945). Univ. Ariz. Bull., 16/4.

Suess, H. E. (1965). J. Geophys. Rev., 70, 5937.

Weitland, J. (1960). Diss. Hamburg.

1.4. OUR 1000-YEAR OAK ANNUAL-RING CHRONOLOGY

B. Huber and V. Giertz-Siebenlist

Translated extracts from Sitzungsberichten der Osterr. Akademie der Wissenschaften, 178 (1969).

Since Huber, v. Jazewitsch, John and Wellenhofer (1949) published their paper entitled Annual-ring chronology of the Spessart Oaks, the mean curve has been considerably strengthened and extended: the oldest recent Spessart oak available to them started with a striking '1421 peak', it has been found regularly since then. The 1530-1540 "saw signature" was shown, but not mentioned in the text because it would have been premature to state whether that sequence of rings was typical.

The chronology was extended by the Ziegenhain oak beams of Pastor Paulus (von Jazewitsch, 1954/55), and with Niess by the annual-ring chronology of Hessian oaks (Huber, Siebenlist and Niess, 1962/3; and Niess 1966). As typical ring-sequences of historic (and even prehistoric) buildings become matched and finally connected to absolutely dated recent sequences, one can produce annual-ring chronologies of increasing length....

A completely new stage of increased certainty was achieved when the sequences for individual samples could be combined into a framework of many parallel sequences. Now that the oak chronology of the last millennium has been covered 100-fold for the most part, a summary report may be made on the state of the research achieved. We might compare it with the position of the 19th century pioneers gradually reducing the blank areas on the map of Africa: a homogeneous phase has been produced out of the scattered phases of individual points. Covering it 100-fold in this way eliminates the earlier danger of having overlooked individual missing rings and makes the absolute dating reliable for each year.

The availability of such extensive material is still in the descriptive phase. To trace the ring-width variation back to its climatological causes (annual-ring climatology) is a further task and we are only at the very start of it in our area...

Parallel to our efforts, Hollstein of Trier has worked out an oak chronology of over 2000 years for West of the Rhine and has already published it in part (Holstein 1965). In its main features it accords with our chronology for East of the Rhine. At present these two chronologies are being tested for the dating of Cistercian buildings in Normandy by Horn and in England by Schove; it has already been established by Bauch, Leise and Eckstein (1967) that the oak chronologies of the North German plain deviate greatly.

The charts (Figs. 1 to 11) show the mean value of every ring width and the number of samples by which the particular year is covered, for ten centuries. When the direction of a line (falling or rising) agrees in at least 75% of the samples, it is shown as a wide line. The charts are self-explanatory, but some 'signatures' should be mentioned.

The 1421 peak has been found in 129 out of 150 (rise), and 145 out of 155 samples (fall). It is a most important aid in dating in the transition from recent to historic dates. The preceding three minima, 1311 (60 out of 61), 1320 and 1326 (51 out of 52) deserve mention.

After 1421, the indicator-years (Weiserjahre) are increasingly well covered: in Budingen the abundant material from the castle, the town and the church of St Remigius were synchronised by a 'double depression' which is dated to 1393/94; the minumum varies from 1393 to 1394. The best signature in the 1000-year sequence is the 1530-1540 "saw"; it consists of five alternate rises and falls in ring-widths and is covered well over a hundred times by our material east of the Rhine and by Hollstein's west of the Rhine. It is even indirectly traceable in Becker's silver fir chronology (1970). Up to the 1534/35 rise, all the fluctuations occur in over 85% of the samples.

The hardest winter of the millennium, 1739/40, caused decreasing ring widths and extensive 'dying' of oak (Hausendorff 1940) and lead, about 1745, to one of the smallest ring widths in the whole 1000-year sequence. The signatures of the last hundred years are taken from Elling's dissertation of 1962/3.

References

Bauch, J., Leise, W. and Eckstein, D., 1967. Holz als Roh- und Werkstoff 25, 285-291.

Hausendorff, W., 1940. Frostschaden an Eichen, aus der Geschichte des Forstamtes Grimmitz/Uckermark. Z. Forst-und Jagdwesen 72, 3-35.

Hollstein, E., 1965. Bonner Jahrbucher 165, 12-27.

Niess, W., 1966. Budinger Geschichtsblatter 6, 24-72.

For details of other references, see this volume, pages 7 to 10.

Ed. Note: The main area from which the samples for this chronology came is shown in the Map on page 32.

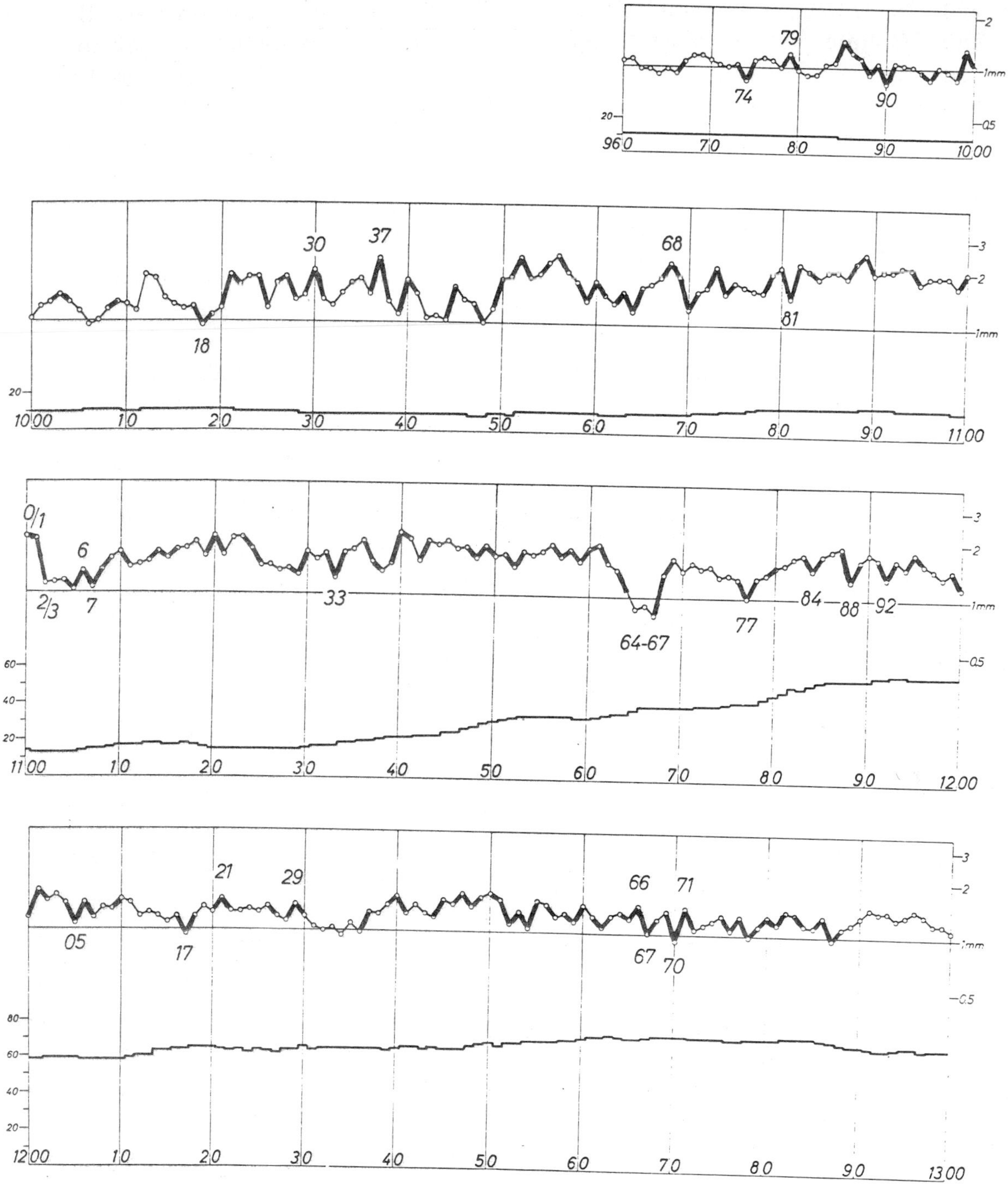

Fig. 1-4. 1000 year Oak Chronology. 960 - 1300

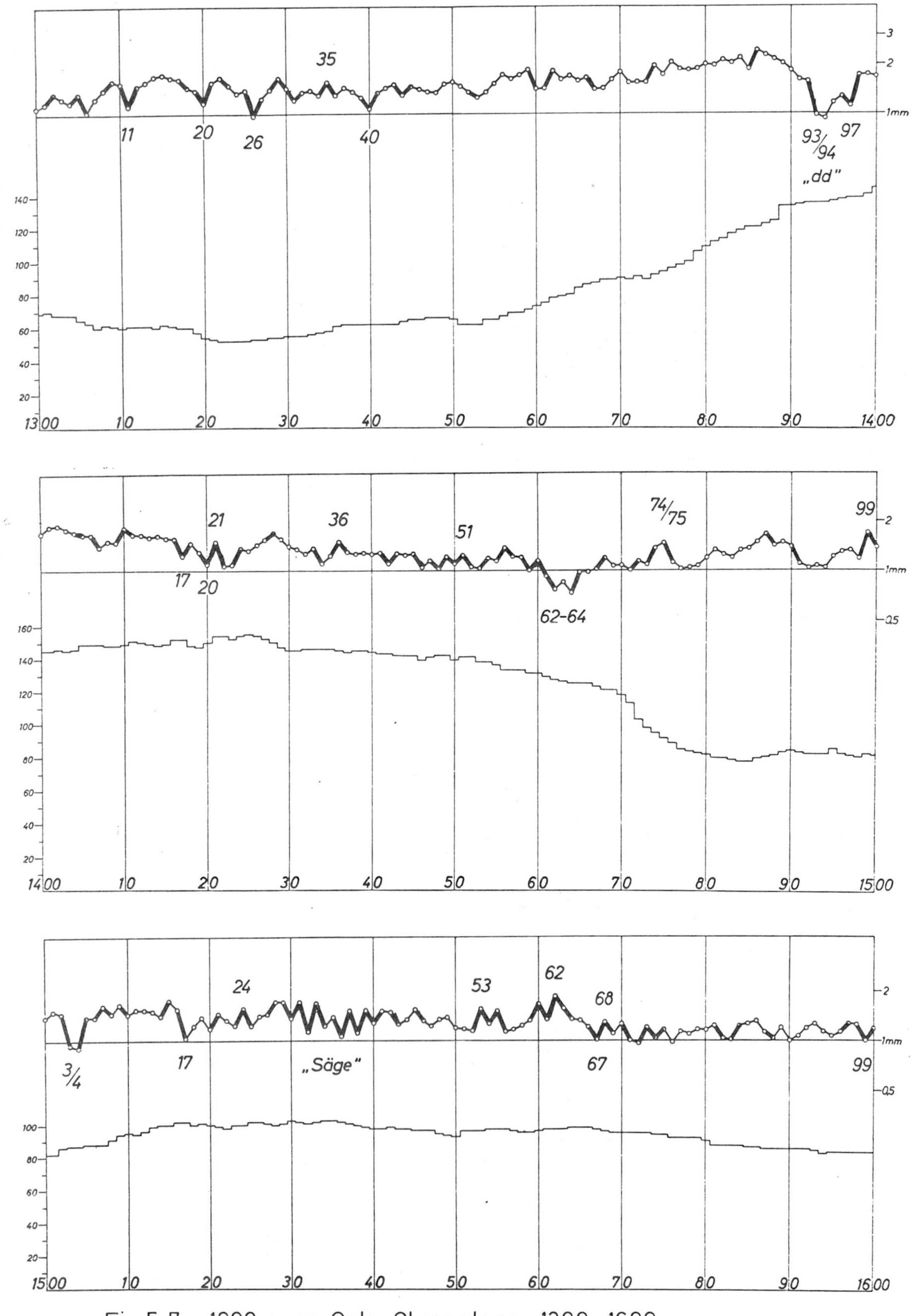

Fig.5-7. 1000 year Oak Chronology. 1300-1600

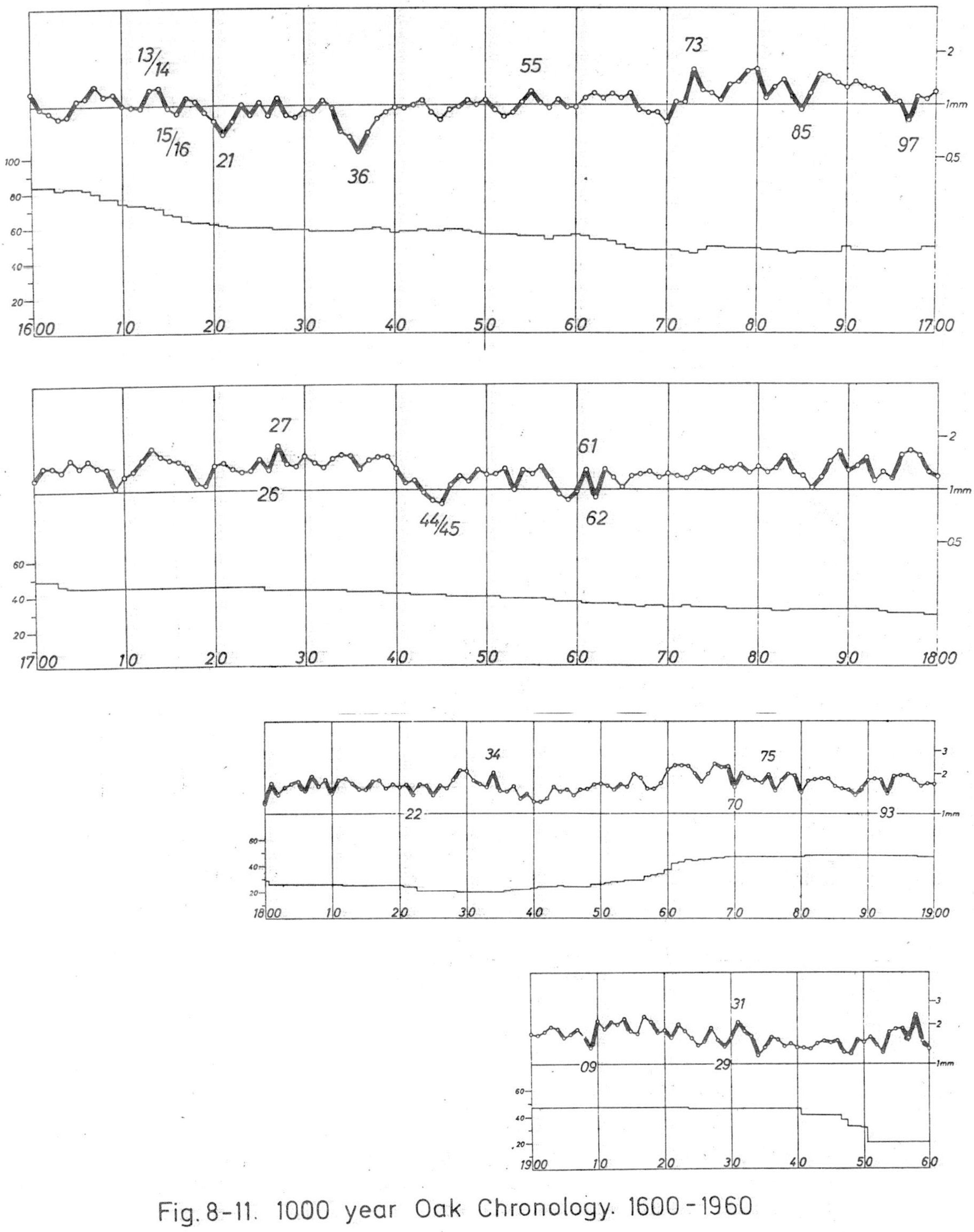

Fig. 8-11. 1000 year Oak Chronology. 1600 - 1960

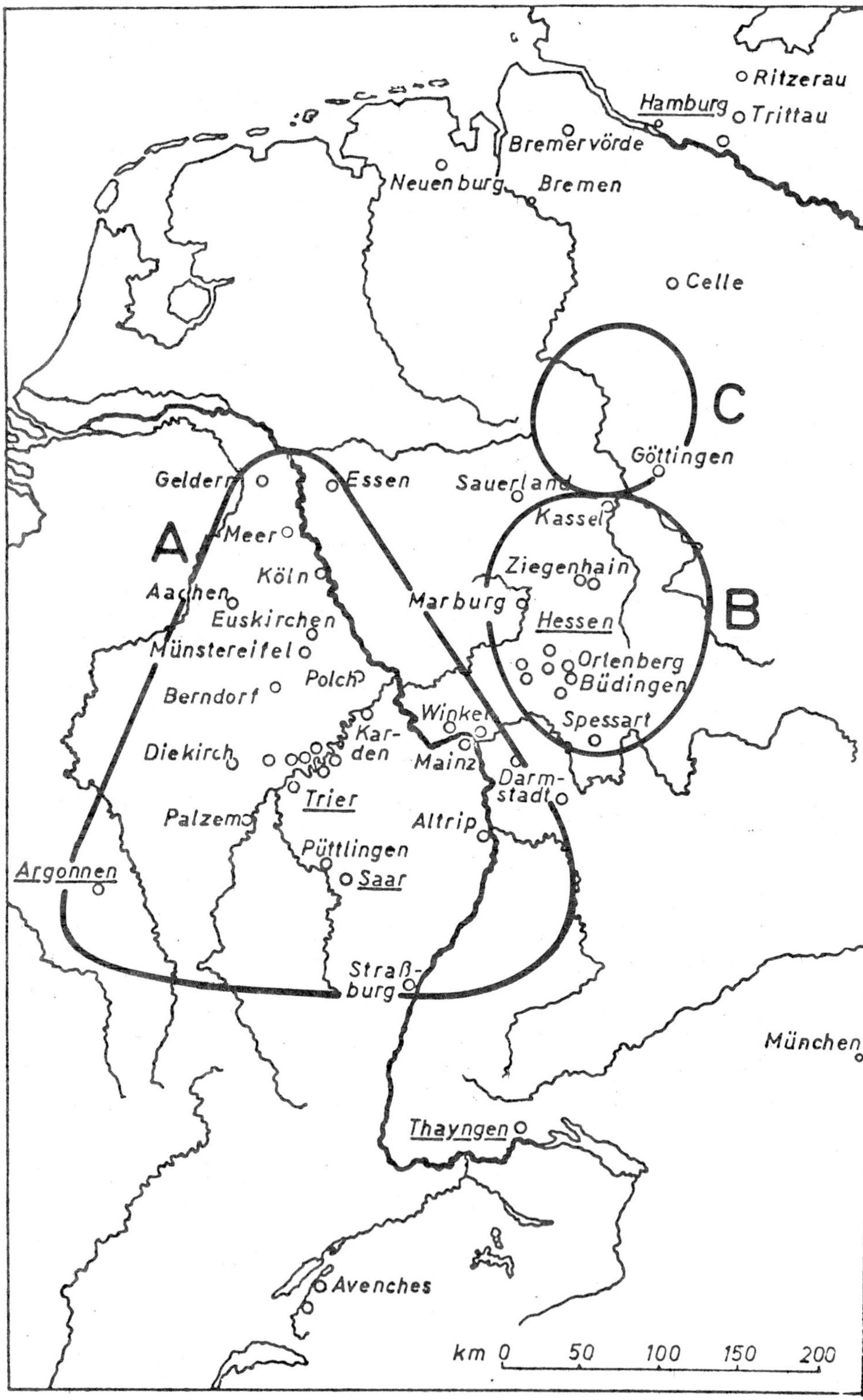

Map (after Hollstein). Origin of samples for Central German Oak Chronologies.

Area A Hollstein (1965). West of the Rhine

B Huber-Giertz (1969). Spessart and Hesse

C Delorme (1972). Weser and Leine Uplands

1.5. WOOD TECHNOLOGY AND THE DATING OF OAK: WEST GERMAN CHRONOLOGIES FOR OAK AND BEECH

E. Hollstein, Trier

Extracts from Translations by Ruth Morgan of papers in Bonner Jahrbücher, 165 (1965) 12-27; Germania, 45 (1967) 70-84; Kunstchronik, 21 (1968) 159-164 and 168-181; with additional illustrations from papers in Rheinische Heimatpflege, 4 (1965) 1-13 and in Mitt. Dendrol. Ges. 66 (1973) 165-172.

Summary

Careful excavation of archaeological timber has sometimes preserved the edge between the sapwood and bark. Observations on such edges show that in earlier times oak timber was usually felled in the spring or early summer. The absence of the cracks and the distortion that occurs on drying show that the wood was used soon after felling, that is while still 'green'. Fortunately this permits exact dating.

Where only the heartwood-sapwood boundary exists, the approximate date of felling can be estimated for oak by allowing for the number of sapwood rings. These, for the trunks of west German oaks have been found to be 20 ± 6 for 100-200 year old trees, and 26 ± 7 for trees over 200 years old.

The West German Oak Chronology has been formed, by Huber's methods, from over 200,000 ring-width measurements on samples (as cores or sawn pieces) from 80 sites. It spans the last 2,500 years and applies over a large area which includes Luxemburg, parts of France and of Switzerland, as well as western Germany. Its ring-widths for the period 822-1964 A.D. are listed (1965). An earlier part of the chronology is based on many samples from remains of Roman and Celtic bridges and from timber used in the salt-mines at Bad Nauheim. It consists of 856 annual rings and is attributed to the period 486 B.C. to 369 A.D.

The German beech chronology has been extended back to 1320 by ring-widths from samples from Trier Cathedral and buildings in the Saar and Moselle regions. Fir was also used in some of these buildings.

Unlike radiocarbon dating which is restricted in its accuracy by a relatively large standard deviation, tree-ring analysis, in principle, can date timber to an exact year if sufficient samples are available. Thousands of years are covered by the oak tree-ring chronologies in Europe. The leading founder and expert Prof. Dr. Bruno Huber together with his colleagues have placed tree-ring research at the service of archaeology, art history and climatology. Oaks of the Spessart Forest were used to form the Central German oak chronology. It covers hundreds of years and with it numerous art monuments in Hesse have been dated (Huber, Siebenlist and Niess, 1964).

Apart from cores or sawn pieces taken from living or recently felled trees, much of the material from which ring-widths were obtained came from the dendrochronology collection in the custody of the archaeologist Dr. Heinz Cuppers at the Rhine Landesmuseum at Trier. This consists of hundreds of oak samples from the Rhineland region between Strasburg and Geldern (see Map). The research has also been helped by the Rhine Landesmuseums of Bonn and Mainz; by the Romano-German museums of Cologne and Mainz; by the Landeskonservatoren of the Federal Lands; and by the dioceses of Aachen, Cologne and Trier.

Oakwood samples with wood edge

The felling year of a wood sample can be determined dendrochronologically, if the last annual increment below the bark is present and measurable. Forest workers and carpenters call this conspicuous point in an aged beam "waney edge". Construction wood which shows the wood edge is distinguishable today from the so-called "sharp-edged" beams. When collecting samples of wood (Niess, 1963; Hollstein, 1964) from historical buildings, one can choose the best places where the wood edge has not fallen victim to the carpenter's axe. Helpers can recognise with practice the suspected wood edge on the cross-sections of beams and piles. In this way Cüppers, in his investigation of the Roman bridge at Trier, rescued numerous piles with their entire wood boundaries. Thanks to this care we are in a position to distinguish the different building periods of the individual bridges, even of the individual piles, to the exact year!

When the sapwood rings of a wood sample are not too decomposed by decay or eaten away by woodworm, the wood edge can be identified under the microscope (Plate 1). In many cases, the remainder of the bast and entire bark fibres are preserved. The outer part of the cross-section of oak, the so-called sapwood, is soft after centuries of immersion in water (Noack, 1965), so that drying causes extreme shrinkage and distortion of the wood structure. The annual ring widths cannot then be measured with the required accuracy and the wood edge is not always identifiable with certainty. The samples are therefore best analysed in damp conditions.

From the wood boundaries of important historical finds, durable sections were prepared as evidence. By curve comparison on a light-table, it has been shown again and again that wood from the same object or building context was felled in the same year. Such samples often allow the estimation of the time of year when the tree was felled. For example in the timber from Cologne Cathedral (Plate 1) the rings for the year 1311 had just started to form. The tree from which it came was felled in the spring, probably May. Similar observations on uninjured wood edges indicate that the main felling time in antiquity and the middle ages lay (contrary to present custom) not in the winter but in spring and early summer.

Oakwood Samples with Sapwood

Despite care, it is not always possible to preserve the wood edge uninjured. Often the sapwood was already removed by the carpenter as useless, and the remainder left to the destruction of insects, fungi and bacteria, while the resistant heartwood—particularly under water—can last for a thousand years. Sapwood is also sometimes lost after rescue and subsequent drying. We first observed this when investigating old museum stock that had not been waxed. The tree-ring researcher who hopes for exact dating endeavours to recognise at least the sapwood boundary, i.e. the transition zone between the dark firm heart-

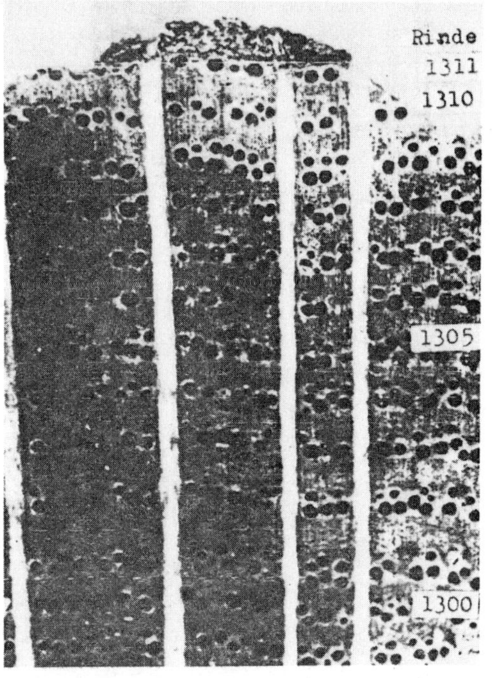

Plate 1. Oak timber from south side of choir of Cologne Cathedral. Bark (Rinde) present. As formation of the ring for 1311 has only just started, the tree was felled in the spring.

wood and the light soft sapwood. He can then estimate the rings lost as sapwood, whereas, in a wood sample without any remaining sapwood, the number of rings is unknown. Indeed among Roman piles several were found in which over 100 annual rings had been lost through treatment or decay. Thus wood samples with a complete wood edge or at least recognizable remaining sapwood are particularly important.

Differences that have been noticed between heartwood and sapwood of oak are:-

(a) The colour; oak sapwood is pale, the heartwood vivid brown, or blackish to deep black after long immersion in water.

(b) The open vessels in the sapwood; tyloses occur in the heartwood, i.e. fine membranes clog the vessels (pores).

(c) The thickness of the sapwood.

The first two are evident with certainty in historic samples, but the third is not. In its place, there is another feature on the basis of our experience. Namely with beams and round wood whose sapwood has been lost, not through human treatment but through insects or decay, the last annual ring of heartwood is preserved around the full circumference or the greater part of the cross-section. One recognises in the regular roundness that the transient sapwood is absent, while the tough heartwood is fully preserved. The change to heartwood does not proceed "annually" and so the boundary line between heartwood and sapwood rarely coincides exactly with one and the same annual ring. In such a case the sapwood boundary can be described as "probable". It is often supported by parallel samples from similar objects having their sapwood boundaries in the same position on the tree-ring curve.

Analysis at Trier has shown that identification of sapwood by pale colour can be deceptive. We possess oak samples in which the sapwood is a darker colour than the heartwood. On the other hand, it sometimes happens that the edge of a cross-section has changed to a paler colour due to lime or other chemical influence and thus appears to be sapwood. However, this is an exception. As a rule the sapwood boundary can be recognised with certainty.

The number of sapwood rings in oak

The estimation of absent sapwood rings is a statistical problem. The number of sapwood rings interests the forester and user of timber less than the width of the sapwood, and for oak the values given by different authors have mainly been based on a single tree. From the abundant historical oakwood collected at Trier, counts on close to 200 trees have been obtained. As Table 1 shows, the values for Anatolia published by Gürsu and Bernhart (1964) correlate well with the Trier results, despite the great distance apart, and the number increases with the age of the tree as noted by various authors.

For the W. German trees about 90% had 12 to 30 sapwood years, while the extreme values (below 10 and above 40 years) are very scarce, perhaps 1% of the total. When wood samples are sent for tree-ring analysis, information on the felling year and perhaps the construction year is expected. When the wood edge is absent, the tree-ring analyst is faced with the alternative of reporting only the date of the last annual ring measured or indicating the probable period of the felling year. With single samples this is particularly delicate, because one may be concerned with a chance repair or reused timber. In addition there is the large overall variation in the number of oak sapwood rings (say 7 to 46 years) if one attempts to be 100% certain. So we are reluctant to date a single sample and point out that a successful match gives no

Table 1

Number of Oak Sapwood Years

	Number of sapwood rings	
Age Group	W. Germany	Anatolia
Less than 100 years	16.0 ± 4.5	-
100 to 200 years	20.4 ± 6.2	20.6 ± 4.1
Over 200 years	25.9 ± 7.5	29.4 ± 7.1

guarantee of the entire construction. The same restriction applies to the dating of single samples by radiocarbon, but in even stronger measure.

The certainty of judgement increases with the number of samples from the same object. By tree-ring analysis of a sufficiently large number of randomly selected samples in a uniform building, we find an accumulation of the heartwood/sapwood boundary in a period for which the standard deviation is about ± 6 years from the mean. This leads to the conclusion that all the wood was felled in the same year, the year calculated from the appropriate value in Table 1 or for comparable values for other areas. The following conditions must be fulfilled:

1. All the wood collected for a tree-ring curve must have been used in a uniform construction, e.g. as beams of a storey, piles of a bridge pier.
2. The sapwood boundary must accumulate at a mean value, with a mean square deviation at the most of ± 6 years; that means that perhaps two-thirds of all sapwood boundary years must lie within a time span of less than 12 years.
3. No sapwood boundary year may be less than 7 or more than 46 years away from the calculated felling year.

If only one of these assumptions is unfulfilled, one cannot assume a uniform felling year.

The calculated felling year lies within a period whose boundaries are set (i) by the frequency distribution of the sapwood years, and (ii) by the quantity of dated samples. With only one sample, the felling year, in 90 out of 100 cases for oak aged 100-200 years, lies 12 to 31 years from the sapwood boundary. If one claims greater certainty one must extend the probability area: e.g. for 95% certainty, from 11 to 34 sapwood years. The more the samples from the same object, the more the probability is increased. Thus from 8 parallel samples, the felling year lies, with 90% certainty, 17 to 22 years from the mean year of the dated sapwood boundary.

Relation between Felling Year and Use

This question has been of particular interest to me. Seasoned wood would be pointless for structures such as piles below water. Maintained forests and woodland arrangements included strict rules that felled wood for building must be used at once, e.g. "The authorised must use the wood for building within one year, otherwise it will be removed from him" (Bauer, 1964). This is verified when tree-ring dating can be checked by documents. Thus I dated beams from a timber-framed house in Mönchen Gladbach to the end of 1606, or the winter 1606/7, without knowing that the house inscription on the door lintel read "Anno 1607".

Wood Technology: That construction wood was usually used at once after felling can be deduced from an examination of its cross-section. In wood used under water—without exception split trunks—the characteristic dryness cracks which develop when wood has been in store for some time after felling

Plate 2. Rhombic distortion of a beam originally right-angled. The angle of deviation proves the beam was used when freshly felled. Sapwood/bark edge at top left for 1779. From Filsch Church, Trier, consecrated 1780 (R. Summa, Trier).

Plate 3. Floor-board from house in Münstereifel with sapwood edge for the year 1167.

are normally absent. Foundations and building piles must have been rammed in while still moist, as in the Roman bridge at Trier of 44 A.D. A few dryness cracks have been recognised by microscopic analysis in the pile-framework in a river-bed since they became filled with sand or mud. If the wood edge was preserved in such cases, dendrochronology showed that the tree was felled one or more years before the others on the site.

Building timbers allow another observation. Beams made with a right-angled cross-section change to being oblique-angled on drying, so those of this shape must have been in a moist, freshly felled state when used. This occurred in the heavy roof beams of 1108 in the Basilika of Münstereifel and for the wood of the church at Filsch of 1780 (Plate 2).

While wooden surfaces that are to be fine planed need to be dry, surfaces that are to be adzed or cut can be in a moist condition. Moist wood, when cut across the fibres, preserves vessel-sharp surfaces—as was found in the ends of the floor boards (Plate 3), cut across the grain, in a Romanesque house in Münstereifel. They explain why the felling-year (1167) is contemporary with the building year.

The above evidence proves the use of wood directly after felling, i.e. dendrochronological dates can as a rule be considered as building dates. It follows that in antiquity and in the Middle Ages construction wood was principally carpentered in a fresh condition, contrary to present building requirements. The reason is certainly to be found in the predominance of fashioning by axe—broad axe, thrust axe and adze being the most important tools for thousands of years and they carve best in moist wood.

The custom with wood for cabinet making was different. Seasoning was needed. The study of furniture, panelling, windows, and doors shows a variable interval between the date from tree-ring analysis and the date the object was made. This also applies to the study of wood sculpture.

In my experience reused wood is rare (it was found only twice) and can be recognised either by the presence of empty peg-holes or now by tree-ring analysis.

THE WEST GERMAN OAK CHRONOLOGY

Our work on this chronology began in 1961. Samples from trees felled recently in Alsace, the Argonne Forest, the Saar, Luxemburg and the North Eifel formed the pattern for recent times. Fig. 1 shows that the pattern for the period 1890 to 1950 is very similar to that for the oaks of the mid-Rhine area of Hesse that forms the main constituent of the Central German Oak Chronology. The same figure also shows the good agreement with oaks grown in North Switzerland and the limited agreement with the pattern derived by Weitland (1960) from 64 oaks for the lowland area in North Germany.

Long-lived oaks have formed the basis for earlier parts of the chronology. Their large number of rings (up to 400) have greatly assisted bridging. Synchronisation is good between the separate mean curves for areas west of the Rhine and in turn between them and mean curves for areas east of the Rhine such as the Odenwald, Spessart, Hesse and even the Wesergebirge. For the

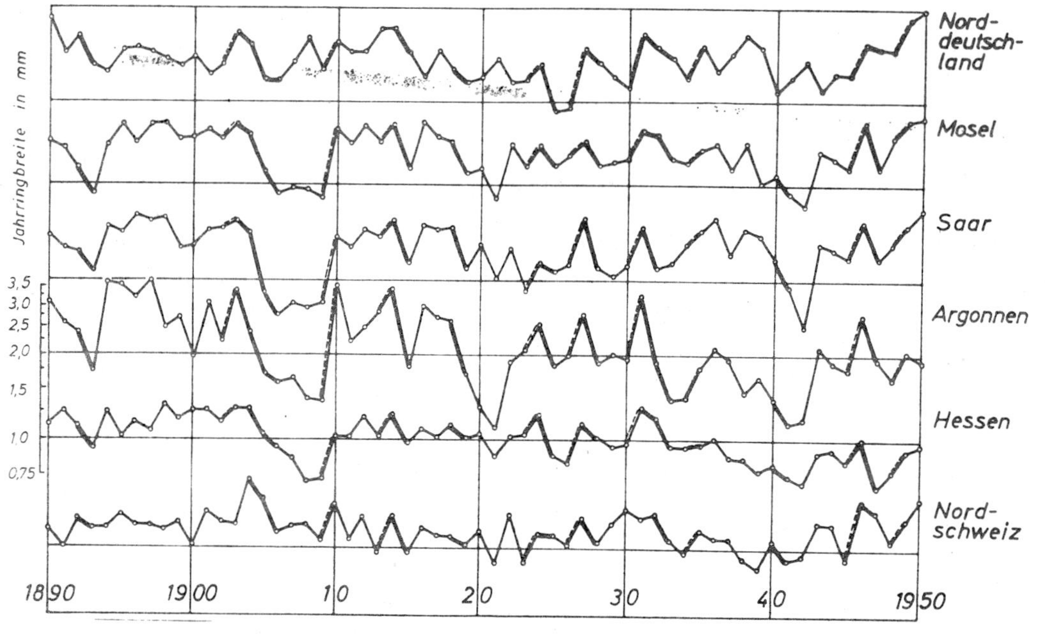

Fig. 1 Ring-Width Curve for Recent Oaks. Logarithmic scale on left applies both to oaks from Hesse and from the Argonne Forest. Thicker connecting lines apply to Indicator Years.

period 1005 to 1950 the "disagreement value" is 24% between the West and Central German chronologies. Important signatures such as the 1530-40 "saw" and the deep depression from 1164 to 1167 are common to both.

The chronology back to 822 A.D. (ring-widths published as Table 4 in Bonner Jahrbücher, 1965) is based on the widths of about 14,000 annual rings, The extension into the 12th, 11th, 10th and 9th centuries was made possible by 48 samples from Romanesque buildings (Fig. 2). In this part of the chronology, the years 874, 883, 905 and 910 showed reproducible minima, while the wide ring for 942 and the subsequent trough up to 946/7 forms an important signature.

The 400 year old oaks of the first Roman bridge at Trier together with other Roman and Celtic bridges and material from the salt-mines of Bad Nauheim (oldest felling date 252 B.C.) form a separate part of the chronology, 856 years long, starting in early La Tène times. On the assumption that the last ring from eight heavy piles from the Roman bridge over the Rhone at Cologne belongs to the year 310 A.D., the chronology overall spans the years 486 B.C. to 369 A.D. It has been derived from 270 samples on which over 200,000 ring-width measurements were made. (Ed. The parts for 244 to 144 B.C. and for 54 B.C. to 47 A.D. appear in Figs. 2 and 3 of the 1967 paper in Germania, 45).

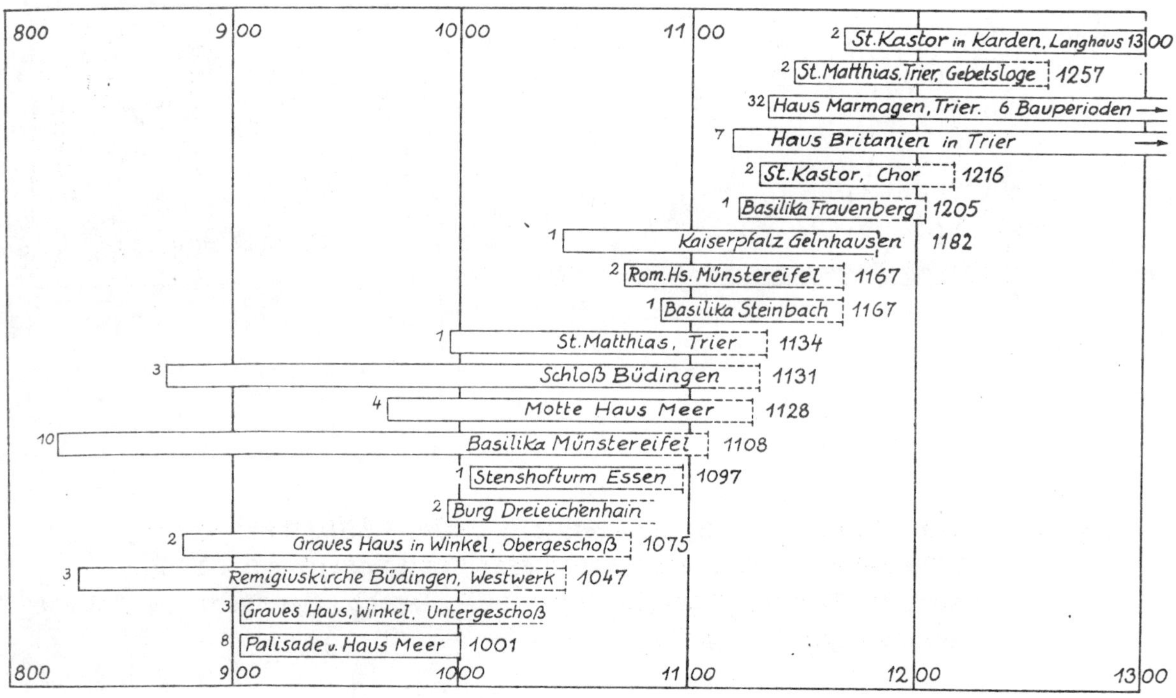

Fig. 2 Bar Chart showing Origin of Samples (numbers at left) for West German Chronology 822-1300 A.D. Felling Year derived from Bark ▭ ; from sapwood/heartwood boundary ⌐ ‐ ‐ ┐

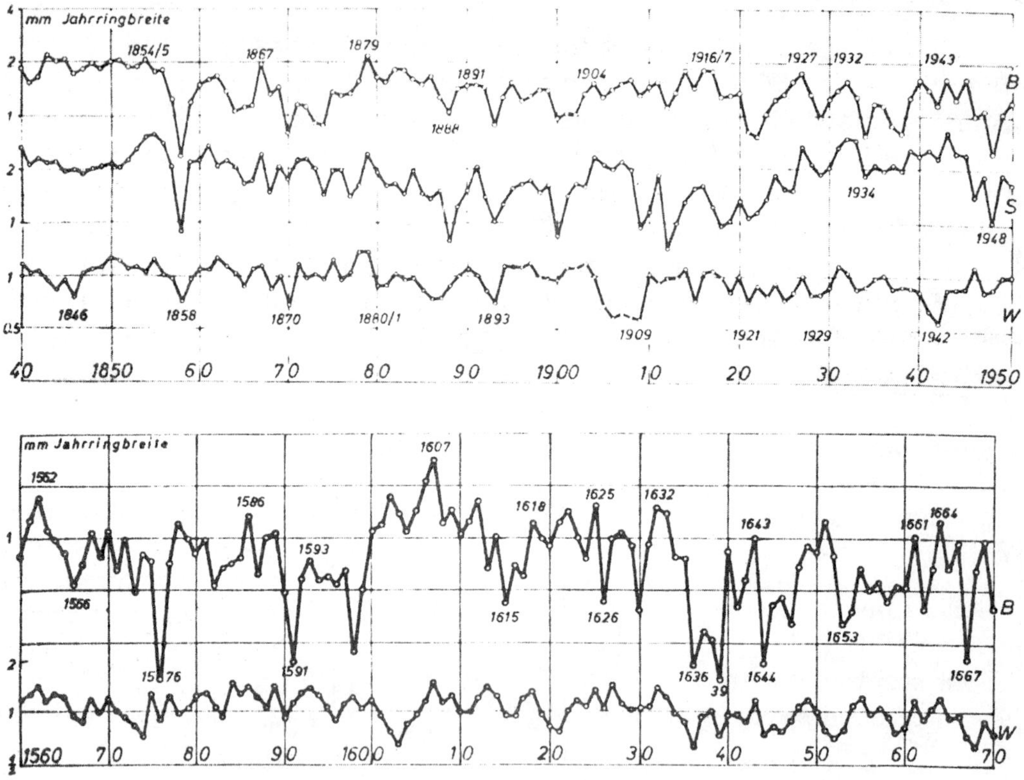

Fig. 3 Ring-Width Curves, logarithmic scale. 1840-1950 above, 1560-1780 below. Notice common signature years, e.g. 1858. Key, B, Beech, Moselle region; S, Beech, Spessart; W, West German Oak Chronology.

Sites that have supplied material for this chronology extend beyond western Germany to Switzerland, France and Luxemburg (Fig. 1). It is apparent that it applies over a large region that covers all the areas adjacent to the Rhine with the exception of the Northern Lowlands. It will probably be advantageous, however, to have a separate curve for oaks of the Lower Rhine in order to improve the transition to the North German lowlands. Work on this has started (Kunstchronik, 1968) with material from:

(a) The Motte Burg Meer in Grevenbroich, 1001;
(b) The Motte Husterknupp, 964;
(c) The Burg Uda in Oedt, 1308;
(d) The abbey of Kornelimünster, various dates;
(e) Aachen Townhall, 1330;
(f) Choir stalls of Cologne cathedral, 1308/1311;
(g) The romanesque Stenshof tower in Essen, 1100.

Their tree-ring curves have been synchronised with those of the Moselle and Hesse, yet they show characteristic disagreements in particular years.

OTHER CHRONOLOGIES

Red Beech: A chronology based on samples from the Saar and Moselle has enabled a chronology back to 1320 to be constructed. The crucial bridging from recent to historical material was made from results on thin split planks with which the roof of the modest village church of Filsch (near Trier) was covered. The felling year of the beech and oak was found to be 1779/80; the building date over the door was 1780! Plate 4 shows part of a polished section from one of the planks. The surface shows one place where the ring for 1639 is virtually absent.

This beech chronology shows good agreement (Fig. 3) with that derived (back to 1684 A.D.) by von Jazewitsch (1953) for beech from the Spessart Forest and the Bavarian Forest. It will be noticed from Fig. 4 that there are, however, considerable differences between the patterns for oak and for beech.

Fir: Although the mid-Rhine is now to the north of the natural growth area for *Abies alba*, in earlier times fir was both used for piles and carpentered for use in buildings. Examples from Roman and early medieval times, mostly of the same felling year as oak carpentered with it, are related to the following centuries:

2nd	Support wall of the Roman bridge at Trier, 13 samples
2nd	Mainz excavation "Am Brand", 6 samples
3rd	Trier, Palace Hall, 2 samples
4th	Trier cathedral, 7 samples
6th	Büttelborn, excavation, 1 sample
11th	Speyer cathedral, 1 sample

The ring-width sequence for the period 970 to 1047 is well matched by that derived by von Hornstein (1964/5) for 10 samples of fir from the Minster at Constance (Fig. 4).

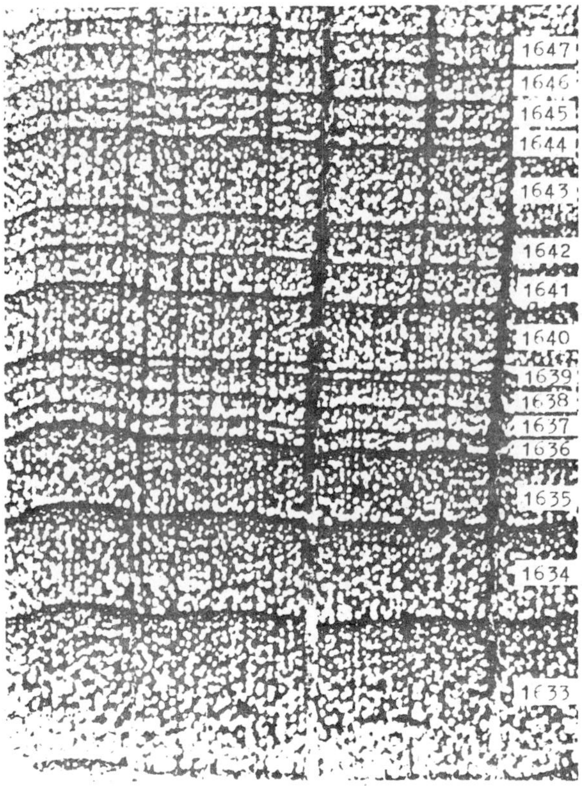

Plate 4 Red beech from Filsch Church 1779/80. Notice the extremely narrow width of the ring for 1639 on the left side.

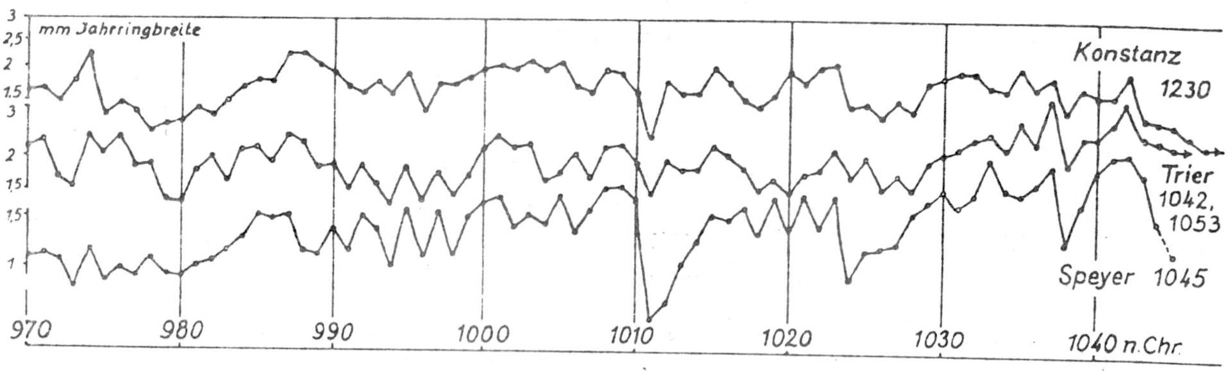

Fig. 4 Ring-Width Curves for Fir, 970-1047 A.D.
For Constance Minster, 10 samples; von Hornstein
For Trier Cathedral, 33 samples; Hollstein

References

Those not listed at the end of the Huber Memorial Lecture (this volume, page 7) are as follows:

Bauer, E. 1964. Trierischer Volksfreund 12, 5-8.

Gürsu, I. and Bernhart, A. 1964. Forstw. Centralblatt 83, 222-247.

Hollstein, E. 1964. Neues Trierisches Jahrb., 26-45.

Niess, W. 1963. Allg. Forstzeitschr. 18, 375.

Noack, D. 1965. Bremisches Jahrb. 50, 20-49.

Weitland, J. 1960. Diss. Hamburg.

1.6. A MEAN CURVE FOR OAKS OF THE SOUTHERN WESER AND LEINE UPLANDS: ITS USEFULNESS AND CHARACTERISTICS

A. Delorme, Göttingen

Extracts from (1) Translation by Wendy Bell (Climatic Research Unit, Norwich) of Dissertation, Forestry Faculty, University of Göttingen, 1972, and (2) from Forstwissenschaftliches Centralblatt 92 (1973), 335-342.

Summary

519 samples of oak of recent or historic origin have been dated and used to form a mean curve from 1004 to 1970 for which the agreement with curves for neighbouring areas has been calculated. Its usefulness has been examined and its qualities (age-trend, mean fluctuation, relation to pith-age and frequency of indicator years) assessed. For forming a mean curve from these oaks, 10 samples are considered sufficient, 30 samples good and 60 optional.

While Huber's oak chronology is applicable to the whole of south Germany, the north is not an homogeneous area for dendrochronology. The part of southern Lower Saxony which includes the Weser and Leine uplands, is a transition zone between the two (Map, page 32), so it seemed advisable to create its own chronology. The Weser upland itself is defined in the Grosses Brockhaus as "a mountain and hill region between Hanoverian Münden and Minden. It is bordered in the east by the Leine valley; in the south it merges with the Hessian upland and in the S.W. with the Rhine Massif. Towards the N and N.W. it forms by far the most prominent range of hills in the N.W. German plain". The chronology also incorporates results for samples from the Leine upland (from the forests of Reinhard and Kaufungen) as well as from N. Hesse and Westphalia.

Further aims were to use the standard mean curve for methodological investigation, and to date archaeological and architectural objects.

Material and Method

A total of 594 different pieces of oak were analysed and from 519 of them 81,000 annual rings were dated. 75 samples, that is about 13%, could not be dated.

Sixty recent samples were collected from Escherode, Münden, Holzhausen, Göttingen, Reinhausen, Bovenden, Sudershausen and Fürstenburg. These sites ensure that the geographical area represented by the historical samples is also covered by the recent samples. Particular care was taken when selecting the tree-stumps.

The historical material (obtained by boring) comes from an area which is bordered by Oberkaufungen in the south, Warburg in the west and Brunswick in the north-east. A large number of pieces of wood from Brunswick and nearby Wendeburg were undatable. Particularly numerous samples originate from Münden, Warburg, Hemeln and Helmarshausen.

Ring widths were measured in 1/100 mm and, in accordance with normal procedure, graphic representation was made on transparent half-log paper. This method (Huber, 1941) has the advantage that relatively similar ring

	Weserbergland 1004-1970	Ziegenhain 1225-1950	Süddeutschland 832-1963	Westdeutschland 822-1964	Wienhausen I 1831-1966	Wienhausen II 1331-1546	Wienhausen III 987-1321	Vierlande 1338-1966	Weitland A 1788-1956	Weitland B 1767-1956	gewogenes Mittel
Weserbergland 1004-1970		1225-1950 GL = 77,2 R = .561 Rho = .601 t = .82	1004-1963 GL = 76,5 R = .417 Rho = .427 t = .82	1004-1964 GL = 71,1 R = .097 Rho = .098 t = .77	1831-1966 GL = 61,5 R = .379 Rho = .438 t = .41	1331-1546 GL = 68,4 R = .262 Rho = .282 t = .57	1004-1321 GL = 63,7 R = -.249 Rho = -.273 t = .55	1338-1966 GL = 63,5 R = .277 Rho = .247 t = .58	1788-1956 GL = 68,5 R = .284 Rho = .263 t = .67	1767-1956 GL = 62,7 R = .137 Rho = .132 t = .56	GL = 70,8 R = .274 Rho = .279 t = .71
Ziegenhain 1225-1950	1225-1950 GL = 77,2 R = .561 Rho = .601 t = .82		1225-1950 GL = 78,0 R = .763 Rho = .779 t = .85	1225-1950 GL = 67,2 R = .121 Rho = .122 t = .62	1831-1950 GL = 59,7 R = .089 Rho = .043 t = .40	1331-1546 GL = 63,7 R = .419 Rho = .475 t = .64	1225-1321 GL = 62,5 R = .117 Rho = .106 t = .65	1338-1950 GL = 64,0 R = -.024 Rho = .006 t = .53	1788-1950 GL = 67,3 R = -.214 Rho = -.265 t = .50	1767-1950 GL = 60,7 R = -.214 Rho = -.207 t = .45	GL = 70,0 R = .307 Rho = .327 t = .67
Süddeutschland 832-1963	1004-1963 GL = 76,5 R = .417 Rho = .427 t = .82	1225-1950 GL = 78,0 R = .763 Rho = .779 t = .85		832-1963 GL = 75,6 R = .509 Rho = .489 t = .89	1831-1963 GL = 59,8 R = .339 Rho = .393 t = .40	1331-1546 GL = 64,2 R = .449 Rho = .475 t = .52	987-1321 GL = 65,7 R = .038 Rho = .208 t = .44	1338-1963 GL = 63,3 R = .142 Rho = .112 t = .52	1788-1956 GL = 59,8 R = .005 Rho = .025 t = .45	1767-1956 GL = 58,5 R = -.073 Rho = -.070 t = .36	GL = 70,5 R = .360 Rho = .380 t = .71
Westdeutschland 822-1964	1004-1964 GL = 71,1 R = .097 Rho = .098 t = .77	1225-1950 GL = 67,2 R = .121 Rho = .122 t = .62	832-1963 GL = 75,6 R = .509 Rho = .489 t = .89		1831-1964 GL = 51,1 R = .164 Rho = .218 t = .24	1331-1546 GL = 61,9 R = -.027 Rho = -.019 t = .39	987-1321 GL = 60,8 R = .070 Rho = .173 t = .36	1338-1964 GL = 60,9 R = .256 Rho = .254 t = .34	1788-1956 GL = 57,4 R = .266 Rho = .233 t = .37	1767-1956 GL = 55,8 R = .194 Rho = .169 t = .16	GL = 67,2 R = .234 Rho = .236 t = .61
Wienhausen I 1831-1966	1831-1966 GL = 61,5 R = .379 Rho = .438 t = .41	1831-1950 GL = 59,7 R = .089 Rho = .043 t = .40	1831-1963 GL = 59,8 R = .339 Rho = .393 t = .40	1831-1964 GL = 51,1 R = .164 Rho = .218 t = .24				1831-1966 GL = 58,5 R = .059 Rho = .125 t = .43	1831-1956 GL = 57,2 R = .134 Rho = .187 t = .32	1767-1956 GL = 54,8 R = .227 Rho = .197 t = .31	GL = 59,7 R = .228 Rho = .303 t = .36
Wienhausen II 1331-1546	1331-1546 GL = 68,4 R = .262 Rho = .282 t = .57	1331-1546 GL = 63,7 R = .419 Rho = .475 t = .64	1331-1546 GL = 64,2 R = .449 Rho = .475 t = .52	1331-1546 GL = 61,9 R = -.027 Rho = -.019 t = .39				1338-1546 GL = 61,1 R = -.226 Rho = -.238 t = .51			GL = 63,9 R = .178 Rho = .199 t = .53
Wienhausen III 987-1321	1004-1321 GL = 63,7 R = -.249 Rho = -.273 t = .55	1225-1321 GL = 62,5 R = .117 Rho = .106 t = .65	987-1321 GL = 65,7 R = .038 Rho = .208 t = .44	987-1321 GL = 60,8 R = .070 Rho = .173 t = .36							GL = 63,3 R = -.029 Rho = .047 t = .47
Vierlande 1338-1966	1338-1966 GL = 63,5 R = .277 Rho = .247 t = .58	1338-1950 GL = 64,0 R = -.024 Rho = .006 t = .53	1338-1963 GL = 63,3 R = .142 Rho = .112 t = .52	1338-1964 GL = 60,9 R = .256 Rho = .254 t = .37	1831-1966 GL = 58,5 R = .059 Rho = .125 t = .43	1338-1546 GL = 61,1 R = -.226 Rho = -.238 t = .51			1788-1956 GL = 74,1 R = .640 Rho = .621 t = .78	1767-1956 GL = 67,5 R = .479 Rho = .483 t = .71	GL = 63,9 R = .200 Rho = .195 t = .52
Weitland A 1788-1956	1788-1956 GL = 68,5 R = .284 Rho = .263 t = .67	1788-1950 GL = 67,3 R = -.281 Rho = -.265 t = .50	1788-1956 GL = 59,8 R = .005 Rho = .025 t = .45	1788-1956 GL = 57,4 R = .266 Rho = .233 t = .37	1831-1956 GL = 57,2 R = .134 Rho = .187 t = .32			1788-1956 GL = 74,1 R = .640 Rho = .621 t = .78		1788-1956 GL = 75,6 R = .623 Rho = .605 t = .82	GL = 66,4 R = .334 Rho = .325 t = .57
Weitland B 1767-1956	1767-1956 GL = 62,7 R = .137 Rho = .132 t = .56	1767-1950 GL = 60,7 R = -.214 Rho = -.207 t = .45	1767-1956 GL = 58,5 R = -.073 Rho = -.070 t = .36	1767-1956 GL = 55,8 R = .194 Rho = .169 t = .16	1831-1956 GL = 54,8 R = .227 Rho = .197 t = .31			1767-1956 GL = 67,5 R = .479 Rho = .483 t = .71	1788-1956 GL = 75,6 R = .623 Rho = .605 t = .82		GL = 62,7 R = .253 Rho = .244 t = .48

Table: Agreement Values between the Weser Upland Oak Chronology (1004 - 1970) and other German Mean Curves for oak. For symbols see text.

width fluctuations can be represented by equal distances independently of the absolute ring width. Both wide and narrow ring samples are easier to compare when on semi-log paper and the 'indicator' value of minima becomes optically prominent.

Synchronisation and Averaging

Dating was done by Eckstein's method (1969) of calculating the agreement value which must be exceeded to keep the error probability under 5%, 1% or 0.1%. Some of the samples were dated immediately by comparison with Huber's South German standard curve, others only relatively to one another in the first place and subsequently via their mean curve by the South German curve. As the work progressed, the dated curves were formed into an independent mean curve with which most of the remaining undated samples could be synchronised.

The FORTRAN computer programme AMIT facilitated the continuous extrapolation of the mean curve. It can draw dated individual curves, print the mean values, the appropriate absolute years, and the number of samples involved.

The year-to-year fluctuations of a mean curve are less pronounced than those of a single curve. The more the samples in the mean curve, the more the fluctuations that reflect the individual process of growth determined by the general climate are equalised. Mean curves from a region that is climatically homogeneous are alike and enable a shorter length to be used for synchronisation.

For the Weser and Leine upland, the mean curve, derived by taking arithmetic means from the 519 dated pieces of oak, extends from 1004 till 1970. On average there are 83 components for each year, the maximum being 226.

Indicator years: According to Eckstein's definition, indicator years are those in which deflection occurs in the same direction such that coincidental agreement can be disregarded with 99% confidence. Changes in direction with statistical confidence of 99%, 95% and 90% were determined and are given, with the values for the mean curve, in the appendix (not reproduced here). For the period 1300 to 1968, a period well covered by samples (from 17 to 226 times), 54% are indicator years with 99 .9% confidence, and 66% with 99% confidence.

SIMILARITY BETWEEN THIS AND OTHER GERMAN CHRONOLOGIES

The ten German oak chronologies listed below were compared (Table 1) by Eckstein's percentage of agreement (GL), by their rank coefficient (RHO) after Spearman, by their correlation coefficient (R), and by their trend coefficient (t) after Glock (1941).

Chronology	Author	Years	Coverage
Weser Upland	Delorme, 1972	1004-1970	1 to 519
Ziegenhain	v. Jazewitsch, 1954/5	1225-1950	1 to 34
South German	Huber and Giertz, unpublished	832-1963	
	" " " , 1969	960-1960	10-150
West German	Hollstein, 1965	822-1964	1 to 36
North German			
Wienhausen I	Bauch, Liese, Eckstein, 1967	1831-1966	1 to 5
Wienhausen II	Eckstein, unpublished	1331-1546	1 to 12
Wienhausen III	Eckstein, unpublished	987-1321	1 to 20
Vierlande (S.E. of Hamburg)	Eckstein et al., 1972	1338-1966	1 to 56
A, East of the Elbe	Weitland, 1960	1778-1956	4 to 33
B, West of the Elbe	Weitland, 1960	1767-1967	4 to 34

Only Weitland's and the North German curves termed Wienhausen I, II and III represent a precise location. The Weser and Leine upland curve and especially the South German and the West German curves incorporate material from considerably larger, overlapping areas and the curves are unequally represented by samples.

The results show that the percentage of agreement (GL) reveals similarities between the ring series better than either of the correlation coefficients. Low agreement values occur between W. Germany and the poorly substantiated Wienhausen I (51%), and between the distantly situated W. Germany and the Weitland B curves. The highest value is that between the S. German and the Ziegenhain curves (the "catchment" area of the former includes that of the latter completely) and other high values occur between adjacent areas.

Some curve pairs with a relatively high agreement value reveal no correlation because at least one of them is weakly substantiated in essential parts or an age-trend effect was present.

The last column in Table 1 lists the mean of the agreement values or of the correlation coefficients, all suitably weighted for the overlap length, for each chronology with the nine others. The highest, that of the Weser and Leine upland curve, 70.8%, shows that this area is indeed transitional between the zone of application of the South and West German chronologies and that of the North German ones, cf. A. Thiess (1966). By contrast Wienhausen I does least well with an average value of only 59.7%. This is not because it represents a marginal area but because it is poorly substantiated and does not adequately represent the mean. The second lowest value, the 62.7% of Weitland B, is attributed to its north-westerly origin.

The agreement values with Holmsgaard's (1955) Danish oak mean curve were also calculated. Values range from around 56.0% with W. Germany, up to 69.3% with Weitland B, i.e. the Danish curve resembles the German chronologies up to a point. M. Jahrig (written communication, 1972) dates several historical chronologies of areas in the D.D.R. by the S. and W. German oak chronologies. The agreement values lie between 54.8% and 69.4%. By contrast Ermich (1960) found no agreement between a Polish and a Bavarian mean curve.

When compared with one another, the North German curves (average agreement 64.9%) are less similar to one another than are the south, west and central German curves to one another (73.8%). Overall, although the similarity between N. German and the other German oak curves is slight, it is not to be ignored.

The usefulness of a mean curve is determined to a large extent by its quality. The agreement values (W) in Table 1 were used to determine the minimum overlap in years required, on average, to date, with a reliability of 99.9%, the 519 single curves from the Weser and Leine upland by the different German chronologies. The formula used (Eckstein, 1969, based on Huber, 1943) is

$$\text{No. of Years} = \left(\frac{154.5}{W-50}\right)^2$$

The value is about 60 years for the Weser-Leine mean curve itself, and about 100 years for the S. German and Ziegenhain chronologies. On the other hand the W. German chronology, which requires an average overlap of 155 years, can only be used with considerable limitations while all the other chronologies (requiring $\nless$ 250 years) are virtually unusable.

CHARACTERISTICS OF THE WESER-LEINE MEAN CURVE

A chronological series can be described by criteria such as average ring width, age-trend, mean ring width fluctuations and length of period. The average annual ring width is dependent on height in the trunk and may be different in dominant and suppressed trees. Sometimes the wind or the steepness of a hill causes eccentricity in the rings on one side of a tree.

Age-Trend: In order to use annual ring curves to investigate climate and yield—though not for dating—one must eliminate the age-trend. The measured widths are 'standardised' to index values that are independent of the age of the tree by relating them to the corresponding values of a smoothed curve reproducing the age-trend. Wellenhofer, Rudakov and Ferguson calculate the index from the running mean, while others use mathematical functions whose coefficients are determined by the method of the least squares. To deal with long-term fluctuations it is important for the tree population to represent as wide an age-spectrum as possible. Then the climatically determined long-wave curve oscillations are distributed over the whole length of the age-trend curve.

One can eliminate the age-trend without standardising the measured annual ring widths, if one forms a standard curve from numerous single curves in such a way that trees of different age are present throughout the mean curve. One can proceed on the assumption that a curve that is well substantiated by recent and historical material will no longer be subject to any significant age-

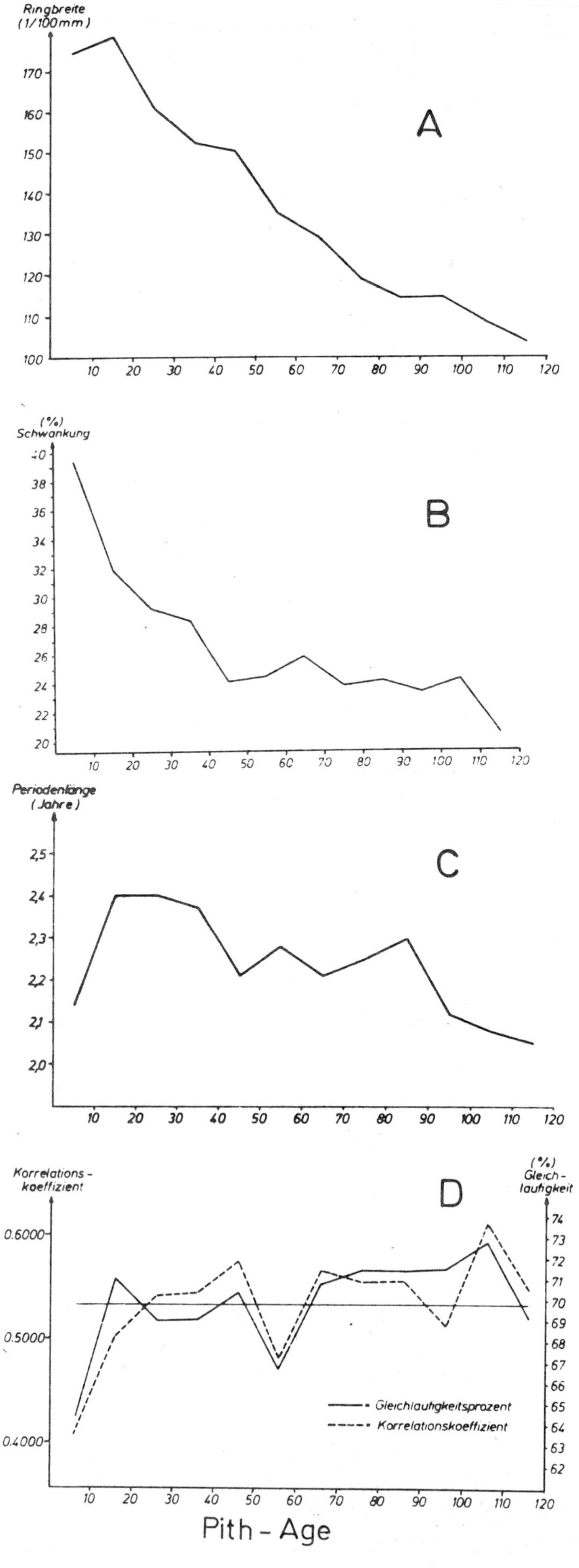

Fig. 1. For 102 oak single curves, dependance on relative pith-age of A. Ring-width; B. % Mean Fluctuation; C. Period Length and D. Agreement Value (G. L. %)

trend. According to Hustich (1949) such a curve is ideal for dendro-climatological purposes.

When establishing the Weser and Leine upland chronology any age-trend was not eliminated, first, because the curve was to be a foundation for dating and not for climatic investigations; secondly, because in oak the age-trend is less pronounced than in other species; and lastly, because the conditions discussed in the previous paragraph are true of extensive parts of this chronology.

Mean Fluctuation of Annual Ring Widths: Various methods have been used, e.g. by Douglass, Fritts and Hustich to calculate the average fluctuations (American "mean sensitivity") of annual ring series. In this work Jazewitsch's (1948) method was adopted. The fluctuation of the ring widths of two successive years is determined from its absolute difference, divided by the smaller of the two values, and the result expressed as a percentage.

The investigations by Wellenhofer (1948) and others have shown that the mean fluctuation (M.F.) of different species of trees in the same locality can differ noticeably from one another. As a particular climatic factor becomes increasingly dominant, trees of the same species exhibit an increasing M.F. Thus Mikola (1956) found that in Finland, where temperature increasingly limits growth, the M.F. in spruces and pines increases from the south to the northern timber-line. Similarly, Fritts et al. (1965) have shown that M.F. increases with increasing aridity. Locations with a single dominant climatic factor are the most suitable for dendrochronologic investigations and for this reason strongly oscillating ring series ("sensitive series") are generally given precedence over weakly fluctuating ("complacent") series.

Curve Qualities and Relative Pith-Age: The dependence of curve quality with the pith-age was studied with all the pith-controlled samples (102 in number) which had more than 120 rings. In Arizona, discs from eight different heights of a ponderosa pine were used to investigate changes in curve qualities with pith-age and the height from the stump. The positions of many of our samples in the tree are unknown but the material, historical and recent, is very comprehensive and our pith-dates extend from the 13th to 19th century with the date only rarely in the same calendar year.

Holmsgaard's method (1955) was used to calculate the average ring width with increasing pith-age grouped into periods of ten years. The results (Fig. 1A) show a clear trend with the ring width reaching a maximum at the age of 10 or 12. Figs. 1B and 1C show respectively how the mean fluctuation and the period length changes with pith-age, while Fig. 1D gives the results for variation in the agreement values with pith-age. For these results the 102 individual samples were compared in ten-year periods with the Weser-Leine mean curve. Overall, it appears that the widths of successive annual rings are determined not by the actual age of the tree but by the age of the ring in relation to its pith.

Intensity of Deflections: The number of the falling deflection was found to be greater than the number of rising ones both in single and in mean curves. However the differences become less significant with increasing substantiation of the mean curve.

The predominance of downward trends in individual curves can be explained by their age-trend. The better a mean curve is substantiated, the more probable it is that particular calendar years will be substantiated by individual curves which represent all phases of tree life and that the influence of the age-trend will be increasingly eliminated. Differences in the averaged ring widths then represent an average reaction of many trees to the climate, which is independent of the life-phase of individual trees. Because of the predominance of falling deflections in the case of well-substantiated mean curves, it is to be expected that falling deflections in the mean are weaker than rising ones.

Indicator Years, Downward and Upward Trends: 73.2% of the years in the period 1300 to 1968 are indicators. Of these 38.3% have a downward and 34.9% an upward trend. In the class with the highest significance ($P < 0.001$), 28.1% fall compared with 26.0% that rise.

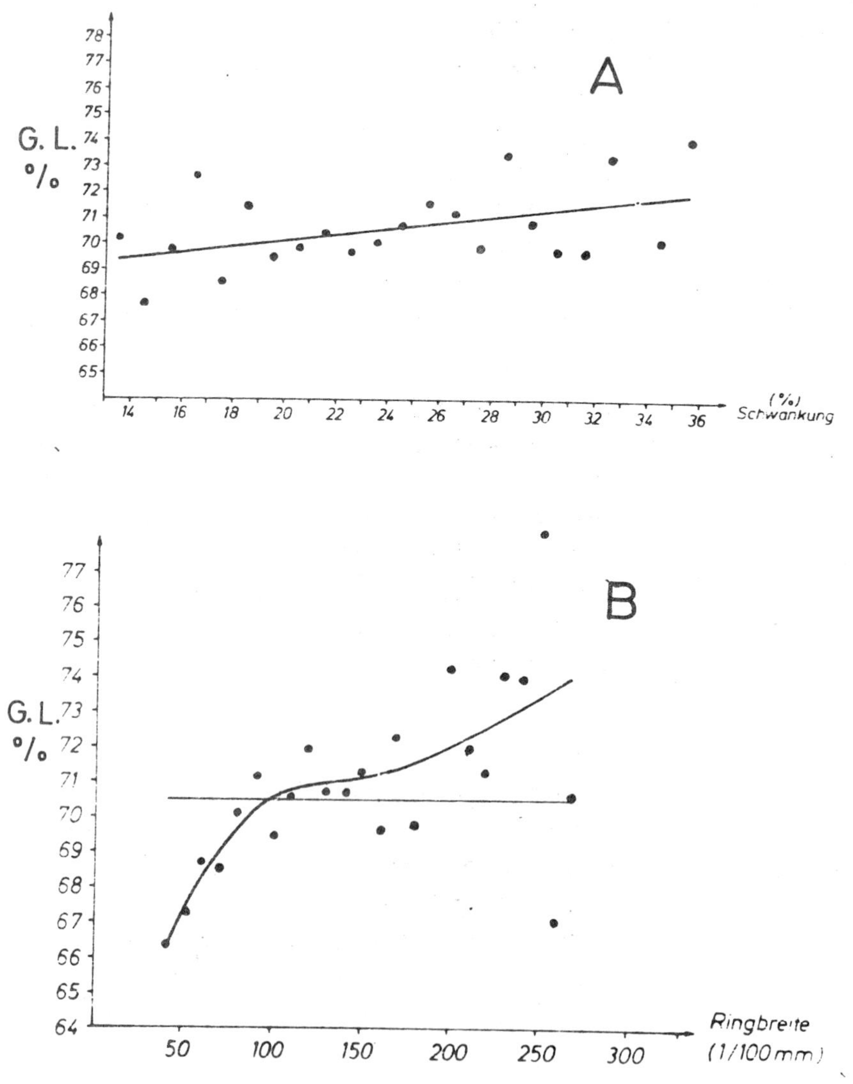

Fig. 2. For 519 single curves, dependance of their agreement (G.L.°/o) with their mean curve as a function of A. Their mean fluctuation and B. Their ring-width.

Variation of Crossdating Quality with the Mean Fluctuation (M.F) and the Mean Ring Width

It is to be expected that single curves with an above average M.F. will also show an above average agreement with their mean curve. This was found with the 519 dated oak curves (Fig. 2A). The points, though scattered, show a slight but definite upward trend. Thus fluctuations of the mean curve (in N. America taken as an index of crossdating quality) can only be used as an index for oak curves to a very limited extent.

The agreement between each of the 519 dated sequences and the standard curve were also analysed as a function of the mean ring width (Fig. 2B). In accordance with Wellenhofer's observations, the wider ringed curves show a higher agreement as well as a higher fluctuation in this area which is climatically homogeneous. Hence climate is not responsible for differences in the mean ring width. By contrast, in Arizona along a climatic gradient with increasing aridity curve fluctuation (M.F.) becomes stronger and agreement values higher when the rings become narrower (Fritts *et al.*, 1965).

References

Those not listed on pages 7-10 are as follows:

Bauch, J., Liese, W. and Eckstein, D. 1967, *Holz als Roh- und Werkstoff* 25, 285-291.

Eckstein, D. 1969, Dissertation, Hamburg.

Fritts, H. C., Smith, D. G., Cardis, J. W. and Budelsky, C. A., 1965, *Ecology* 46, 393-401.

Glock, W. S. (1941), *Botanical Review* 7, 649-713.

Holmsgaard, E. 1955, *Det forstlige Forsogsvaesen i Danmark* 22, 1-246.

Hustich, I. 1949 , *Geogr. Annaler*, 90-105.

Mikola, P. 1956, *Tree-Ring Bull.* 21, 16-20.

Rudakov, V. E. 1964, *Sovetskaja Archeologija*, H.2, 79-86.

Thiess, A. 1966, Diplomarbeit Hamburg.

Weitland, J. 1960, Mitteilungen der BFA für Forst- und Holzwirtschaft Reinbek, 48.

1.7 USSR PAPERS TO XII INTERNATIONAL BOTANICAL CONGRESS, LENINGRAD, 1975

Titles taken from
Biolecological Fundamentals of Dendrochronology

Published Vilnius-Leningrad, 1975. Editor T. Bitvinskas

	Pages
T. T. Bitvinskas, B. A. Dergachev, L. A. Kairiukstis, G. E. Kocharov, K. K. Jankevicius, Elaboration of Bioecological Fundamentals of Dendrochronology in the USSR.	5-11
English Translation	86-90
V. S. Nesterov, M. I. Rozanov, G. I. Kirienko. Dynamics of trees increment of different types its development.	17-20
B. A. Kolchin, N. B. Chernih. Absolute dendrochronological scales of northern regions of Eastern Europe for the space of 12 centuries.	27-31
E. D. Labzhanidze. The effect of light schedule change on the formation and structure of annual wood-rings in the same-aged stands.	33-37
V. G. Kolischuk, O. N. Petriv, R. I. Polovnikov. Fir increment, its structure, dynamics and conditionality.	39-42
M. I. Rozanov, V. G. Nesterov, G. I. Kirienko. Dynamics peculiarities of trees increment and their evaluation by tabulating indices.	43-46
G. G. Schiatov. The long-term cycle of growth indices of Larix sibirica at the northern timber line (1103-1969 A.D.).	47-53
R. Pakalnis. On the synchronism of water level fluctuations in closed lakes of Baltys and Jaskutis.	55-59
E. V. Dmitrieva. Influence of climate on tree increments in different habitats of the northern area of the Southern Taiga.	61-63
E. P. Shpalte. Dendrochronological and dendroclimatological investigations in the Latvian SSR.	65-67
T. T. Bitvinskas, J. J. Kairaitis. The radial growth variations of oak stands and their relationships to environmental conditions, climate and the solar activity in the Lithuanian SSR.	69-74

1.8. DENDROCLIMATIC AND DENDROCHRONOLOGIC INVESTIGATIONS IN THE USSR

L. A. Kairiukstis, Chairman of the Commission for Dendrochronological Investigations of the USSR Academy of Sciences

Summary

There is increasing interest in deriving information from woody plants about the biosphere in the past. In many institutions of the USSR, dendroclimatology and dendrochronology are flourishing. Regularities (11 and 22 year cycles) in the growth patterns and their dependence on climatic indices have been determined.

Dendroscales based on old trees have been formed and long-term forecasts of meteorological processes based on tree-ring data and astrophysical phenomena have proved to be correct in regions of the European part in the USSR.

The Commission for dendroclimatic investigations at the Academy of Sciences of the USSR coordinates the research and is directing it to the solving of local, regional and global problems in climatology as well as in other spheres of man's activity.

Scientists of many countries are interested in plant reactions to environmental changes. One aspect is the study of woody plants as reflections of the past biosphere. With the help of dendrochronology, reliable long-term prognosis of biosphere changes can be made, while the exact dating of past climatic and astrophysical phenomena may be possible. In addition, many biological, forestry, archaeological as well as agricultural problems can be aided. Investigations along these lines are being carried out in the USA, West Germany, Scandinavia and other countries.

In the Soviet Union, dendrochronology and dendroclimatology have been studied since the last century. Such famous naturalists as D. I. Mendeleev and I. I. Vavilov paid attention to these subjects. The principals were formulated by F. N. Shvedov (1892) and developed by V. A. Rudakov (1951), A. Zvedris (1962), A. A. Molchanov (1976), T. T. Bitvinskas (1974) and others. The investigations were summarized at All-Union Conferences (in 1958 in Vilnius and in 1972 in Kaunas, see References) organised by the Botanical Institute of the Academy of Sciences of the Lithuanian SSR; in a special symposium Bitvinskas *et al.*, 1975) at the 12th International Botanical Congress in Leningrad; and in the All-Union Conference "Astrophysical phenomena and radiocarbon" (Tbilisi, 1976).

Summary of Results

1. The dynamics of growth variation of tree and woody coenoses, and the width changes of annual rings of many species over the last 100-300 years have been studied. Pine (*Pinus silvestris*), spruce (*Picea abies*), oak (*Quercus robur*), black alder (*Alnus glutinosa*), birch (*Betula verrucosa et B. pubescens*), Siberian larch (*Larix sibirica*), dwarf stone pine (*Pinus pumila*) and juniper (*Juniperus turkestanica*) are the main species investigated. The areas cover the Baltic (Lithuania, Latvia), the south 'taiga' zone (Leningrad, Moscow, Kalinin region), the Carpathians and the Urals and include the upper limits of forest growth.

Great differences in growth depending on the soil moisture regime for different sites have been found for the same geographical region. Pine and spruce under conditions where temperature or moisture limit growth reflect the 11 and 22-year-old rhythms of solar activity and climatic variations. The amplitude of the variation of growth as a result of climate amount to -70 to +100%. A number of hydrothermal indices, reflecting the influence of environmental conditions on forest growth, have been found. Linear dependence of growth with amplitudes of solar activity in 22-year-old cycles have been determined. Original methods of studying width change of annual rings of creeping woody species have been developed.

2. Seasonal growth of the main forest species was studied and the dependence of growth on separate climatic factors in the year itself and in previous years determined. Inertness indices, that is the independence of climatic factors of the current year, were observed. This inertness varies with the age of the tree. For spruce at needle age of 6-7 years, 60 to 70% of current growth is already predetermined by the climate of previous years whilst the growth which will begin in 2 years is predetermined to the extent of 10 to 30%. Total growth of the following year may be ascertained by the bud size in autumn because there is linear dependency between bud size and growth (Kairiukstis, 1973).

The relative effects of factors such as light, moisture, temperature have been assessed (Kairiukstis and Juodvalkis, 1972, and this vol. page 173 . This makes it possible to decipher climatic factors of the past from the annual rings, particularly those of trees of stable growth rhythm and those less subject to variation from class to class.

3. To provide an exhaustive answer about climate changes during the past we need dendrochronological data on the radial growth of trees and stands of similar ecology over a wide area. Dendroclimatic profiles from the north to south and from the west to the east are being created. The profile from the Kola Peninsula through the Baltic Area to the Carpathians has been formed. Other profiles, from the Baltic through Moscow, the Urals to Vladivostok are being obtained.

4. The possibilities of re-creating the climatic fluctuations during the last 500-8000 years in separate parts of the USSR depend on the presence of old trees and stands; of wood in archaeological, ethnographical and historical monuments; and of wood remains in peaty and fluvial deep-water deposits and in the layers of permafrost.

A number of dendroscales based on old trees have been created. The Novgorod chronology reflects climatic variations of the southern taiga during a period of 1300 years (Kolchin, 1965), while that of Western Lithuania, based on excavations of pine in peaty zones, covers 2200 years. The Smorgan oak scale for the southern Baltic area is being formed for a period of 6000 years. These dendroscales are based on cross dating by dendrochronology and by radiocarbon measurements (Bitvinskas, 1978).

Recommendations from resolutions at conferences are already being used in astrophysics, hydrology, climatology, forest management, archaeology and in other spheres of science. Long-term climatic forecasts based on dendroclimatic investigations on ecosystems are proving correct in a number of regions of the European part of the USSR (Nesterov and Rozanov, 1975).

Present Research

This is being carried out at the laboratory of dendroclimatology and dendrochonology at the Botanical Institute of the Academy of Sciences of the Lithuanian SSR, at the botanical and dendrochronological laboratory of the Limmological Institute of the Siberian department of the Academy of Sciences of the USSR, while specialised groups are investigating dendrochronology at the Komarov Botanical Institute at the Academy of Sciences of the USSR; at the Institute of Plants and Animal Ecology of the Ural Scientific centre at the USSR Academy of Sciences; at the Botanical Institute of the Ukrainian SSR Academy of Sciences; at V. P. Sukachiov Institute of Forest and Wood at the Siberian Department of the USSR Academy of Sciences; at the Lithuanian Research Institute of Forestry; at the Forestry laboratory of the USSR Academy of Sciences; at the Timiriazev Agricultural Academy; at Moscow State University; and at the Biology-Soil Institute Far East Scientific Centre of the USSR Academy of Sciences.

In 1974, the USSR Academy of Sciences set up a Commission for dendroclimatological investigations. The Commission not only coordinates the research work but organises All-Union conferences and symposia; deals with problems and methods; issues scientific publications; and represents this sphere of science in the USSR and abroad.

Note by Editor

Ed. The 'taiga' zone is defined as the drier, colder, continental zone of the northern coniferous forests.

References

Bitvinskas, T. T., 1974. Dendroclimatic investigations. Gidrometizdat, Leningrad. pp. 172.

Bitvinskas, T. T., 1978. On the problem of the creation of super-long-term dendroscales in the Southern Baltic area. B. Environmental conditions and radial growth of trees. Kaunas, 1978.

Bitvinskas T. T., Dergachev, V. A., Kairiukstis, L. A., Kocharev G. E. and Jankevicius K. K., 1975. Elaboration of biological fundamentals of dendrochronology in the Papers to the 12th International Botanical Congress. Leningrad, July 1975. Vilnius/Leningrad.

Kairiukstis, L. A. and Yuodvalkis, A. I., 1972. Peculiarities of the seasonal formation of annual rings in connection with climatic conditions. Papers of the Second All-Union Conference on Dendrochronology and Dendroclimatology, Kaunas.

Kairiukstis, L., 1973. Mixed spruce stand formation and cutting. Mintis publishing house, Vilnius. pp. 358.

Kolchin, B. A. 1965. Dendrochronology of Eastern Europe. Moscow. Nauka. pp. 62-85.

Molchanov, A. A. 1976. Dendroclimatic fundamentals of weather forecasts. Moscow. Nauka publishing house.

Nesterov, V. T. and Rozanov, M. I. 1975. Bioecos theory and its application to superlong-term prognosis of nature. Doklady TSKhA No. 208, pp. 183-188.

Rudakov, V. E. 1951. Method of studying climate variation by the width of annual rings. Doklady AN Arm SSR pp. 75-79.

Shvedov, F. N. 1892. Trees as chroniclers of droughts. Meteorologicheskii Vestnik No. 5.

Zviedris, A., Matizanis, Ya. 1962. Patterns in the variation of ring width of woody species in the Latvian SSR. Izvestiya AN LatvSSr, No. 8, pp. 16-18.

2.1. OAK CHRONOLOGIES FOR CENTRAL EUROPE. THEIR EXTENSION FROM MEDIEVAL TO PREHISTORIC TIMES

B. Becker, Stuttgart-Hohenheim and A. Delorme, Göttingen

The Southern German master chronology for oak was developed by Huber and co-workers at Munich (this volume, page 27). It was first extended by Hollstein (1965) at the Rheinische Landesmuseum in Trier. During recent years contributions for this chronology have been made by Hollstein (Trier), Becker (Stuttgart-Hohenheim), Delorme (Göttingen), Schmidt (Cologne), Lambert and Orcel (Neuchâtel, Switzerland). The geographical area now covered by various series extends over many parts of Central Europe.

Series derived recently from very numerous sub-fossil oaks embedded in Holocene gravel terraces of river valleys and excavated during dredging have improved the early medieval and Roman-Iron Age parts (Fig. 1) and also extended the chronologies to much earlier times (Figs. 2 and 3).

Details of much of this work formed the subject of papers to the Mainz Conference in 1974 and are published in its Proceedings (Frenzel, 1977). In southern Germany, oaks from deposits along valleys of the Danube, the upper Rhine and the Main have been evaluated (Becker and Frenzel, 1977; Becker and Schirmer 1978). In central Germany many oaks from the Rhine between Mainz and Karlsruhe as well as from the upper Weser and its tributaries, in particular the Fulda and Werra (Fig. 4) have been measured (Delorme, 1976 and 1977). These with others to the north and west are included in the more extensive map of Fig. 5. This includes the main area (Rhineland) for the oaks studied at Cologne by Schmidt (1977).

1. The Early Medieval-Roman-Iron Age Chronologies (Fig. 1)

According to Hollstein (1977) the chronology for West Germany covers the past 2690 years (1973 A.D. to 717 B.C.). Nevertheless, the ring-widths are not yet published completely. Hollstein commented at the Mainz Symposium that in this chronology problems still exist between 400-300 A.D., that is to say in linking together Parts B and C of the Master Chronology shown in the lower part of Fig. 1.

On the other hand, the hitherto published parts of the Hollstein chronology could be reduplicated by many successful crossdatings. In general, these cross-correlations have been achieved by the construction of two large floating chronologies from tree-ring analyses on sub-fossil oaks (Becker and Frenzel 1977, Delorme 1976). The two series (Becker, Delorme and Schmidt 1977), date, on the Hollstein time scale, from 755 A.D. to 217 A.D. (Master Chronology B in Fig. 1) and from 261 A.D. to 397 B.C. (Master Chronology C in Fig. 1). Both series were sent to the Trier laboratory in 1975, but no comment is available and we are still waiting for the exchange of the Hollstein data.

For the Roman and Iron Age periods, the cooperation between the laboratories at Stuttgart, Göttingen, Neuchâtel and Cologne has provided many tree-ring dates. The existing series are plotted in Sections 2 and 3 of Fig. 1, together with the published parts of the Hollstein Master Chronology in Section 1 and the chronologies of the late Prof. Huber in Section 3. The last were

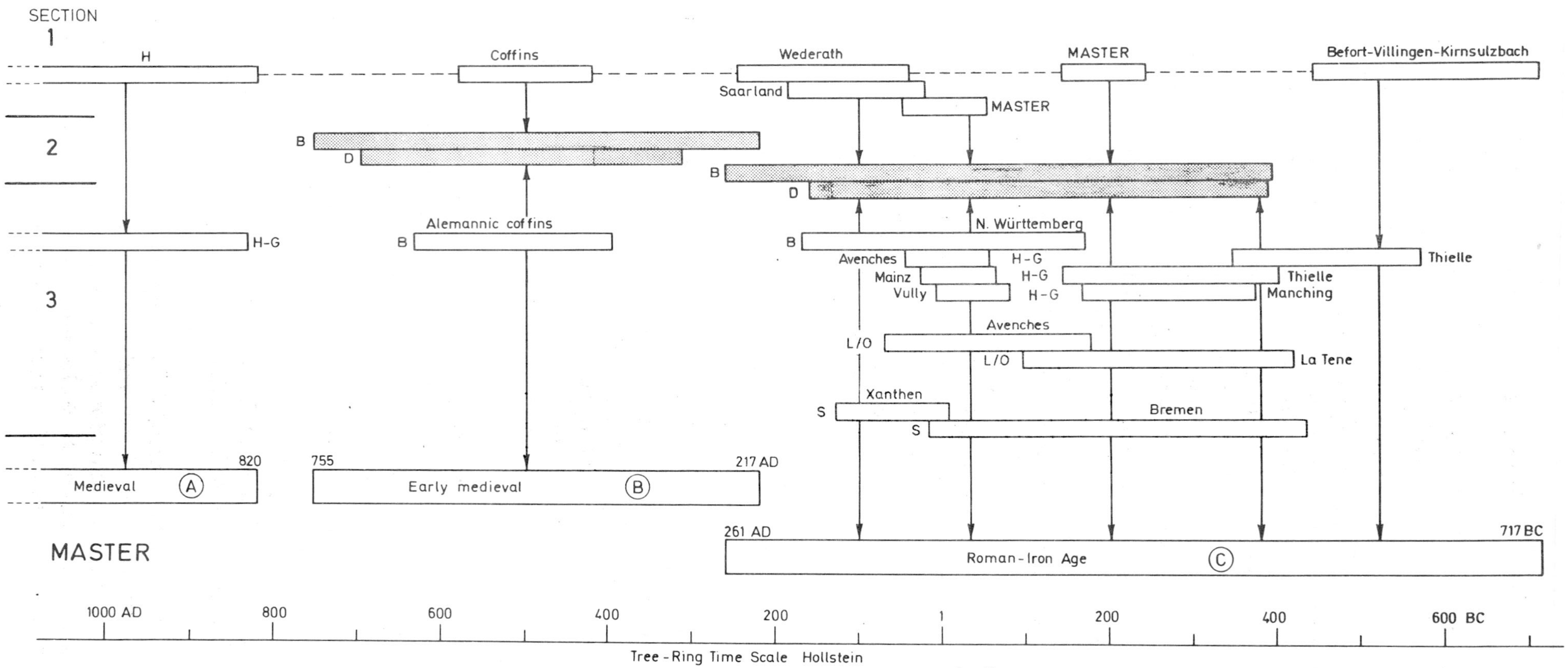

Fig. 1. The Early Medieval / Roman / Iron Age Oak Chronologies for Central Europe 1978.

MASTER Present position at Stuttgart-Hohenheim, Parts (A), (B) and (C)

Section.

1. Cathedrals, coffins, bridges etc. Hollstein (H), Trier. MASTER, parts published
2. Sub-fossil oak. Becker (B) and Delorme (D)
3. Archeological sites (Germany and Switzerland). Huber/Giertz (H-G); Becker (B); Lambert/Orcel (L/O); Schmidt (S).

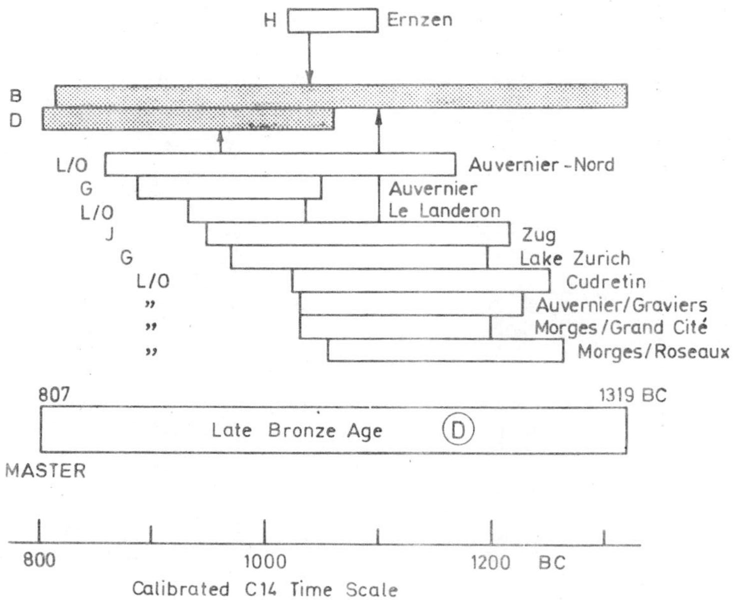

Fig. 2. Late Bronze Age Chronologies for subfossil river oaks (shaded) of the Danube, Rhine, Main and Werra crossdated with chronologies for oaks (unshaded) from 10 Swiss and W. German sites.

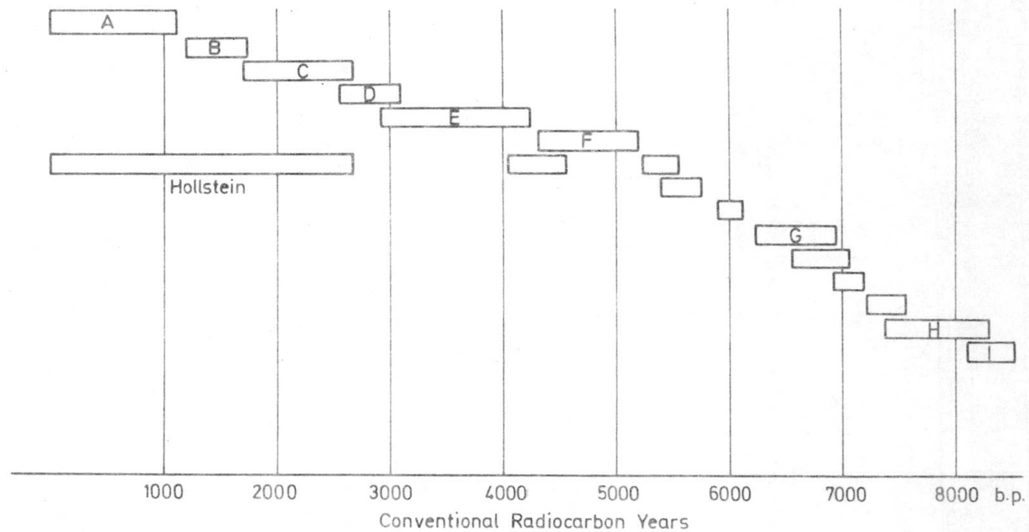

Fig. 3. The Postglacial Oak Tree-Ring Chronologies of Central Europe. A, B, C, D. As in Figs 1 and 2.

measured by V. Giertz-Kerner and were used by Hollstein as well for the construction of his chronology. The following series have now been dated:

(a) Chronology of oak coffins of the Alemannic Period: 629-398 A.D. (Becker 1974).

(b) Chronologies of Roman and Celtic bridges and settlements from Switzerland, derived by Huber and Giertz-Kerner, dated by Hollstein and Becker: 44 A.D.-572 B.C. (Avenches, Mainz, Vully-le-Haut, Thielle).

(c) Chronology of Roman settlements in Northern Würtemberg: 168 A.D.-173 B.C. (Becker, not yet published).

(d) Roman chronologies from Switzerland: 69 A.D.-419 B.C. (Avenches, La Tène, analysed by Lambert and Orcel).

(e) Roman and Iron Age chronologies from Xanthen and Bremen: 127 A.D.-436 B.C. (established by Schmidt).

The tree-ring datings listed above demonstrate the accuracy of our two subfossil oak chronologies (Section 2 of Fig. 1). Nevertheless it must be stressed, that at present the correlation of Master chronologies A and B is still lacking due to the gap in our data between 820-755 A.D. A final and definite check of the Hollstein time scale therefore is not yet achieved. According to the existing crossdatings with Hollstein's published material the master chronologies B and C available in S-Central Europe date from 755 A.D.-217 A.D. and 261 A.D.-717 B.C.

2. The Late Bronze Age Chronology (Fig. 2)

Recent progress has been made with the Late Bronze Age chronology (Series D, Fig. 2). The archaeological series consists of Swiss lake dwellings from Zug/Sumpf (Huber and v. Jazewitsch 1958), Lake Zurich and Auvernier (Giertz-Kerner, unpublished) and from six sites on the shore of Lake Neuchâtel (cf. Lambert and Orcel, this volume,). In addition, a series from Ernzen is involved. It was analysed by Hollstein and cross-dated by him with the Zug/Sumpf series (Hollstein 1970).

The various cross-correlations have been achieved after a successful construction of the master curve (shaded in Fig. 2) derived from subfossil oaks of the Danube-Rhine-Main-Werra valleys by us. The overall master chronology, according to radiocarbon calibration by H. Suess (Suess and Becker 1977), is dated 807-1319 B.C.

3. Bronze Age and Neolithic Chronologies (Fig. 3)

A further extension has been achieved in the Bronze Age and Neolithic subfossil oak chronologies. Our 1307 years Bronze Age series (Chronology E, Fig. 3) is dated, according to ^{14}C calibration by Suess, to 1619-2925 B.C., and two Neolithic series to 2750-3286 B.C. and 2982-4024 B.C. (Chronology F, Fig. 3). In turn, archaeological oak chronologies of Swiss lake dwellings at Thayngen-Weier, Burgäschisee-Süd and -Südwest (Huber and Mertz, 1963) and Niederwil (Munaut, unpublished) are now crossdated with the river-oaks, extending the 1076 years master-curve to 4057 B.C. (Becker, Proceedings of the 9th International Radiocarbon Conference, held at Los Angeles and La Jolla in 1976, in course of issue).

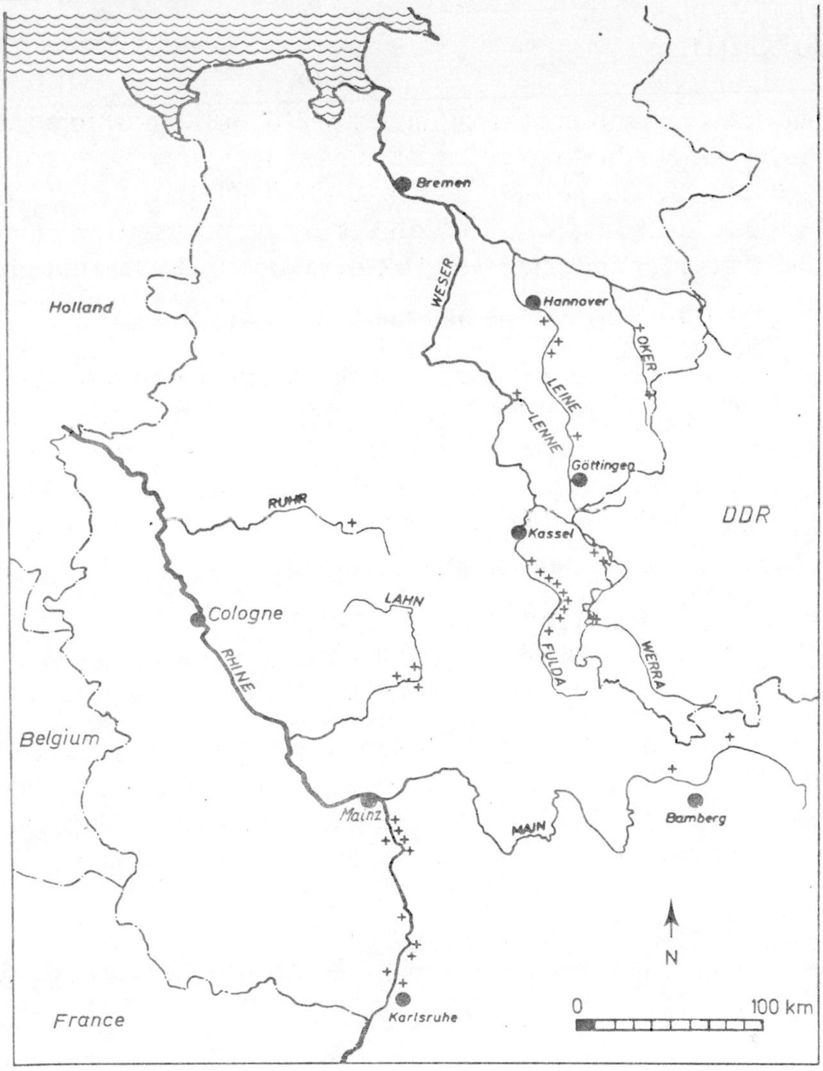

Fig.4. The 35 Localities of Subfossil Oaks examined at Göttingen.

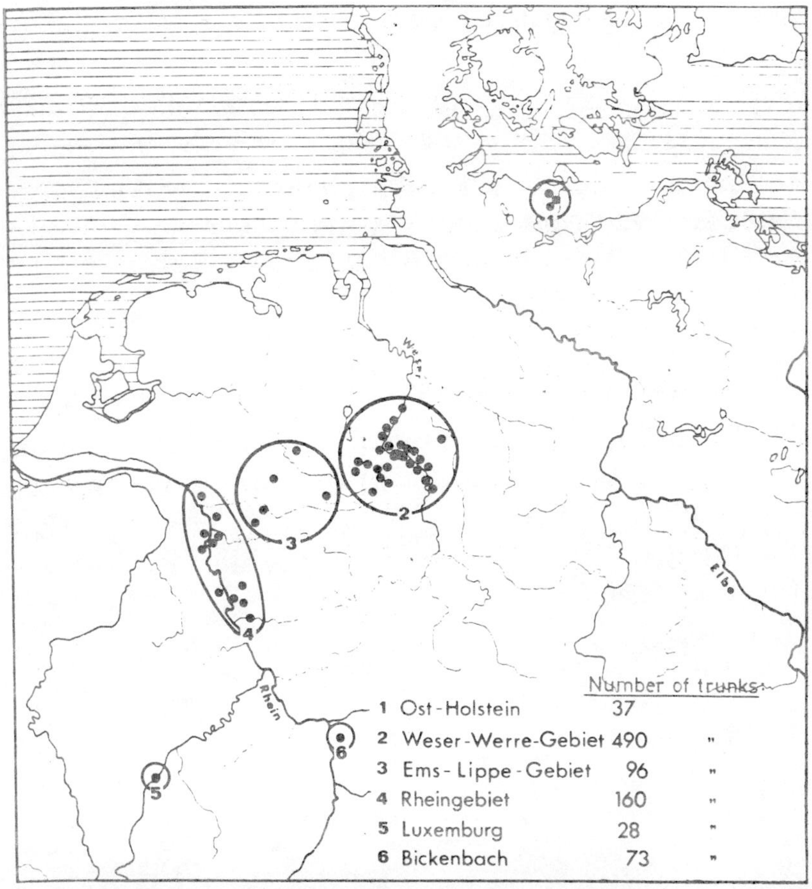

Fig.5. Localities of Subfossil Oaks in Northern Central Europe that are being

At present, the absolute tree-ring dates or ^{14}C calibrated part of the Central European oak chronology covers the past 6000 years. Gaps or doubts still exist at 400-300 A.D., 700-800 B.C., 1400-1600 B.C. and 2700-2800 B.C. Hopefully the still existing gaps will be closed by a continuation of our cooperation between the tree-ring laboratories in Germany and Switzerland.

4. The postglacial subfossil oak chronology (Fig. 3)

Fig. 3 shows a schematical diagram of the present stage of the postglacial oak chronology. To avoid problems, which occur in comparing absolute tree-ring dates, calibrated ^{14}C dates and conventional ^{14}C dates, the series are plotted against the conventional ^{14}C time scale before present. The oldest part of the chronology, not published before, consists now in three large floating subfossil oak chronologies of Becker and Delorme:

Chronology G: 707 tree-rings 6235-6941 B.P.
Chronology H: 917 tree-rings 7388-8304 B.P.
Chronology I : 399 tree-rings 8120-8518 B.P.

The exchange of data between Stuttgart-Hohenheim, Göttingen and Cologne (Becker, Delorme and Schmidt, 1977) based on the continuing analyses of subfossil oaks together with the archaeological oak series that is being constructed in Switzerland should provide the chance to construct a continuous oak tree-ring chronology for Central Europe.

We wish to thank the Editor for arranging for Figs. 1-3 to be plotted.

References

Becker, B. 1974. Fundberichte aus Baden-Württemberg, 1, 545-564.

Delorme, A. 1976. Neue Ausgrabungen und Forschungen in Niedersachsen, 10, 225-237.

Frenzel, B. (ed.) 1977. Dendrochronologie und Postglaziale Klimaschwankungen in Europa, Erdwiss, Forschung, 13, Steiner Verlag, Wiesbaden:-
Becker, B., Delorme, A. and Schmidt, B. Pages 143-146;
Becker, B. and Frenzel, B. Pages 43-61;
Delorme, A. Pages 62-67;
Suess, H. and Becker, B. Pages 156-165.
Schmidt, B. Pages 91-98.

Hollstein, E. 1965. Bonner Jahrbücher, 165, 12-27.

Hollstein, E. 1970. Kurtrierisches Jahrbuch, 10, 15-20.

Huber, B. and v. Jazewitsch, W. 1958. Flora, 146, 445-471.

Huber, B. and Merz, W. 1963. Germania, 41, 1-9.

2.2. GEOGRAPHICAL AND TEMPORAL DIFFERENCES IN THE FLUCTUATIONS OF RING WIDTHS IN CENTRAL EUROPE: CLIMATIC IMPLICATIONS

by O. Fürst, Munich, 1963.
Extracts from Translation by W. Bell, Climatic Research Unit, Norwich, of paper in Flora, 153, 469-508.

Summary

For 164 curves of annual ring widths (138 individual curves and 26 means of 5 curves) of different species, regions and periods, we have calculated the average interannual fluctuation of the individual curves (S_E) and of the mean curves (S_M), and their ratio ($R = S_M : S_E$). For the first time the index formulated by Douglass and his team in Arizona has been applied to Central Europe.

Both for deciduous trees and for conifers the average values of S increase from west to east (Spessart-Bavarian Forest-Beskides) that is, as the climate becomes more continental. The apparent exception, trees growing on the eastern edge of the Vosges, is explained by the continental character of the Colmar "dry island" in the rain shadow at the ridge of the Vosges.

The fluctuations shown by random samples from the present, the High Middle Ages, the Bronze and the Neolithic Ages indicate that in the Neolithic the climate in northern Switzerland was definitely more extreme, more continental than at present; on the other hand, the climate of the late Bronze Age was more equable. In the High Middle Ages, there was a marked change about 1330 from a period of high fluctuations (with many cold winters) to one of lower fluctuation. In the mid-17th century the change about 1650 from a period of low fluctuation to one of high fluctuation reflects the onset of the Little Ice Age.

INTRODUCTION

Ring widths in general decrease with age due to the fact that the new growth is distributed over an increasingly large circumference as well as an increasingly tall trunk. In addition however, one observes year to year fluctuations in the ring widths. Sometimes the change from wide to narrow annual rings is gradual, sometimes it is sudden. This investigation is concerned exclusively with these inter-annual fluctuations and their influence by the following factors:

(a) the conditions of temperature and precipitation and water which directly or indirectly affect growth;
(b) the light conditions, which are of the greatest importance for the formation of new organic substances;
(c) all factors which are very damaging to trees e.g. rupture due to wind or snow, and animal or plant pests.

Leonardo da Vinci realised that in Italy — in the Mediterranean part of the northern subtropics — the width of the annual rings is directly proportionate to the annual precipitation. Ugo Buli was able to demonstrate and confirm his observations in an investigation on the Ravenna pine-groves. The most extensive investigations in this field were conducted on the Douglas firs of the Colorado Basin in the U.S.A. Douglass and Huntingdon, and also Schulman were able to establish a close correlation between precipitation and annual ring growth. These and other investigations were conducted in areas with a minimum of one of the two main factors that determine growth, namely precipitation.

In Europe, a series of investigations was concerned with the question how much the annual ring width is dependent upon annual temperature. Here too locations with a minimum of the relevant factor were selected. For instance, Erlandson (1935) and Ording (1941) worked on pines from Scandinavia. Müller-Stoll was able to discover in yews from the Tirolean Karwendel a direct relation between annual ring width and summer warmth at Munich. Artmann studied the annual ring curves of cembra pines and larches from the Wetterstein Alps and found that fluctuations in the widths are directly dependent on the duration of sunshine and summer warmth at the Zugspitz observatory. Brehme drew similar conclusions about larch at high altitudes.

In areas in which there is no single climatic factor likely to be minimal, as in most parts of Central Europe where the ratio of warmth and humidity is essentially favourable, dendroclimatological analysis of annual rings is more difficult. All the same, here too one could find locations where a decisive factor for the annual ring widths is dominant (von Jazewitsch 1961). Indeed, there may be the possibility of deriving information about European climates of the past from annual rings. An observation reported in 1942 by Huber and Holdheide is significant. For Bronze Age piles of pine from the Wasserburg Buchau on the Federsee (Upper Swabia) they established that the ring widths fluctuated more widely and uniformly than recent pines from the same location. Huber and von Jazewitsch made the same observation about wood from oak pile dwellings from excavations of Neolithic settlements at Thayngen-Weier, near Schaffhausen and inferred that the climate at that prehistoric period must have been somewhat less equable than at present.

The purpose of this investigation has been therefore to enquire how much the character of annual ring curves has changed since the Neolithic Age. To understand a possible reason for changes, a comparison has been made between recent annual ring curves of various regions in Europe, especially between those in a more oceanic climate (to the west) and those in a more continental one (to the east).

Ring Width Curves Available and Method

The curves (from the dendroclimatological department of the Forstbotanisches Institut in Munich) were for oak, beech, fir and spruce. They had already been analysed dendrochronologically and the results published (Huber, von Jazewitsch, John and Wellenhofer 1949, Müller-Stoll 1951, von Jazewitsch 1953 and 1955; Huber and von Jazewitsch 1958). I also included freshly-bored cores from the Bavarian Alps, for which the curves had been synchronised and used in the Alpine fir and Alpine spruce chronologies.

The average fluctuation (S) in the annual ring widths over a period of n years is calculated by the expression first used by Douglass and then used in 1942 in Central Europe by Huber and Holdheide: the sum of the differences from year to year in the annual ring widths (regardless of the sign) is divided by the sum of the annual ring widths themselves and multiplied by 100w/(w-1). It is thus, throughout this paper expressed as a percent. The Americans call this index "sensitivity". However, Huber has repeatedly emphasized, e.g. in Geol. Rundschau 1961, that the average fluctuation is a measure of the sensitivity of different species only if their curves fluctuate differently at the same location and the same time (e.g. beech > oak, larch > cembra pine). If on the other hand a species fluctuates differently at different times, or in different areas, one may assume that it is not the "sensitivity" of the species that has changed, but that the variations in the fluctuations result from other external conditions.

One obtains a value for S which is the fluctuation of a definite number of annual ring widths. Theoretically it could be 200%, but in fact, in America and in several prehistoric curves for Central Europe (Huber and Holdheide 1942), it is at most 70-80%, and generally in the range 20-40%.

The equation was first used to calculate the average fluctuation of every 10-year period of all the annual ring curves used in the investigation. Next, sets of 5 curves were synchronised and formed into mean curves and their 10-year fluctuations calculated. Due to the smoothing effect of averaging, they fluctuate less than individual curves.

The quotient $S_M : S_E$ (average fluctuation of the mean curves in relation to the individual curves), here called R, serves as a measure of the agreement of the individual curves with the mean curves (American "crossdating"). In Arizona the Americans often reach 80%; in Central Europe the figure is generally around 50%, but occasionally 75-80%.

RESULTS

Dependence of S on the age of the tree: I kept noticing that the fluctuations of the annual ring widths vary during the different stages of the tree's life, i.e. with the so-called "age-trend". The dependence of the average fluctuation, S, on age was therefore examined closer. 10-year means of the average fluctuation of the annual ring widths were calculated for 6 Spessart oaks. The values show that the individual trees have a higher average fluctuation during youth than with increasing age. They also show that within the first 30-50 years of life, S adjusts from a maximum to a mean value and then oscillates about the mean. In this investigation we have therefore dispensed with the innermost 50 annual rings of the individual slices. Care was then taken to analyse annual ring series from trees of much the same age.

Dependence of S on the species of tree: In the publication "Dendroclimatic Changes in Semiarid America" of 1956, Schulman investigated many species of trees from the semi-arid American West, and established that these have distinctly different average fluctuations. Examples, each the mean value for four trees, he gave were:

Pseudotsuga taxifolia	53%
Pinus flexilis	21-25%
Pinus ponderosa	24%

Because of its high value the first is particularly suitable for dendroclimatology. Using this species he obtained high correlations (0.85-0.9) between the fluctuations of increase in width and the fluctuations in precipitation. The Sequoia gigantea from semi-humid locations, with an average S value of 16-19%, displays considerably less variability.

In Central Europe, Artmann was able to establish that the change from narrow to wide rings is much more abrupt in larch than in cembra pine. Brehme gives the average fluctuation of the annual ring widths of larches from the Berchtesgaden ibex reserve as 25%. Artmann reckoned an average fluctuation of 19% for the cembra pine, and Wellenhofer calculated 29-23% for the Spessart oak. Weitland gives fluctuations of width of 23% for oaks from areas in the western North Germany plain, and his values are similar to Wellenhofer's.

Müller-Stoll gives average fluctuations of 16%, 17% and 28% for spruces, firs and beeches from the northern foothills of the Western Beskides. Mayer-Wegelin and Möhring-Schulz-Brüggemann have established, for the Eastern Alps, that fir in a mixed population of a virgin forest has an average value of 11%.

The curves on which this article is based were examined species by species, using five curves for each. The beech and oak come from the Spessart, the spruce and fir from the Bavarian Forest. Beech (31%) was found to have a greater average fluctuation than the others — oak (18%), fir (18%) and spruce (17%).

Decrease in S in mean curves: Next we examined how much the average fluctuation decreases in the mean curve and the agreement between the individual and the mean curves (from five single curves). This revealed a significant fact: in mean curves the average fluctuation is lower (Fig. 1).

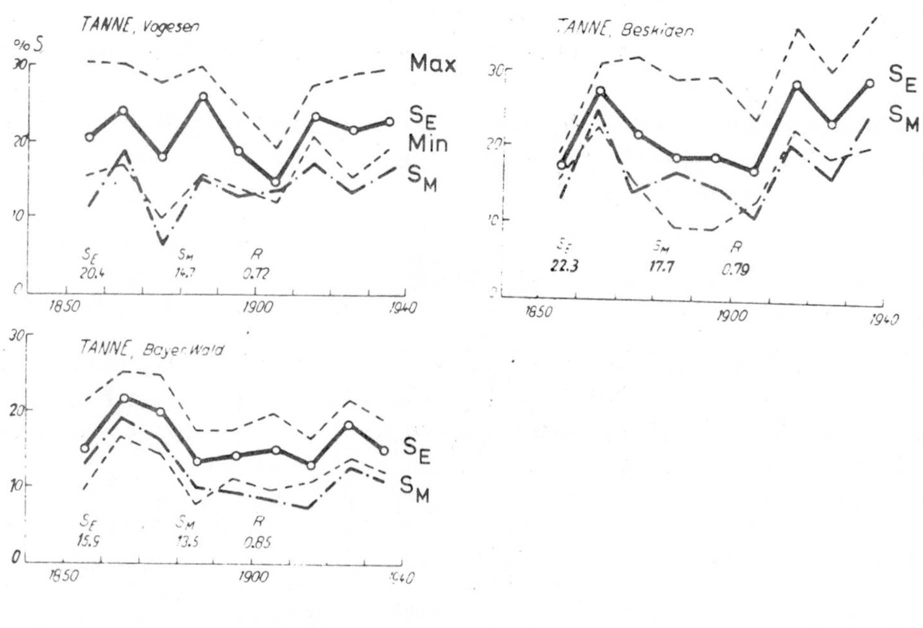

Fig. 1. Fluctuations of Silver Fir in different Localties, 1850 -1940
Values at 10 year intervals.

Key: Heavy lines; mean fluctuation (S_E) of the five component individual curves.
Chain lines; fluctuation (S_M) of mean curve (from 5 single curves)
Lines of dashes; max. and min. fluctuations of the individual curves.

The quotient R measures the agreement between the individual and mean curves. Schulman gives average values of 0.81 for Douglas fir from the Colorado Basin, but for Sequoia gigantea and Pinus ponderosa he often obtained values of 0.98 and 0.96. Values for Central Europe are given below.

Choice of intervals: In order to determine a convenient period for calculating the average interannual fluctuation the average fluctuation was calculated for 5 Spessart oaks for increasing intervals of time for the period from 1740 to 1940.

TABLE 1

Values of the Fluctuations of Individual Curves and of Mean Curves (made from 5 individual curves) for various species and locations, 1850 to 1940.

Species	Location	S_E	S_M	R
Oak	Spessart	19.8	12.9	0.65
	Rimpar	16.0	11.6	0.72
	Thayngen	18.5	13.4	0.72
Beech	Vosges	-	26.3	-
	Spessart	30.9	22.4	0.72
	Bavarian Forest	32.7	23.2	0.70
	Beskides	39.0	28.9	0.74
Silver Fir	Vosges	20.4	14.7	0.72
	Bavarian Forest	15.9	13.5	0.85
	Beskides	22.3	17.7	0.79
	Bavarian Alps	16.3	11.4	0.69
Spruce	Bavarian Forest	16.1	13.3	0.82
	Beskides	21.6	17.8	0.82
	Bavarian Alps	16.0	11.7	0.78

Intervals greater than ten years produce proportionately smoother fluctuations. For a 10-year period the results are scattered by ± 2% around the mean and for any longer interval the average fluctuation coincides with the mean. For this reason the curve-data were divided into 10-year intervals for calculating their average fluctuations. This interval has the advantage that it is often chosen for calculating variations in temperature and precipitation.

I GEOGRAPHICAL VARIATION OF THE AVERAGE FLUCTUATION IN THE PERIOD FROM 1850 TO 1940

Average fluctuations in deciduous trees and conifers at several locations

We must now discover how much trees from different known locations with distinctly different climates differ in their average fluctuations. We can then use these results as a basis for possible statements about past climates.

In the Forstbotanisches Institut, Munich, annual ring curves from the following locations were available:

(a) beech and fir curves from the eastern edge of the Vosges towards the Rhine Rift on the Colmar heights;

(b) oak and beech from the southern Spessart;

(c) beech, fir and spruce from the Bavarian Forest (Mauth-Ost, Buchenau); and
(d) beech, fir and spruce from the Ostrawitza valley in the West Beskides.

I was also able to make comparisons with a mean curve of 7 oak trunks from the Cracow area which Dr. Karol Ermich kindly made available.

Five indi idual curves were selected for each species and for each location. These were about the same length, but above all, the same age. The values calculated for S were plotted. Those obtained for silver fir are shown in Fig. 1. Results were limited to those for 1850 to 1940 since the climatic data for the relevant locations only goes back to 1850 and some of the annual ring curves end in the 1940s. My values for the mean curves of beech, fir and spruce growing in the Beskides is consistent with Müller-Stoll's.

The locations are listed from west to east in Table 2. With the exception of sites on the eastern edge of the Vosges, it is seen that S_M for all the species increases towards the east. This is particularly noticeable for beech in moving from the Spessart to the Beskides.

TABLE 2

Fluctuations of mean of five individual curves, 1850 - 1940.

	Vosges		Spessart		Bavarian Forest		Beskides	
	S_M	R	S_M	R	S_M	R	S_M	R
Oak			12.9	0.65			17.2 (Ermich)	
Beech	26.3		22.4	0.72	23.2	0.70	28.9	0.74
Fir	14.2	0.72			13.5	0.85	17.7	0.79
Spruce					13.3	0.82	17.8	0.82

The climatic conditions at the locations

It is known that climatic factors undergo considerable variations with increasing distance from the sea. A brief outline of the conditions in Europe reveals that precipitation decreases towards the east, with the amount of summer precipitation increasing relatively. Most especially, the annual amplitude of temperature becomes much greater, i.e. in comparison with the equable rather oceanic climate at the coast, the climate becomes increasingly extreme, increasingly continental towards the European interior. The climatic data (Table 3) for the locations comes from the following weather stations or are calculated from neighbouring ones:

Vosges:	Colmar (200m above reference level) adjusted for Barr (400m above reference level)
Spessart:	Königstuhl (563m above reference level)
Bavarian Forest:	Buchenau (750m above reference level)

Beskides: Moravian-Ostrau (253m above reference level) adjusted for the locations Moravka and Mohelnitz (600m above reference level).

TABLE 3

ANNUAL TEMPERATURES AND PRECIPITATION (1850 to 1930).

Location	Temperature °C			Precipitation mm		
	Annual Mean	Jan	July	Annual Mean	Max.	Min.
Vosges	9.2	-0.1	18.0	480	969	389
Spessart	8.0	-1.9	15.8	990	1234	579
Bavarian	5.4	-3.8	14.5	1350	1917	829
Beskides	5.8	-4.5	15.2	860	1191	433

(a) Annual mean temperature: A decrease towards the east is noticeable in the annual temperatures. Thus the average temperature of the Spessart locations is 8°C; it drops to 5.4°C in the Beskides.

(b) Annual amplitude of temperature: The values of 17.7°C, 18.4°C and 19.7°C for the locations in the Spessart, the Bavarian Forest and the Beskides reveal a definite trend towards continental conditions. The amplitude of temperature of 22°C in the Cracow area fits this picture.

(c) Precipitation: In the Spessart and Bavarian Forests the annual precipitation is high, 980-1000mm and 1350mm. A decrease to 860mm a year occurs in the Beskides and to 700mm in the Cracow area. One sees that there are significant differences. The climates of the Spessart, the Bavarian Forest and the western Beskides have progressively a more continental character.

The most western locations, those on the edge of the Vosges towards the Upper Rhine Rift, also show certain continental features as they have a high mean annual temperature of 9.2°C (the winters are quite mild with a January mean of only -0.1°C, whilst the summers are very warm with a July mean of 18.0°C, which gives an amplitude of temperature of 18.1°C). Also, the precipitation is low, only 480mm, due to their position in the rain shadow of the main ridge of the Vosges. Indeed, around Colmar there is the famous "dry island" where individual steppe plants occur as outposts.

Comparison of values of S with the climatic data

The S values show certain agreement with the climatic data for the respective locations, especially with the annual amplitude of the temperatures (represented in Table 3 by the January and July Temperatures) and with the fluctuation quotients of the precipitation.

For beech, we have curves for all the locations. From the Spessart to the Bavarian Forest there is only a relatively small increase in the amplitude of temperature but a marked fall in the average annual temperature: S shows a small increase (Table 2). By contrast, there is a considerable increase in S with the greater continentality of climate that occurs between the Bavarian Forest and the western Beskides. If we turn to the west and compare the data for the Spessart beech with those of the eastern side of the Vosges, here too the increase in S for beech is related to the more extreme climatic conditions.

The comparisons are similar for fir and spruce. For oak, there is a steep rise in S on moving to the east.

II TEMPORAL VARIATIONS FOR THE SAME LOCATION OF THE AVERAGE FLUCTUATIONS OF DECIDUOUS TREES AND OF CONIFERS

Variations in historical time

Brehme (1951) found that larches on high mountains in the Berchtesgaden Land had different average fluctuations at different periods. Especially low values occurred for the period before 1650 (1551-1650, S_M 18%) whereas in the next 100 years the average fluctuation (23%) was much higher. He concluded that these differences were caused by climatic variations.

I have tried to find out whether other species revealed this phenomenon and to determine whether the variations in S depend on the climate. I have taken oak as representative of deciduous species and have investigated recent Spessart oaks and also oak beams in historical buildings in Büdinger and Gelnhausen (dated in the Forstbotanisches Institut to the High Middle Ages and to the 16th to 18th centuries).

The conifers considered were firs and spruces, and these were all recent trees from the forestry district of Kreuth near Tegersee which are growing in the woodland area of Reineck (1100-1200 m high). Unfortunately, chronologies for these two species do not go back as far as those of the oak, so I had to limit this investigation to recent material and that of the 17th and 18th centuries.

TABLE 4

Average Fluctuations of Ring Widths of Oak Curves for Spessart, Büdinger etc. at Different Periods.

Period	S_E	S_M	R
1280 to 1380	18.8	11.5	0.61
1550 to 1650	16.5	11.1	0.67
1650 to 1750	21.4	16.1	0.76
1850 to 1950	19.8	12.9	0.65

For oak, the periods compared were limited for various reasons to hundred year periods as follows:

High Middle Ages	1280 - 1380
Adjacent to 'Little Ice Age'	1550 - 1650 and 1650 - 1750
Recent times	1850 - 1950

The curves do not fluctuate equally over the whole period (Fig. 2 and Table 4). Wide fluctuations give way to weak phases. This is very evident in the High Middle Ages when for the years from 1280 to 1330 there was the high fluctuation of 21% in the individual and 14.2% in the mean curves. This was superseded by a period from 1330 to 1380 with less curve oscillation, the values for the individual and mean curves being 13.5% and 8.9%.

For the period 1550 to 1650 the fluctuations are considerably lower than in the following 100 years (Table 4).

For the period 1850 to 1950, curves for a great many locations are available. In more than half of them the average oscillation is continually decreasing, especially for the Spessart oak. The decadal trends for individual locations show many similarities. Thus all the curves rise from 1850 to about 1870 and fall again in the following decade. The uniformity disappears somewhat from 1880 to 1910, especially in the period from 1890 to 1900 in which 50% of all curves fall, whilst 50% rise. A distinct peak is present, however, in many curves (71%) between 1920 and 1930.

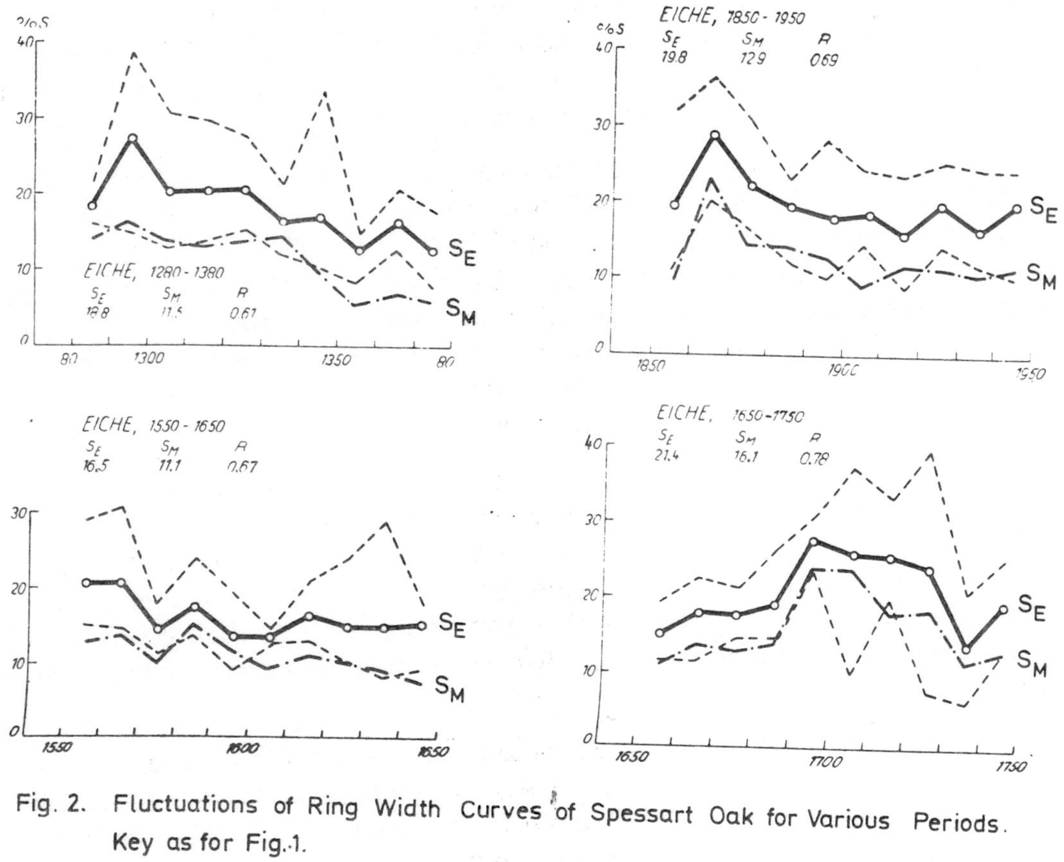

Fig. 2. Fluctuations of Ring Width Curves of Spessart Oak for Various Periods. Key as for Fig. 1.

Over the whole 100-year period in the High Middle Ages, despite the period of strong fluctuations from 1280 to 1330, oaks have relatively low fluctuations (Table 4). In the period 1550 to 1650 the values drop only to rise very sharply between 1650 and 1750. The recent Spessart oaks also fluctuate less than those of the period from 1650 to 1750; however their values are still higher than those of the High Middle Ages. The response of conifers, such as fir and spruce, also reveal an agreement with Brehme's results and those for oak outlined above. Fir and spruce also have a consistently higher average fluctuation at the turn of the 18th century than have recent trees at the same location.

Comparisons with climatic information: To explain the temporal variations in the average fluctuation in deciduous trees and conifers we must look at the history of climate in Central Europe as well as the history of forests and pollen analysis.

High Middle Ages: H. Flohn and A. Wagner agree that this must have been comparatively warm, especially around 1200 when there were many warm summers and mild winters. A temporary period of deteriorating climate, according to Flohn from 1250 to 1330, coincides with the period for which we can establish increased curve oscillation. This is followed by a period of mild oceanic climate which results in a low average fluctuation of the annual ring curves. In the more favourable climate the limit of viniculture reached East Prussia and in the Black Forest above 700 m. The upper limit of settlement was raised in the Alps and highlands, and there was a bonanza in mining in the Tauern when the glaciers retreated to far above their present level. Overbeck was able to establish a distinct increase in Ulmus-pollen between the two recurrence surfaces in the upland moors of the Rhone. The higher position of the tree-limit in the upper highland, established by Firbas and Losert through pollen analysis, confirms the rise in temperature.

16th to 19th centuries: After a short deterioration the climatic optimum of the High Middle Ages once more returned, though according to Flohn by 1500 the summers were already cooler, and about 1540 there was a definite change to cold winters and cool summers. This deterioration is not reflected in our curves, however, as with values of 15.6% and 11.1% they show a weak response, even below that of the High Middle Ages. By contrast, the glacial advance in the first half of the 17th century and its recurrence in the period from 1820 to 1855 stand out clearly in our curves.

During this period, as Firbas suggests, the tree-line dropped at least 100-200 m, and spruce reached its present level in the upper Harz and the Sudetes. Overbeck puts the formation of the recurrence Surface I in the upland moors of the Rhone in this period and sees the climatic change as responsible for its emergence. Similarly, during this period the area available for viniculture declined. Glacial advances also forced mining at high altitudes to be abandoned.

Late 19th century to Present: A noticeable change occurred around 1870. Since then air-temperatures have been slowly increasing, and since about 1900 this warming has occurred on a global scale. In addition, precipitation has increased. Paschinger observed that the Alpine glaciers are again retreating quickly, and as a result the snow-line in the Gross Glockner area has now risen about 150 m. Scherhag believed there was exceptionally heavy precipitation and very even temperatures in the decade from 1920 to 1930. Maurer alleges that a high percentage of the glaciers in Switzerland retreated at this time.

The increasing oceanity of the climate in the last 100 years (i.e. increase in precipitation and decrease in annual amplitude together with warming) is revealed in a decrease in the average fluctuation in comparison with the period before 1800. It is particularly pronounced between 1920 and 1930.

The results give us yet further information. Meteorology tells us that the present climate is still not as warm, and its temperatures are still less than they were in the High Middle Ages. The results produced here are in agreement with the climatological literature.

Conclusion: The results both of the geographical variations and of the temporal variations show that a moderate oceanic climate always corresponds to a low value of the average fluctuation, and that by contrast a period of extreme climatic conditions always results in a high average fluctuation of the annual increase. Both severe cold and great drought are reflected in the same way, i.e. by large fluctuations in the annual ring widths. It is therefore impossible to identify from the fluctuations which of the two is responsible. One can only assume that the weather conditions were extreme.

Variations in Prehistoric Times

Huber and von Jazewitsch (1958) found that the fluctuations in the ring widths of oaks from the Neolithic excavation in the Swiss lake village of Thayngen-Weier (see also this volume, page 21) were surprisingly large. The values for a group of five individual curves and comparable ones for recent oaks in the same area are:

	S_E	S_M	R
Recent	18.4 (24.8-15.7)	15.0 (23.4-10.3)	0.81
Neolithic	22.4 (27.9-18.4)	20.8 (27.0-14.7)	0.92

In addition, the S values for five oaks from neolithic finds at Thayngen-Weier and Burgäschi-Süd show the same trend. However, Bronze Age samples from Zug-Sumpf gave fluctuations (over an 80 year period) that were lower than both recent values and even more so than the Neolithic ones.

This suggests that more extreme climatic features prevailed in our latitudes during the 100-year period investigated in the neolithic than now. Also that during the 80-year period investigated in the Bronze Age the climate must have been significantly more equable, with more regular precipitation and a smaller amplitude of temperature than now.

The trends in temperature during the post-glacial period are well-known, for example, there is abundant evidence about the warm optimum of the period of mixed oak forest in which the neolithic occurs. Yet very different views are held about climatic conditions in the Bronze Age.

The transition from the warm period to the post-warm period in the Neolithic is apparent from the advance of the beech and the fir at a time when the trees of the mixed oak forest were becoming increasingly less important. This can be seen clearly from a list of tree species which Huber and von Jazewitsch compiled from archaeological finds for Thayngen-Weier (neolithic) and Zug-Sumpf (Bronze Age). Remains of tree trunks in the neolithic settlement at Thayngen-Weier consist of 28% oak and 35% ash; in the Zug-Sumpf these drop to 20% oak and 15% ash. Firbas does not deny that a direct deterioration in climate could be the cause of the rapid decline in the ash, which runs remarkably parallel with the advance of the beech.

Gams and Nordhagen assume that the climate of the Alpine foreland in the Bronze Age was warm-dry, which does not however mean like a steppe climate; nevertheless, Keller and Rytz believe that because the beech was widely distributed (confirmed by pollen diagrams) the climate of the Bronze Age was relatively wet and cool without extreme temperatures. Our fluctuation values, although only an 80-year period, agrees with the views of Keller, Rytz and Groschopf.

Note by Editor

Table 1 of the original paper lists the species, origin, number of rings, and the persons responsible for producing the 138 individual curves used. The average number of rings (the first 50 were disregarded) per curve is about 150 (range 49 to 354).

References

See this volume, pages 7 to 10.

2.3. THE CONTRIBUTION OF WOOD DENSITY TO DENDROCHRONOLOGY AND DENDROCLIMATOLOGY

by H. Polge, Nancy, France.

Summary

Variations in wood density, particularly in the maximum annual density, differentiate between the years of a tree's growth. They tend to be more efficient than ring widths for dating purposes.

They are related to climatic factors (such as rainfall) and through them to phenomena such as the melting of glaciers, but they are also sensitive to other factors. For example, two progenies of Maritime Pine differ significantly in their variations of maximum density; and the correlation of maximum density with summer rainfall is always significant, but is negative with Douglas fir and grand fir yet positive with Scots Pine.

The maximum density varies more than ringwidth or other density criteria with the soil and with the height in the trunk of the tree.

The Wood Quality Research Station aims to study the biological aspects of wood quality, i.e. to analyse the influence on wood quality of ecological factors (soil, climate); of sylvicultural factors (competition between trees); of genetic factors; and of artificial alterations of tree biology (fertilization, pruning). An important defect of wood is its heterogeneity due to the fact that its structure changes strongly within and between annual rings. Many years have been devoted at Nancy to solving the problem of quantifying wood homogeneity.

X-ray densitometry of 5 mm cores gives a continuous record of wood density variations from pith to bark. This is satisfactory for defining the heterogeneity as it is easily quantified by measuring the differences between maximum and minimum annual densities, or the slope of the charts at the transition between earlywood and latewood.

After having investigated some hundreds of charts, it appeared that the densities could, by chance, be useful in the field of dendrochronology. It was soon shown (Polge, 1966):

(a) That the shape of the annual density chart was often typical of the year;
(b) That certain quantitative data that could be measured on density records, particularly the maximum and minimum annual densities, were more discriminant than were ringwidths;
(c) That these maximum and minimum annual densities were closely related to climatic factors, specially to rainfall and to temperatures.

Subsequent work as follows has corroborated the usefulness of density for dating purposes:-

1. A study of the effects of industrial smokes on Norway spruce growth has proved (Polge and Keller, 1969):

(a) That air pollution clearly changes the shape and values of density curves;

(b) That the characteristic years (years for which a parameter varies in the same direction in at least 80% of the trees) are much more numerous for maximum and minimum annual densities than for ring-widths, so that it is easier to find a missing ring with the help of density than without;

(c) That densities are less dependant on the age of the tree than is the ringwidth; therefore it is not so necessary to have a reference plot of the same age as the trees to be dated;

(d) That density and ringwidth do not always characterize the same years, so when both parameters are available, dating is easier and more reliable.

2. Another experiment concerning Silver fir and Norway spruce (Keller and Millier, 1970) has confirmed the last point. In the population of fir there were four characteristic years for ringwidth and ten for density criteria, but with only one year common to both parameters. In the spruce population there was also only one year in common amongst the two years that were characteristic for width, and the ten years characteristic for density.

From a main components analysis it appeared that in the spruce the first component, accounting for 54% of the total variation between years, is only related to a density factor, namely to the difference between the maximum and the minimum annual densities.

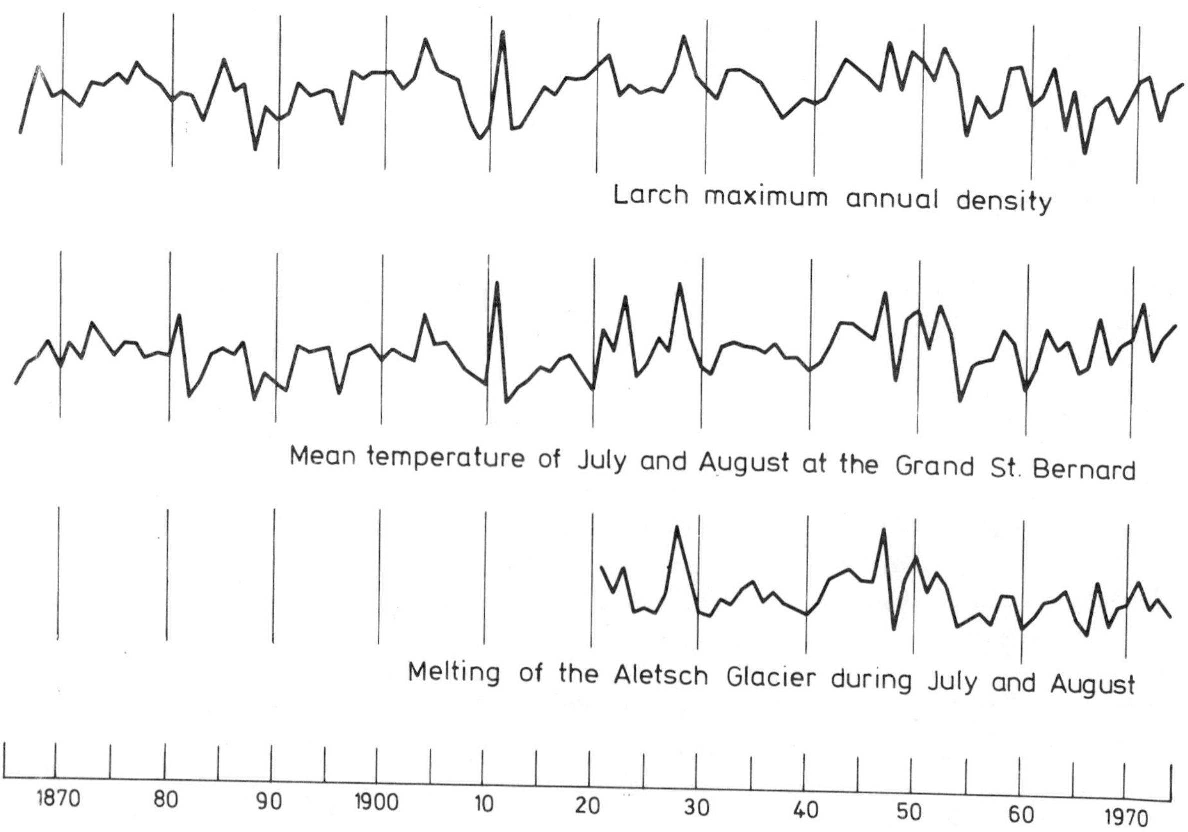

Fig. 1. (After Röthlisberger)

3. Fletcher and Hughes (1970) have used densitometric records to date some medieval oak beams. From both ringwidths and density they have been able to suggest a precise dating for the specimens. They have related high density latewood to high yields of corn, and also described a particular type of density pattern in which the density substantially decreases later in the year due to a monsoon type weather with high rainfall in the late summer.

4. Parker and Henoch (1971) have found that Engelmann spruce latewood density was closely linked to August mean maximum temperature (r = 0.85). They also have shown that during a 117 years interval of time (1850-1966), the maximum latewood density indices gave higher average cross correlation, mean sensitivity and percentage of variance in common, as well as lower first-order serial correlation than the ringwidth indices. This proves the higher dendrochronological quality of density for each of these 4 factors.

5. Röthlisberger (1976) has measured the maximum density of European Larch from 1560 to 1974 and found (Fig. 1) a wonderful parallelism with the melting of a glacier during July and August (results for 54 years) and with the mean temperature of July and August (results for up to 277 years).

FACTORS RESPONSIBLE FOR DENSITY VARIATIONS

Although many examples show the potentialities of density variations as an adjunct or in substitution for ringwidths can we infer that wood density is a panacea in the field of dendrochronology?

Probably not, for if density is very sensitive to the general influence of climatic factors, we also know that it is under a strong genetic control and that it varies widely with the soil characteristics and even with the distance to the living crown in which photosynthates and most of the growth substances are formed. We must therefore expect some interaction between climate and genotype (within or between species), between climate and soil, and between climate and position within the tree. Some recent experiments have shown such interactions.

Interaction between Climate and Genotype

Within Species: From two families used in a progeny test on maritime pine, ten seedlings were randomly sampled per family. Their synthetic, xylochronologic profiles are shown in Fig. 2. These diagrams record, for each year, in abscissa, the mean ringwidth, and in ordinates, the mean maximum and minimum wood densities and these extreme values of density are joined together by segments of straight lines.

We can see that the effect of annual climatic variations is not the same on wood density of both families, particularly at the end of the year 1970 when compared to 1969 and 1971. The maximum annual density again is more sensitive than minimum density or ringwidth. The "t" test is highly significant for all the years from 1967 to 1971, but not for 1970. If we subtract the 1970 maximum density from the mean between the 1969 and 1971 maximum densities, we find a negative value for Family 1 and a positive one for Family 2, and the difference between the families is significant.

The explanation lies partly in the phenology of both families: the increment cores used in this study were collected in June 1972 and at this time the transition from earlywood to latewood was just starting for one family, but very much ahead for the other family. The difference between families is again

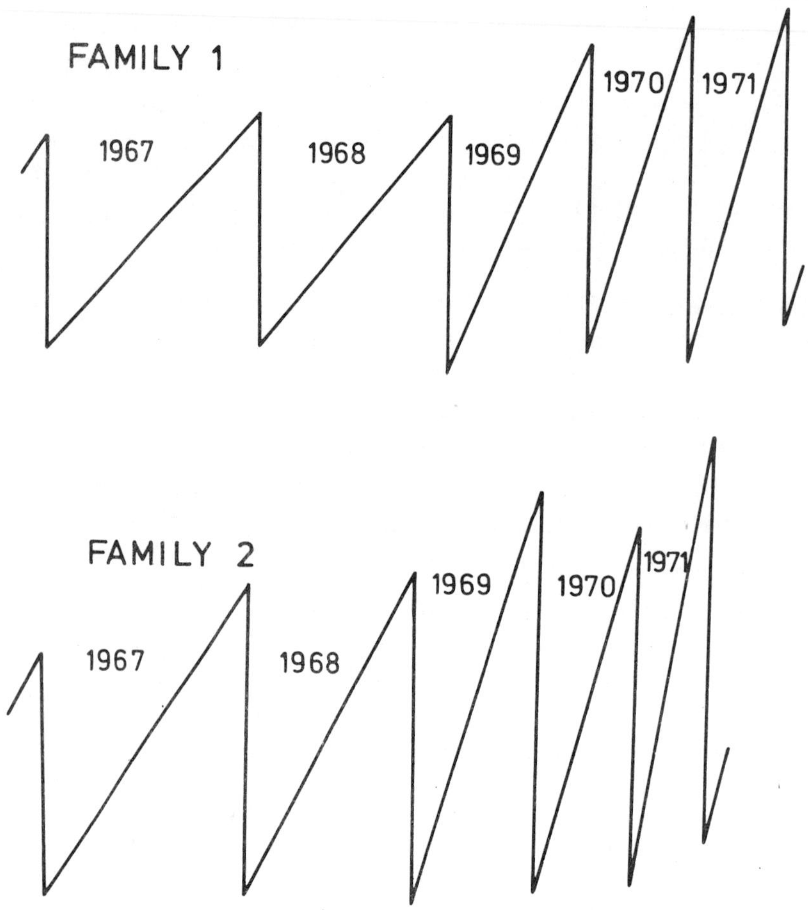

Fig. 2. Maritime pine. Xylochronologic profiles.

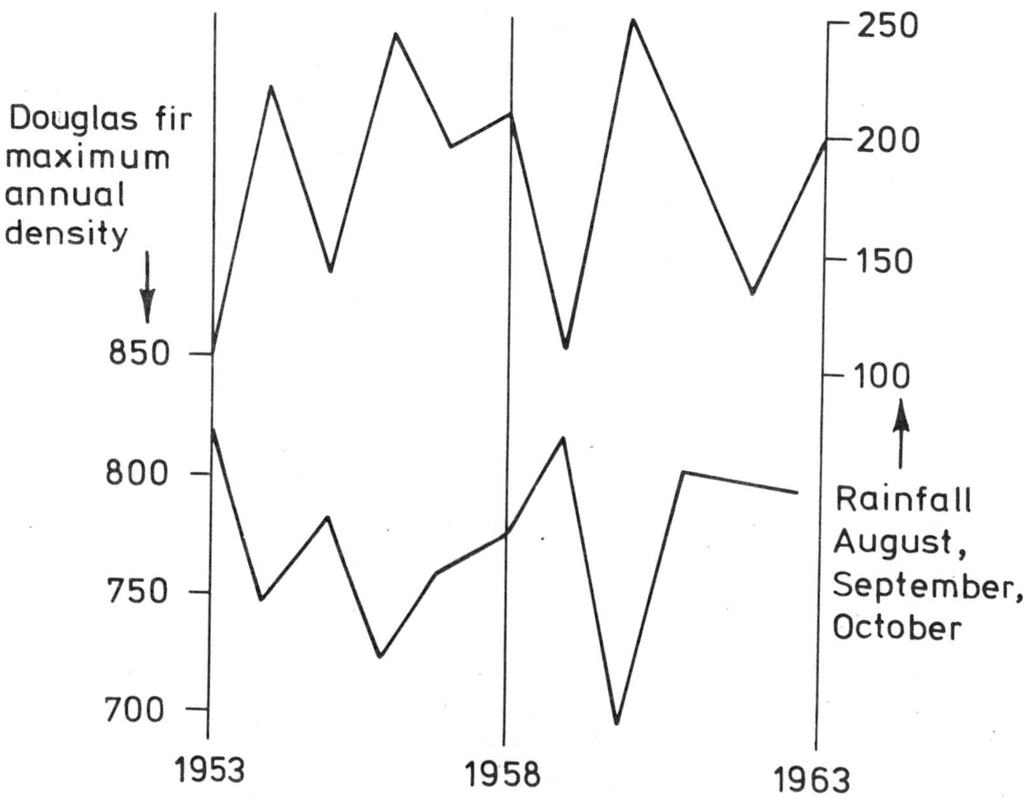

Fig. 3. Correlation of maximum density with rainfall.

very significant (t=4.07). That means that height growth and needle elongation stops earlier for the second family. If the different phases of the growth process do not occur at the same time, it is not surprising that a given climatic factor does not affect equally the wood density of both families.

Between Species: The first study of the relationship between densitometric data and climatic factors (Polge, 1966) dealt with two exotic conifers — grand fir and Douglas fir. Many significant correlations were found relating the maximum or the minimum densities to the rainfall or the insolation. The relationship between maximum densities and rainfall provide an interesting example of the variation from species to species.

For Douglas fir there was a correlation coefficient of -0.89 for 11 years with the rainfall of August, September and October (Fig. 3). For grand fir, the highest coefficient was -0.77 with the rainfall of September and October. Thus in both species low rainfall causes the formation of dense latewood and vice versa. This result seems very logical after the experience of Larson (1963). By artificially modifying the watersupply of Pinus resinosa he proved that in dry conditions the tree produces latewood-type fibres with very narrow lumina and therefore dense wood.

Another investigation with Scots Pine (Bazerque, 1974) gave however, quite opposite results. All the correlations were positive between the maximum density and the monthly rainfalls of June, July, August, September and October, and a highly significant coefficient was found with the total rainfall from June to October (r=+0.623 for 50 years). Fig. 4 shows that generally speaking, high maximum densities are related to high rainfalls. This is difficult to understand, but the exceptions, shown by the years 1938 and 1939, perhaps give at least a part of the explanation. Their rainfalls are the highest

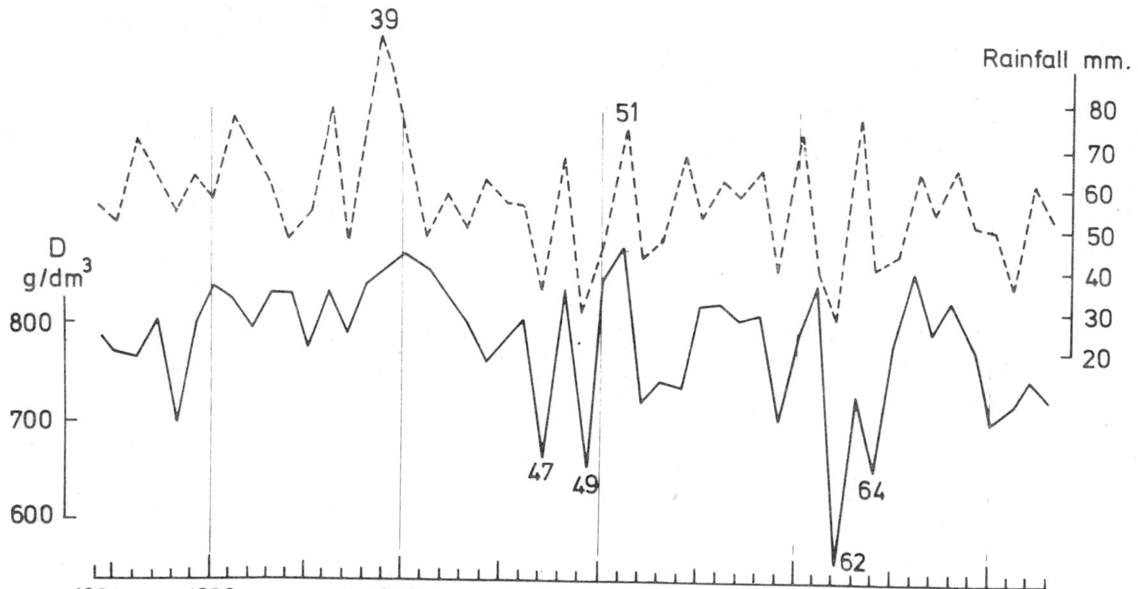

Fig. 4. Relationship between the mean rainfall of June to October and maximum density. FESSENHEIM, SCOTS PINE.

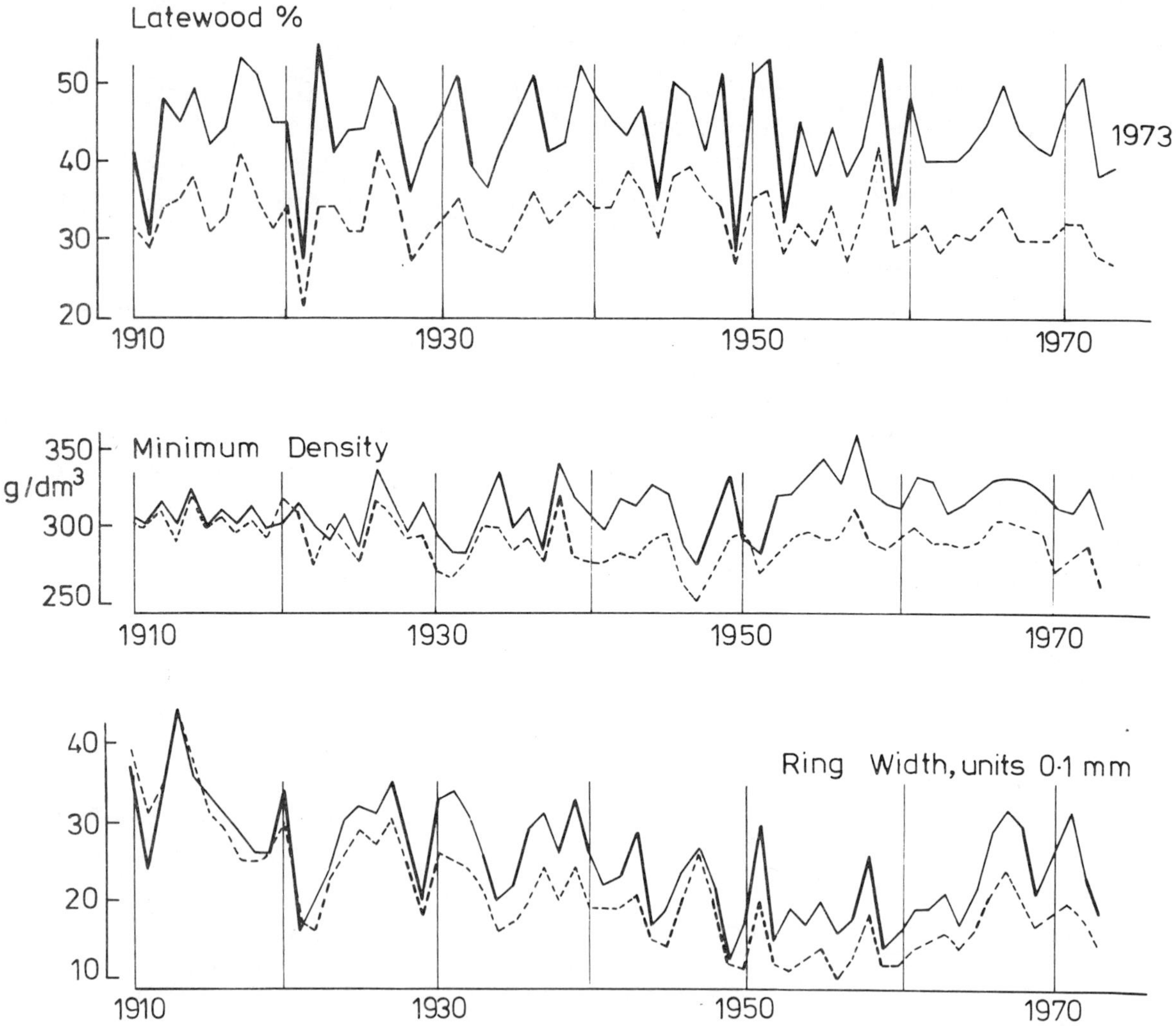

Fig. 5. Comparisons between samples at breast height, 1·3 m (unbroken lines), and at base of the living crown, 10 m (broken lines) for Scots Pine.

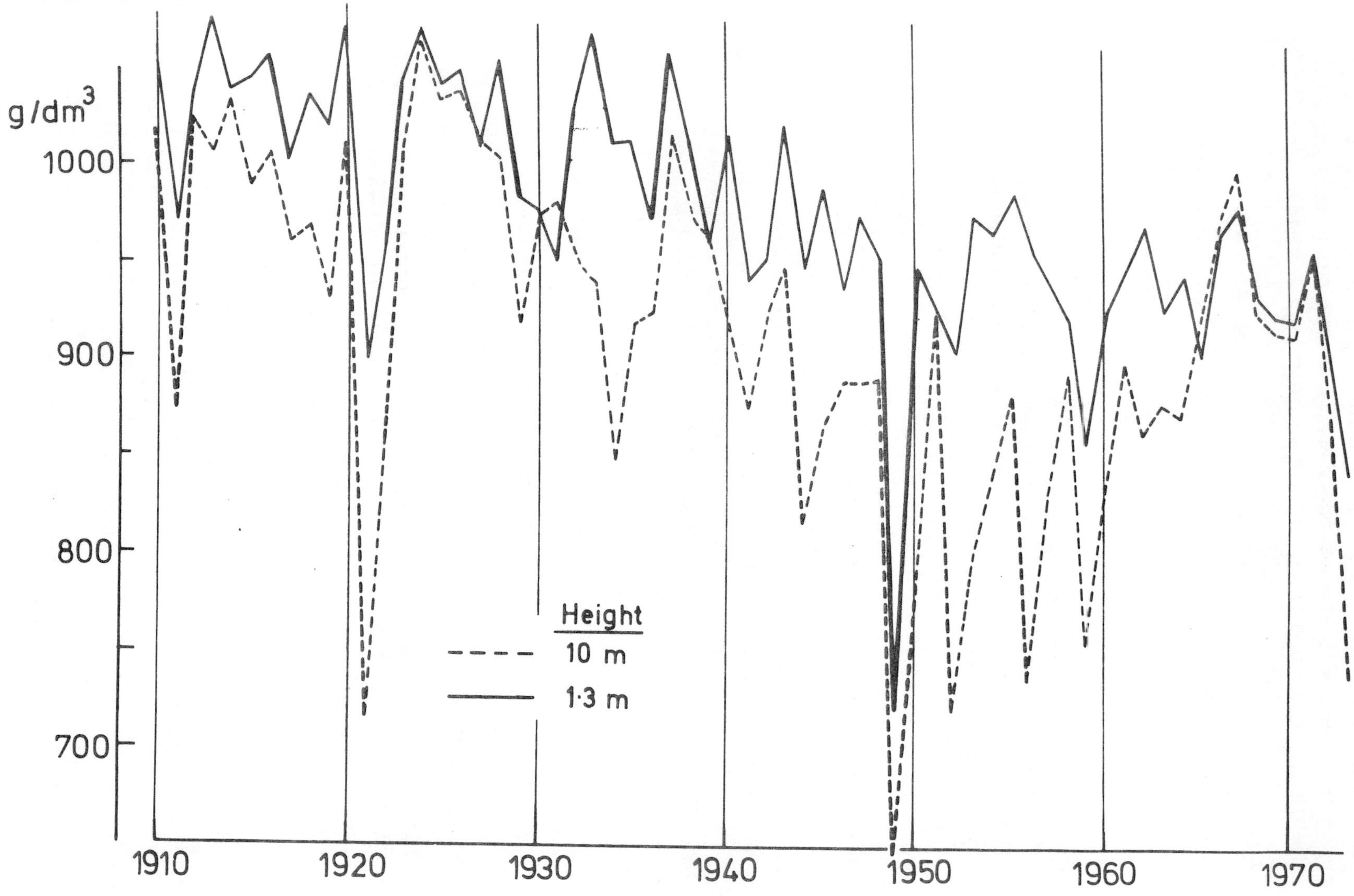

Fig. 6. Comparison between samples at two heights for Scots Pine. Maximum Density.

of the period 1923-1973 but their maximum annual densities although high are similar to many other less wet years. This observation leads one to think of a threshold phenomenon. The Scots Pine stand grows on a poor sandy soil at Fessenheim in the Rhine valley and the water supply may be a limiting factor for all the physiological functions and even for the photosynthesis which plays an important part in the formation of dense wood. Thus below the threshold level the maximum annual density could be related positively to the rainfall.

Interaction between Climate and Soil

The pine stand of Fessenheim, mentioned above, has been compared to another pine stand growing at Charmes in the plain of Vosges, at about the same latitude and at a distance of 110 km. There are climatic differences between the sites: the total rainfall at Charmes is 802 mm, regularly distributed during the year, and at Fessenheim 524 mm with a minimum in winter and a maximum in summer. But they chiefly differ in their soil: contrary to that of Fessenheim, already described, the soil of Charmes is a leached soil with pseudo-gley with a better water supply but with some risk of roots' asphyxiation when the underground water level is high.

Table 1 gives the characteristic years of both stands for ringwidth, maximum and minimum annual densities. It will be noticed that more years are characterized by ringwidth at Charmes and more by maximum density at Fessenheim. At Charmes, 10 years are characterized by ringwidth alone, 10 years by density alone, and 10 both by ringwidth and density; while at Fessenheim, we have only 4 characteristic years for ringwidth alone, 22 for density alone and 12 for both ringwidth and density.

A more important observation concerns the characteristic years common to both stands. They are 8 for the ringwidth, 5 for the minimum annual density, and 5 for the maximum density, but for three of these, 1935, 1937 and 1960, the directions of the variation are opposite.

Effect of Position within the Stem

When dendrochronologists have to date a piece of wood they may not know from which part of the bole the piece had been cut. It was therefore interesting to compare ringwidth and other parameters of ring structure at different levels within the stem. This has been done in a Scots Pine stand near Versailles on 63 annual increments of 28 trees at 2 levels: at 1.30 m (breast height) and 10 m (base of the living crown).

Figs. 5 and 6 show the variation of ringwidth, minimum density, latewood percentage and maximum density at both levels. The characteristic years are indicated by thicker lines.

For the ringwidths, we have 19 characteristic years at breast height. At 10 m there are 13 amongst which only 8 are common with those at breast height. Contrary to what is reported for other species in the literature, the breast height ringwidths tend to be larger than at the base of the living crown, but the variations from year to year are roughly similar.

For the minimum annual densities, we have 5 characteristic years at breast height and 7 at 10 m, with only one in common.

For the latewood percentages, there are 17 characteristic years at 1.30 m and 5 at 10 m. Of these 4 are in common with the lower level. Here again the 2 graphs are generally, as we would expect, parallel with higher percentages at breast height.

Lastly for the maximum annual densities, we have 10 characteristic years at 1.30 m; 9 at 10 m with only 3 common to both levels. For the characteristic years, we find opposite directions for the variations in certain years and the parallelism between the graphs is not good, if we except the very typical years 1911, 1921, 1949 and the last 8 years.

Conclusions

Broadly speaking, I am convinced that the maximum annual density is the best parameter for wood dating. This superiority is a consequence of its high sensitivity to climatic factors. But density is also very sensitive to other sources of variation: genotype, soil and position within the stem.

We have to avoid wrong genalisations, but with improved knowledge of these interactions we should, in due course, be able to identify one year by low maximum density in one species and by a high one in another species; or, for the same species, by a high density on one site and a low one on another site. In that way we shall get to know not merely the broad outline of a year's climate but also the temperature, rainfall or insolation during a limited period of the growing season.

The studies done so far show that the information given by density does not duplicate that given by ringwidths. Together they will make the dating of wood and the reconstitution of past climates more and more reliable and easy.

References

Bazerque, F., 1974, Etude dendrochronologique et dendroclimatique de deux peuplements de pins sylvestres. Thèse de spécialité, Université de Nancy I, U.E.R. Physique, Chimie, Biologie.

Fletcher, J. M., and Hughes, J. F., 1970, Uses of X-rays for density determinations and dendrochronology. The University of British Columbia, Faculty of Forestry, Bulletin, 1, 41-54.

Keller, R., and Millier, C., 1970, Utilisation des composantes de la densité en xylochronologie. Annales des Sciences Forestières, 27, 151-196.

Larson, P. H., 1963, The indirect effect of drought on tracheid diameter in Red Pine. Forest Science, 9, 53-62.

Parker, M. L., Henoch, W. E. S., 1971, The use of Engelmann Spruce latewood density for dendrochronological purpose. Canadian Journal of Forest Research, 1, 90-98.

Polge, H., 1966, Etablissement des courbes de variation de la densité du bois par exploration densitométrique d'échantillons prélevés à la tarière sur des arbres vivants. Applications dans les domaines technologique et physiologique. Thèse Docteur ès-Sciences Appliquées, Faculté Sciences, Université de Nancy.

Polge, H., Keller, R., 1969, La xylochronologie, perfectionnement logique de la dendrochronologie. Annales des Sciences Forestières, 26, 225-256.

Röthlisberger, 1976, Gletscher- und Klima-schwankungen im Raum Zermatt, Ferpèche und Arolla. Die Alpen, 3/4, 59-152.

TABLE 1

Characteristic Years for Pine at Charmes and Fessenheim

Years	Ringwidths		Maximum Annual Densities		Minimum Annual Densities	
	Charmes	Fessen-heim	Charmes	Fessen-heim	Charmes	Fessen-heim
1925	-				+	+
1926	-				+	+
1927		+				
1928		-		-		
1929		+		+		-
1930	+	+				
1931					-	
1932	+			-		
1933						+
1934	-	-			+	+
1935			+	-	-	-
1936	+		-			
1937	+		+	-	-	
1938				+	+	
1939			-			
1940	-		+	+		+
1941		+			-	-
1942	-	-		-	+	
1943	+					
1944	-	-	-			
1945	+		+			
1946				+		
1947	-			-		
1948	+			+		-
1949	-			-		+
1950	+	+		+		
1951	+	+		+		
1952		-		-	+	
1953				+		
1954						

POLGE						Wood Density
1955				+		
1956		−				−
1957	+		+			+
1958					−	−
1959				−		
1960			−	+		
1961			+	+		
1962	−	−		−	+	
1963				+		
1964						
1965	+			+	−	
1966	+	+		+		
1967				−		
1968						
1969						
1970		−		−		
1971			−			
1972	+					+
1973		−				
Number of characteristic years	21	16	11	26	12	13

+ Values higher than in the preceding year: - Values lower than in the preceding year.

2.4. THE DENDROCLIMATOLOGICAL VALUE OF EUROPEAN LARCH IN THE FRENCH MARITIME ALPS

Françoise Serre, Marseille

In order to develop dendroclimatological work in the Mediterranean zone of Europe, thirty eight larches (Larix decidua Mill.) growing in a nearly pure stand situated north of Nice (France), at 2100 m altitude, were sampled. Sections of seven stumps were also collected at this site.

The site is located in the south-west corner of the larch distribution area and is situated at the upper altitudinal forest border. The trees are mostly gigantic and vigorous but, nevertheless rather affected by weather. All these facts were promising as concerns the age and the sensitivity of the trees and therefore for their climatological value.

Actually, millennia-long chronologies are available (the oldest stump sectioned dates from 930 A.D.) and the analysis of the material reveals effective tree-ring width variations. The skeleton plot technique has been used for cross-dating.

The comparison of ring width variations with the meteorological data for the region by means of a Benzecri's analysis of correspondences and a Fritts' response function shows the influence of June and July precipitations (high precipitations correspond to narrow rings, low precipitations to wide rings) and the influence of June to September temperatures (high temperatures lead to narrow rings, temperature below the general mean to wide rings).

These relations still apply when the mean ring width variations during the thousand years available are compared with the record of noteworthy climatic and economic events in France and even in Europe since the year 1000 A.D. (Sanson, 1945).

Moreover, the annual variations of narrow and wide rings are in good agreement with the annual variations of the advance and retreat of some alpine glaciers probably on account of the peculiar sensitivity of the two systems to moisture.

The so-called "medieval warm epoch" and "little ice age" can be detected in the variation of the mean curve for all the samples.

The surprising reverse relationship of the larch to the early summer precipitation can be explained by its intrinsic high transpiration rate which needs high light intensity and a dry atmosphere. The European larch of the alpine mediterranean zone proves to be of real dendroclimatological interest.

Benzecri, J.-P. et al., 1973—l'analyse des donnees. Il L'analyse des correspondances. Dunod éd., II, 624p.

Fritts, H. C., Blasing, T. J., Hayden, B. P. and Kutzbach, J. E., 1971—Multivariate techniques for specifying tree-growth and climate relationships and for reconstructing anomalies in paleoclimate. Journal of Applied Meteorology, vol. 10, no. 5, 845-864.

Fritts, H. C., 1976—Tree Rings and Climate. Academic Press, 567p.

Sanson, J., 1945—Recueil de données statistiques relatives à la climatologie de la France. Mémorial de la Météorologie Nationale, no. 30, 148p., 1 carte, Paris (1961).

2.5. X-RAY DENSITOMETRIC RESULTS FOR SUBALPINE CONIFERS AND THEIR RELATIONSHIP TO CLIMATE

F. H. Schweingruber, O. U. Bräker and E. Schär,
Birmensdorf, Switzerland

Summary

An automatic data recording system for X-ray densitometric equipment has been established. Maximum density in subalpine trees is more suitable than the width of the earlywood, latewood or whole ring for chronological as well as for dendroclimatological investigations. All conifers investigated (Picea abies, Abies alba, Pinus montana, Pinus cembra, Larix decidua) are well synchronised. Apart from samples that come from Mediterranean sites, spruce and fir give mean curves that show high agreement for the entire Alpine region (from Raxalpe near Vienna to Mt. Cenis, France).

Maximum density in spruce is highly correlated with summer temperatures (July, August, September) and with run-off from selected glaciers. It appears that by using X-ray densitometry, the fluctuations of summer temperatures during past centuries can be reconstructed.

Defoliation produced by Larch bud moth is visible in the maximum density and in the ring-width curves.

As voluminous bibliographies and the results presented herein demonstrate, the best density structure diagrams can be obtained at present with the X-ray method. At the Swiss Forest Research Institute the technique developed by Polge has been tested and partially improved; the data acquisition and processing steps have been improved so as to use technological and electronic possibilities (Lenz, Schär, Schweingruber 1976). The aim of the study is to interpret the climatic information that is contained in certain annual ring parameters, in particular the maximum density and growth ring width, for samples collected in the alpine region (Fig. 1).

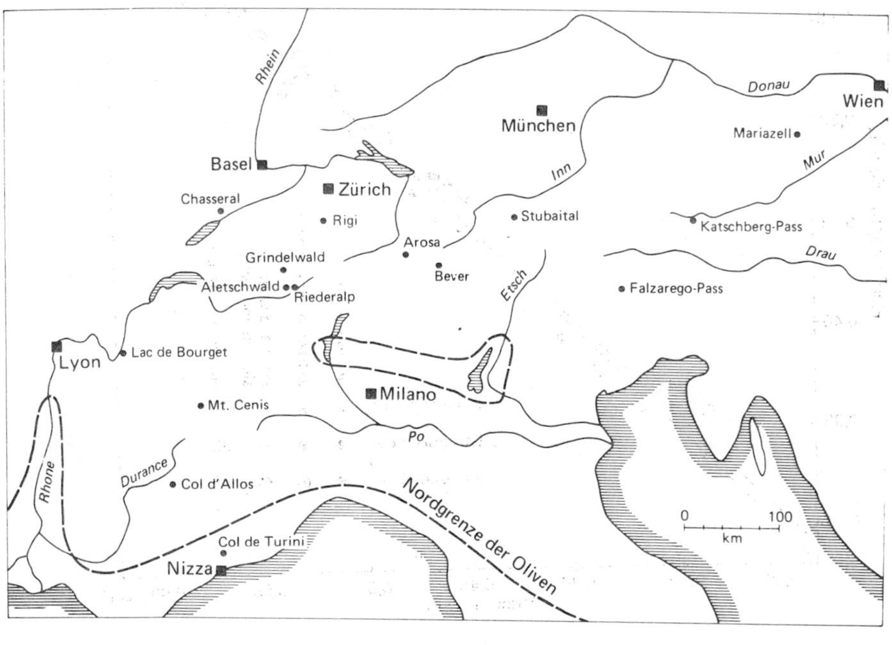

Fig. 1. Origin of core samples.
■ Geographic orientation • Core sampling sites

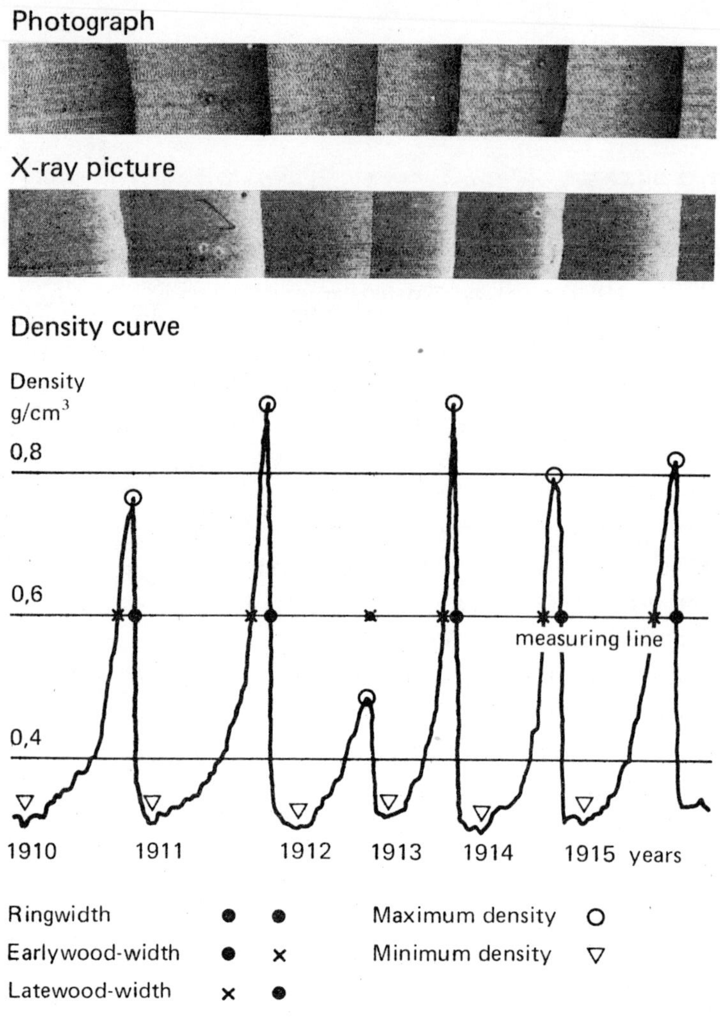

Fig. 2. Density diagram with wood parameters.

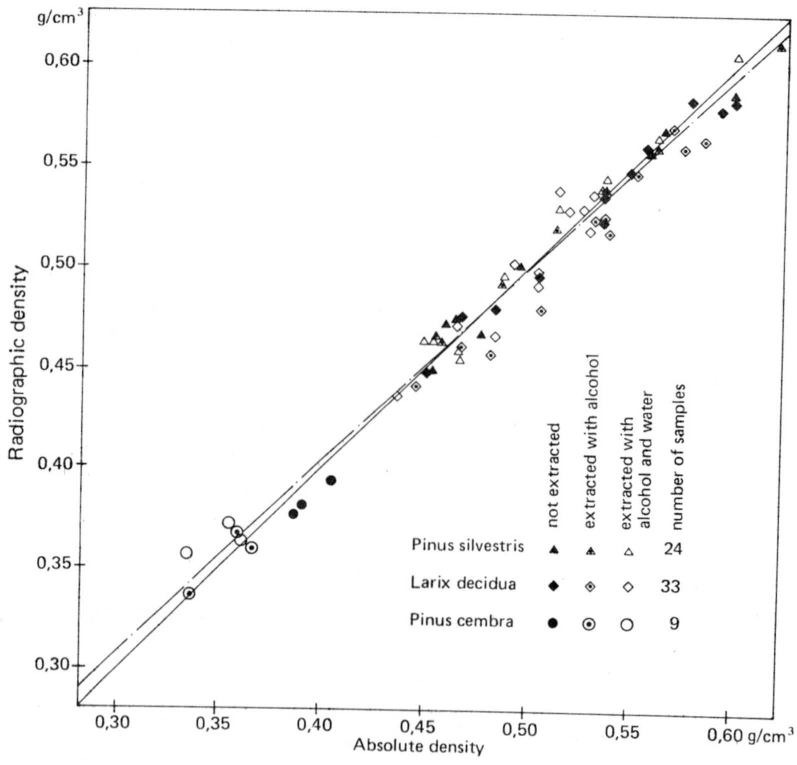

Fig. 3. Our radiographic densities (corrected by x 0·933) versus absolute densities.

Sources of error in data acquisition

The density diagrams (Fig. 2) were drawn with a linear plotter. An electric data-acquisition system is connected to the microdensitometer and the linear plotter selects values which represent anatomic characteristics. These values are recorded on punch tape. The values were checked, corrected, presented in pointplot form (computer CDC 6500/6400, Polytechnical School of Zurich, satellite at Swiss Forest Research Institute, Birmensdorf), and subsequently synchronized. The statistical analysis was made in part with programmes developed by H. C. Fritts.

Because the movable, and sometimes irregularly distributed substances in the cells and the cell walls influence the density, they are extracted, the resins with alcohol, and the heartwood substances with water. The resins are 20-30% lower and the heartwood substances in larch > 100% higher in density. Our average radiographic densities are higher than the true densities (Fig. 3). For *Larix decidua*, *Pinus silvestris* and *Pinus cembra* it is 7.2% higher; for *Picea abies*, 12.9%. The corrected radiographic values are practically equal to the absolute density.

In addition to technical errors in density recording, anatomical damage may falsify the density diagrams. In the subalpine region large bud moths attack larch trees more or less periodically (Fig. 4). Reduction of photosynthetic activity is indicated by a decrease in the width of the growth rings, and a decrease in the maximum density in the late wood

It was observed that moth infestation has a greater detrimental effect on cambium activity than on the development of cell walls in the late wood; only in years in which the trees are intensively infested is the maximum density reduced. After a year of moth infestation a tree requires 3-4 years before

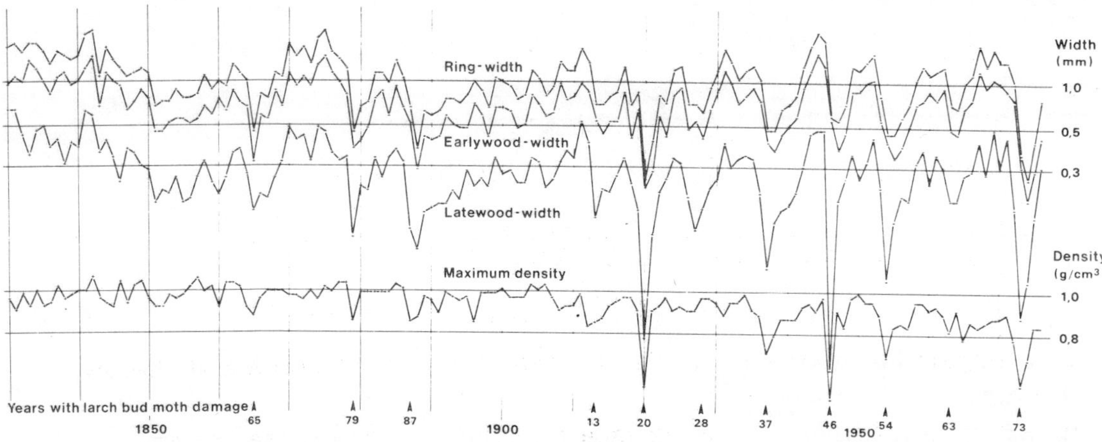

Fig. 4. Effect of infestations of the larch bud moth on width of growth rings and on maximum density.

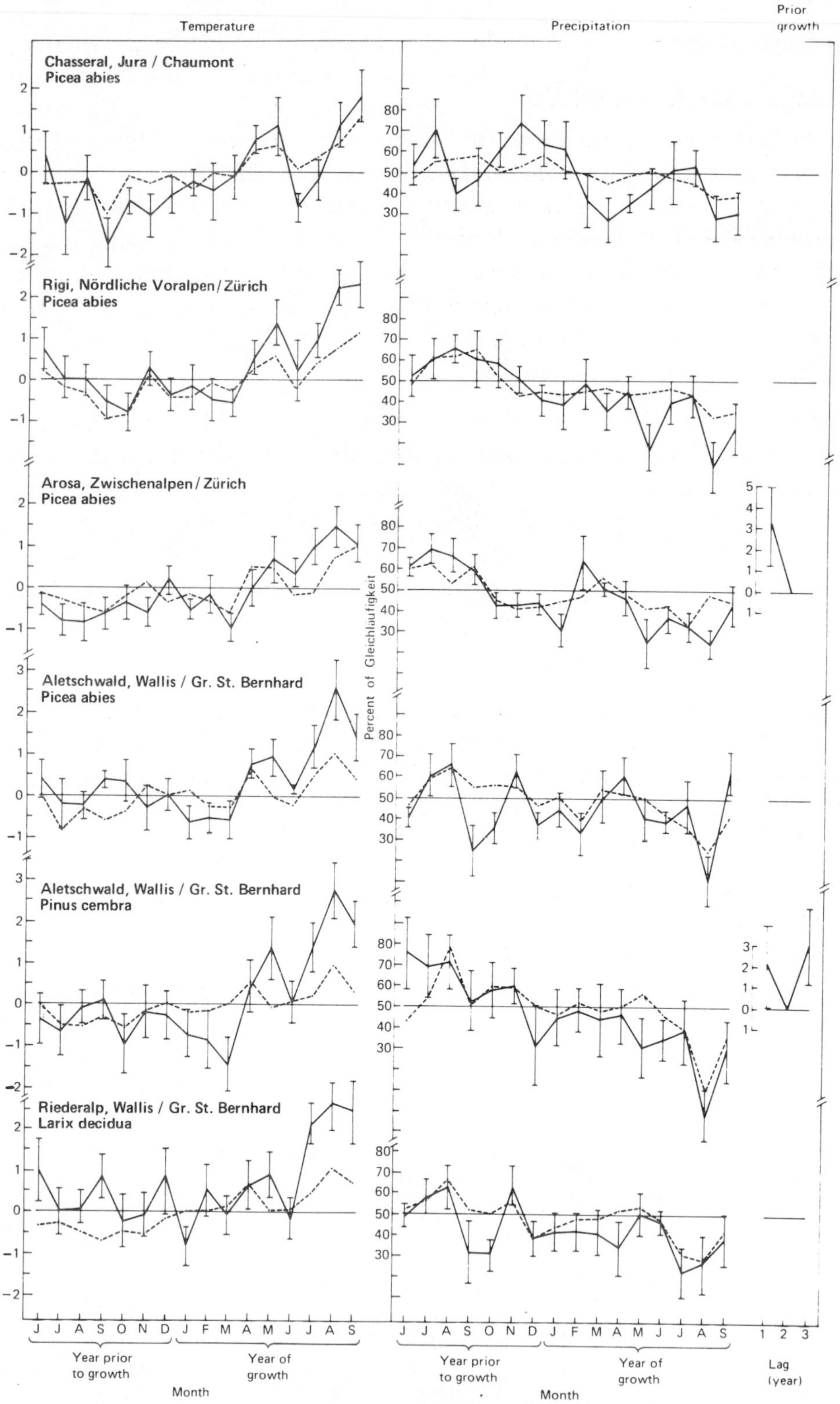

Fig. 5 Headings on left give site of trees, meteorological station and, below, species.

A. Joined by solid lines: indices (scale at left) for response of annual maximum density to mean monthly temperature (left) and to mean monthly precipitation (right). Vertical lines give 95% confidence limits.

B. Joined by broken lines: agreement (gleichläufigkeit) as percentage correlation (central scale) of year to year changes in direction of the annual maximum density curves with mean monthly temperatures and the mean monthly precipitation. For A and B, period is 1900-70.

normal ring widths are again produced. The maximum density is reestablished after only 1-2 years. In modern times as in late Roman and Medieval times, periods of intensive bud moth attack alternate with years of normal growth (H. N. Müller, personal communication; F. Röthlisberger 1976). This phenomenon probably has climatic causes.

Situation in Subalpine zones of trees sampled

The samples were taken from subalpine zones: Eastern Alps (Austria), Central Alps (Switzerland), Southwestern Alps (France) and in the Jura Mountains (Fig. 1).

The sites are dissimilar in their geology, pedology, phytosociology and topography e.g.

Site	Chasseral	Aletschwald
Geology	Chalk	Gneis
Soil	Rendzina	Podzol
Phytosociological unit	Abieti-Fagetum	Rhododendro-Vaccinietum
Exposition	south	north
Altitude	1400 m	200 m

All sites in the regions north of Mt.Cenis have a temperate, humid, central European climate. The Jura and the Wallis are influenced by the western European climate, while Cortina d'Ampezzo is influenced by the insubrian one. The sites with low summer precipitation south of Grenoble are in the Mediterranean climatic region. Typical of all sites is high annual precipitation; above 1800 m the annual precipitation is higher than 900-1000 mm.

Tree Ring Parameters and Climate (cf. Parker and Henoch, 1971: Polge, 1970)

From visual comparison it can be concluded that curves of maximum density correlate better with one another than do curves of early wood width, late wood width, and growth ring width. Therefore we hypothesized that cell wall growth is principally influenced by macroclimatic factors. The agreement (Gleichlaufigkeits Prozent) and response functions (Fritts 1976) were calculated for maximum density with monthly temperature and precipitation (Fig. 5).

In late wood, high amounts of cell wall substances are deposited if the climatologic conditions are the following: high temperatures in spring (April and May) and summer (August and September), below mean precipitation in the current year (January to September), and cool temperatures during the previous vegetation period. Also low temperatures in winter (January and February) favour cell wall deposition. It is noteworthy that high temperatures in spring and full summer favour cell wall growth, but not those occurring in early summer (June). The climate of the previous year does not seem to have any major influence on cell wall development. The most effective ecological factor for physiological processes in the late wood cells seems to be the summer temperatures, especially those in August and September. Based on the similarity of the sequence in curves from climatically different regions in the Alps and their corresponding response functions, it could be proved that regional temperature patterns essentially control the maximum density. As essentially only one climatic factor of the current vegetation period regulates the physiology of cell wall growth, the maximum density is particularly suitable for the reconstruction of summer temperatures in past times.

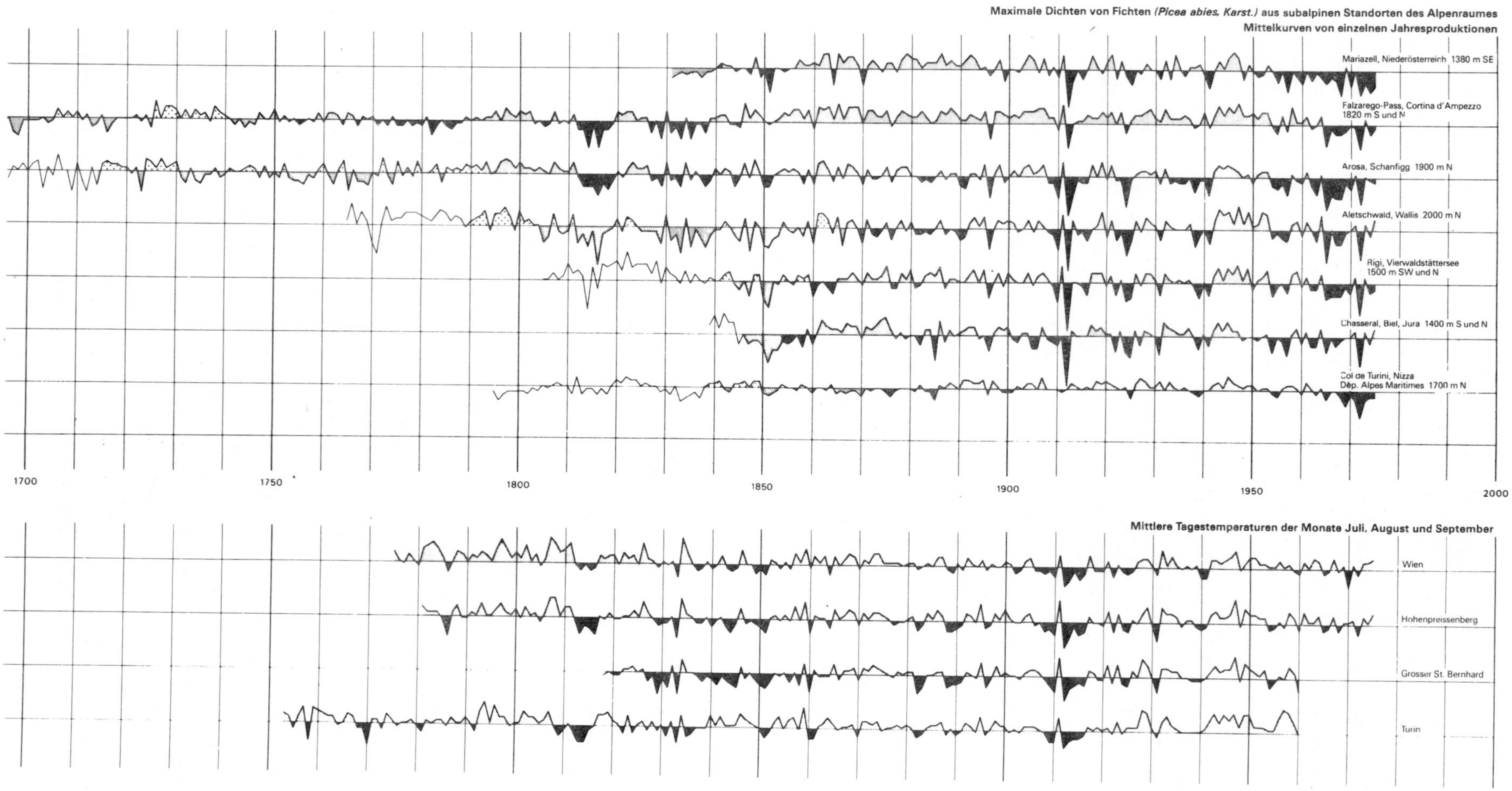

Fig. 6. Above: Spruce, Maximum density. Annual changes of site means.
Below: Mean temperatures for July, August and September. Annual changes.

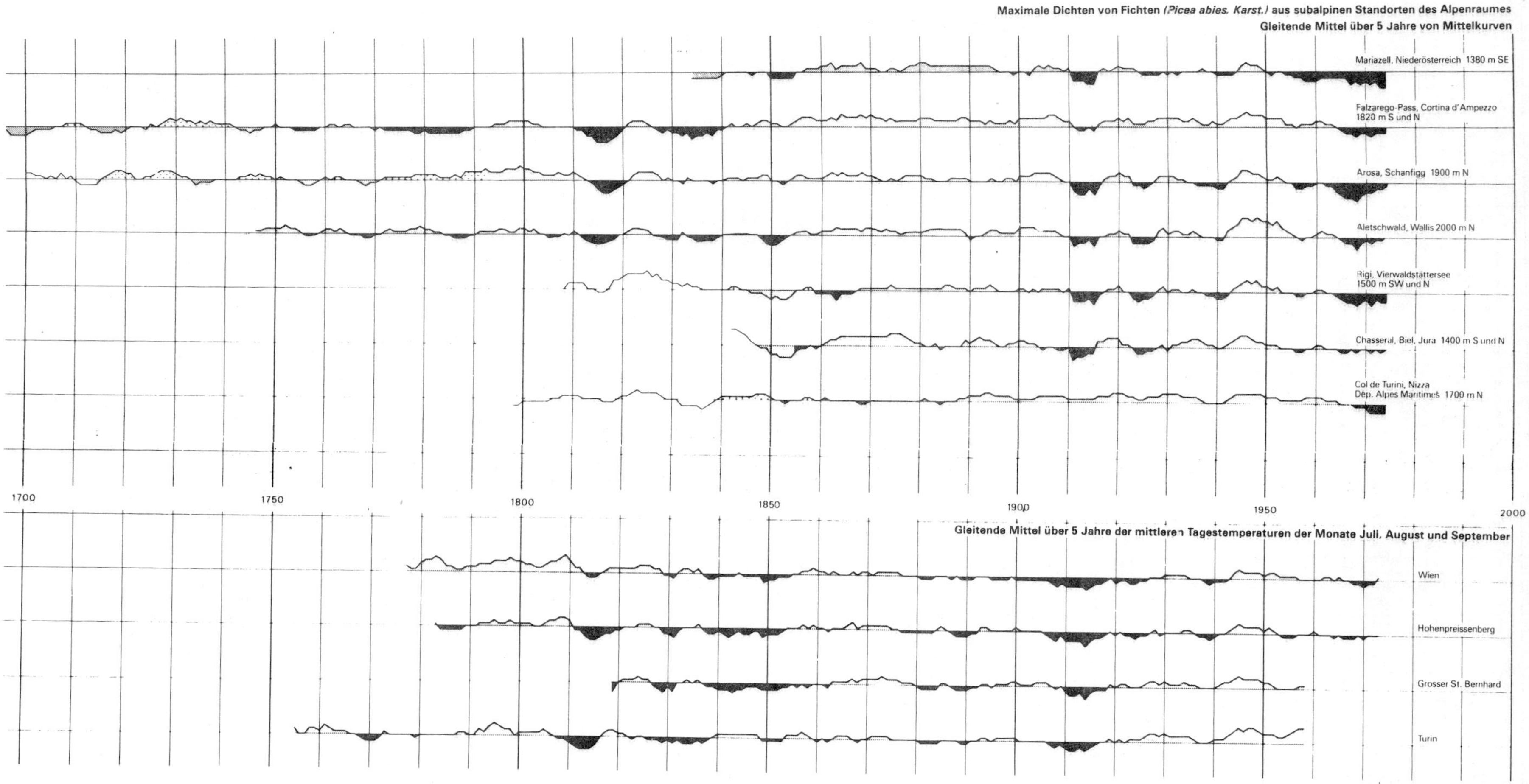

Fig. 7. Above: Spruce, Maximum density. Changes in 5 year running means.
Below: Mean temperatures for July, August and September. Changes in 5 year running means.

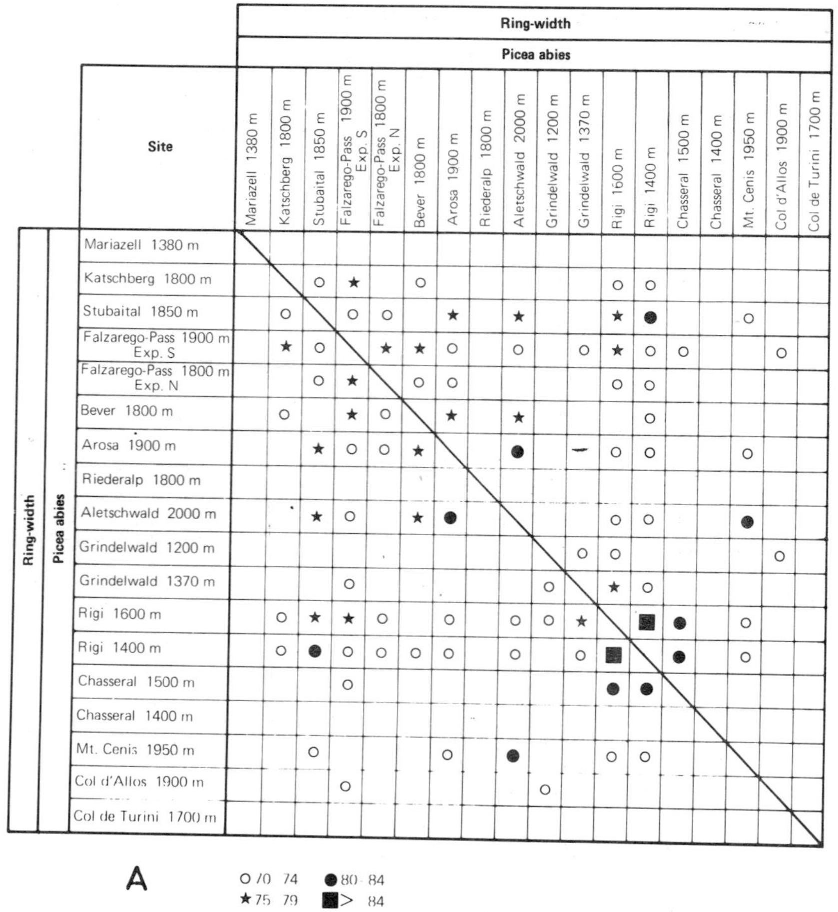

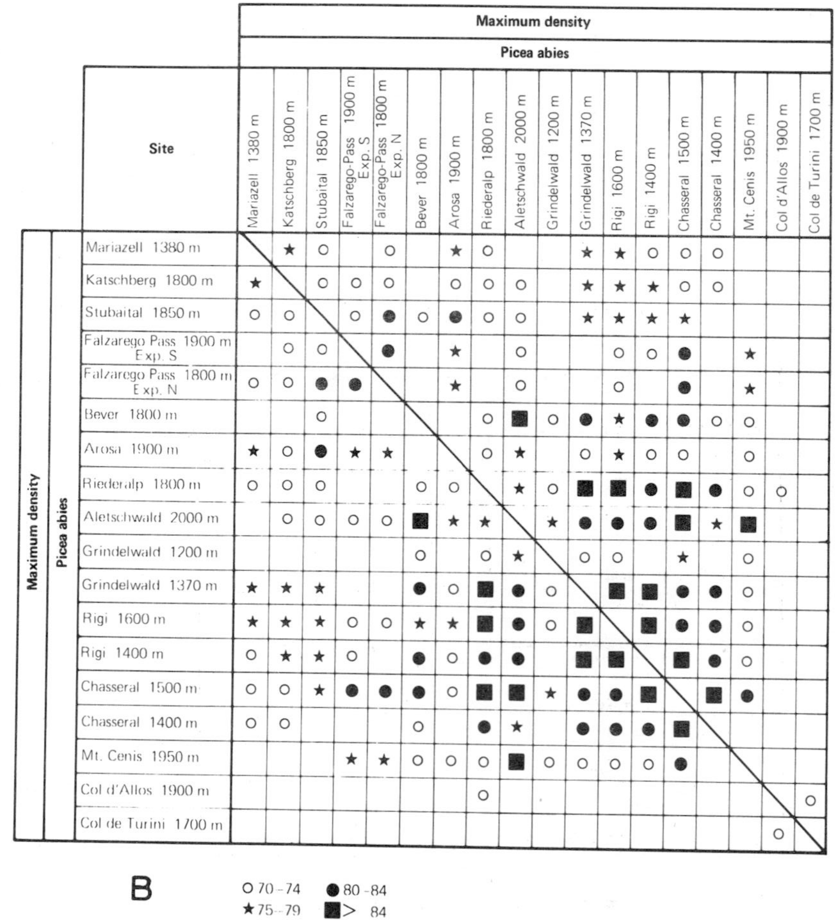

Fig. 8. % agreement for site to site mean curves for alpine spruce.
A. For ring-width. B. For maximum density.

The highest agreement between maximum density and weather curves is to be found between the daily average temperature of July, August, and September. They vary between 70-80 percent. The agreement between the maximum density and average daily minimum or maximum temperature vary between 55-63 percent. Due to the presence of characteristic peaks, the correlation of maximum density and summer temperature curves is evident to the eye (Fig. 6) as periodic trends, above and below the average values, become conspicuous. Yearly values were smoothed by 5 year moving averages in order to bring out long phase fluctuations (Fig. 7). Although there are some problems relating to the curves (especially from statistical imperfections), the fluctuations of the summer temperature during the last 200 years can be characterised as follows:-

Warm phases		Cool phases	
warm	1780-1811	1772-1779	moderately cool
warm	1820-1828	1812-1819	cool
moderately warm	1841-1849	1829-1840	moderately cool
moderately warm to average	1854-1908	1850-1859	moderately cool
moderately warm	1927-1938	1909-1916	cool
very warm	1941-1953	1939-1940	moderately cool to average
moderately warm to average	1958-1963		
		Central Alps	
		1954-1957	moderately cool
		Eastern Alps	
		1954-1975	cool
		Central Alps	
		1964-1975	cool

The observation that calculations with other tree ring parameters are much less homogeneous might be explained by the different ecological conditions on the sites. Late wood width seems to follow the same general rules as maximum density. The early wood width is above all increased by high temperature in June and July. The analyses show that low precipitation in the subalpine zone during the vegetation period does not limit the cells' physiological processes. However, X-ray studies on cores from extremely dry sites in the hill zone in the Swiss midlands, the Valais (Schweingruber in process) and the Rhine valley (F. Huber 1976) show that low precipitation in summer limits the tree ring growth. Together with reconstruction of summer temperature it will be possible to reconstruct the principal climatic factors in limited areas during the wood's growing season.

GEOGRAPHICAL LIMITS OF MEAN CURVES

Mean curves were calculated from 8-14 tree curves from each of the sampling sites. The confidence limit of 99.9% was used; thus only curves with an overlap of at least 80 intervals and an agreement percentage of above 70% were considered. The following discussion is based on Figs. 8 and 9.

Limits of Applicability of Mean Curves for Spruce (Fig. 8)

Within the alpine area the maximum densities show higher agreement than do ring width curves. The following table lists the sites whose mean curves for ring width and/or maximal density have an agreement with the 15 other site curves higher than 70%. The sites Col d'Allos and Col de Turini have been excluded.

A — Ring-width

Ring-width		Site	Picea abies: Mariazell 1380 m	Katschberg 1800 m	Stubaital 1850 m	Falzarego-Pass 1900 m Exp. S	Falzarego-Pass 1800 m Exp. N	Bever 1800 m	Arosa 1900 m	Riederalp 1800 m	Aletschwald 2000 m	Grindelwald 1200 m	Grindelwald 1370 m	Rigi 1600 m	Rigi 1400 m	Chasseral 1500 m	Chasseral 1400 m	Mt. Cenis 1950 m	Col d'Allos 1900 m	Col de Turini 1700 m
	Abies alba	Lac de Bourget 1340 m																		
		Col d'Allos 1900 m				○														
		Col de Turini 1700 m																		
	Larix decidua	Riederalp 1800 m			○					■	○									
		Aletschwald 2000 m																		
		Mt. Cenis 1950 m																		
	Pinus cembra	Stubaital 1850 m		★		○		★	○		○									
		Bever 1800 m																		
		Aletschwald 2000 m				★		★			★									

A ○ 70–74 ★ 75–79 ● 80–84 ■ > 84

B — Maximum density

Maximum density		Site	Picea abies: Mariazell 1380 m	Katschberg 1800 m	Stubaital 1850 m	Falzarego-Pass 1900 m Exp. S	Falzarego-Pass 1800 m Exp. N	Bever 1800 m	Arosa 1900 m	Riederalp 1800 m	Aletschwald 2000 m	Grindelwald 1200 m	Grindelwald 1370 m	Rigi 1600 m	Rigi 1400 m	Chasseral 1500 m	Chasseral 1400 m	Mt. Cenis 1950 m	Col d'Allos 1900 m	Col de Turini 1700 m
	Abies alba	Lac de Bourget 1340 m					○	★		★	●	○	★	★	★	■	■	★		
		Col d'Allos 1900 m					★				★					★		○	★	
		Col de Turini 1700 m																	●	●
	Larix decidua	Riederalp 1800 m	○	○	○			★	○	■	●	★	●	■	●	●	●	●	○	
		Aletschwald 2000 m					○	★	○	★	●	○	★	●	★	■	○	○		
		Mt. Cenis 1950 m						○		★	★	○		○	○	●			○	
	Pinus cembra	Stubaital 1850 m	●				○		★		○			★		○				
		Bever 1800 m			○	★	○	○			★		○			●				
		Aletschwald 2000 m					★	○		●	■		●	●	★	●	○	■		

B ○ 70–74 ★ 75–79 ● 80–84 ■ > 84

Fig. 9. ‰ agreement for site to site mean curves between alpine spruce and other species - silver fir, larch and Swiss stone pine.

A. For ring-width. B. For maximum density.

Site curves which agree with 15 other site curves by>70%

No. of sites	Ring Width Curves	Maximum Density Curves
14-15	none	Chasseral 1500 m, Rigi 1600 m Riederalp, Aletschwald
12-13	none	Rigi 1400 m, Grindelwald 1370 m Arosa, Stubaital, Katschberg
10-11	Rigi 1600 and Rigi 1400 m, Falzarego-Pass, Exp. S	Mt. Cenis, Bever, Mariazell
8-9	Stubaital, Arosa	Falzarego-Pass, Exp. S und N Chasseral 1400 m
6-7	Falzarego-Pass Exp. N, Bever, Aletschwald	Grindelwald 1200 m
< 6 sites	8 sites	None, as all in above categories

Our data showed that agreement does not decrease continuously with increasing geographic distance, but that mean curves from the subalpine zone in the Mediterranean region (south of Mt. Cenis) are definitely different.

From two sites on Rigi mountain, mean curves were obtained which give high agreement when either the ring width curves or the maximum density curves are compared with curves from all other sites. Curves from Chasseral (1500 m), Grindelwald (1370 m) and Aletschwald (2000 m) show high agreement only for maximum density curves. The ring width curves from Grindelwald (1370 m), Chasseral (1400 m and 1500 m) are only slightly similar to the curves from all other sites.

The geographical region over which the maximum density is significantly similar extends from the Wiener-Schneeberge in Austria to the Cottic Alps in southwestern France. Boundaries were found at the following sites:

Boundary	Ring Width Curves	Maximum Density Curves
SW	Up to Mt. Cenis. Approximately on the northern border of the Mediterranean flora region and slightly north of the olive tree extension zone. For the curves from Col d'Allos and Col de Turini there exist only weak similarities.	
East	Katschberg, near the western boundary of the Pontic flora region	Not found. Probably east of the Alpine zone
North	Not found. North of the Swiss Jura mountains, probably north of the Black Forest	Not found
West	Not found	Not found
Altitude	Approximately 1200 m	Approx. 1200 m

The results from spruce and silver fir from the Emmental (700-900 m) show that the curve to curve agreement decreases for both ring width and maximum density curves at lower altitudes. There, the agreement of the maximum density curves is no higher than those of the ring width curves.

From the agreement of single year intervals from all subalpine curves (exclusive of Grindelwald 1200 m, Col d'Allos, Col de Turini) it can be seen that they correspond more or less to the curves from the entire Alpine region (Fig. 10). For 73 intervals, 40 possess agreement between 90-100%. Conspicuous, for example, are the high agreements for 1941-1945 and the low or missing agreements for 1914-1920.

Interrelationship of Site Mean Curves for Various Species (Fig. 9)

From the low coincidence of growth ring curves it can be concluded that the response of cambium activity to environmental factors is species specific. In contrast, cell wall development in late wood (maximum density) is species independent; in all the species examined the temperature appears to have a similar influence on the development of cell walls.

Due to the limited number of curves, it is not possible at present to suggest the curve agreement limits for silver fir, larch and and Cembra pine.

References

Fritts, H. C., 1976. Tree rings and climate. London, Academic Press.

Huber, F., 1976. Problèmes d'interdation chez le Pin sylvestre et l'influence du climat sur la structure de ses accroissements annuelles. Annales des sciences forêstières 33:61-86.

Lenz, O., Schär, E., Schweingruber, F. H., 1976. Methodische Probleme bei der radiographisch-densitometrischen Bestimmung der Dichte und Jahrringbreiten von Holz. Holzforschung 30:114-123.

Parker, M. L., Henoch, W. E. S., 1971. Canadian Journal of Forest Research 1:90-98.

Polge, H., 1970. Tree Ring Bulletin, 30:1-10.

Röthlisberger, F., 1976. Die Alpen, 52:59-152.

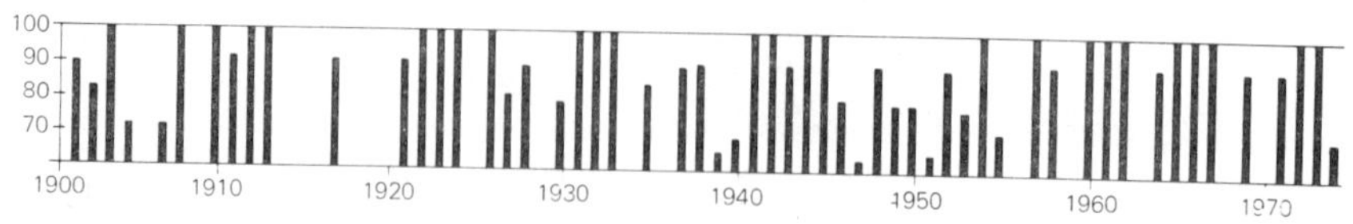

Fig. 10 % agreement of 12 alpine curves Mariazell to Mt. Cenis.

2.6. DENDROECOLOGICAL ZONES OF CENTRAL EUROPEAN FOREST COMMUNITIES

B. Becker, Stuttgart-Hohenheim

Summary

The uniformity of the tree-ring pattern over large distances in the hilly areas of central Europe for beech, oak, fir and spruce respectively are interpreted in terms of ecological zones. They stretch from south of the Alps to the north German plain and from the foothills of eastern France to the Carpathians. The climatic factors in the various zones are discussed.

Introduction

The natural forests of Central Europe discussed here may be divided into those at moderate altitudes (up to 600-800 m in elevation) and those in the lower mountain regions (from 800 to 1300 m high). The former are mainly deciduous forests in which beech (Fagus sylvatica) and oak (Quercus petraea and Q. robur) predominate. The latter, except in the Central Alps, consist of a wide belt of fir (Abies alba) and beech mixed with spruce (Picea abies).

These species—beech, oak, fir and spruce—were used as constructional timber in medieval times, their different amounts depending on the location (Becker, 1977). They therefore form the base for tree-ring chronologies and dating in Central Europe. This paper deals with their ring-width patterns and relates the geographical range of similarities in tree-ring formation to a macroclimatic zonation. In each zone, the species respond in various ways according to their ecological adaptability.

Fig. 1 shows the species and the locations of living trees in Central European forests sampled either by co-workers of the late B. Huber at Munich, or recently by the author. The ring-widths are stored in the International Tree-Ring Data Bank at Tucson, Arizona, U.S.A.

Zones of Uniform Tree-Ring Pattern and Their Climate

Already in 1950 Müller-Stoll (1951) was deriving ring-width patterns which cross-dated over large distances in Central Europe. During the past 25 years, further long-distance correlations have been obtained, e.g. by Delorme (1973b) and Hollstein (1974) for living oaks. Cross-dating was found, not through any systematic study, but by comparing results for historical or prehistorical samples with existing master chronologies.

Oak

A great deal of information is available about the geographical range of the master chronologies for oak in Western, Northern and Central Europe. The most important (known as the Southern German chronology) was developed by Huber and his co-workers (Huber and Giertz-Siebenlist, 1969, and this volume, page 27), its central area being the hillsides of the Spessart Forest, situated about 60 km to the east of Frankfurt-am-Main. The oak stands of that area are situated on ridges and south-facing slopes that receive a high intensity of radiation; the summers are warm and the summer moisture is

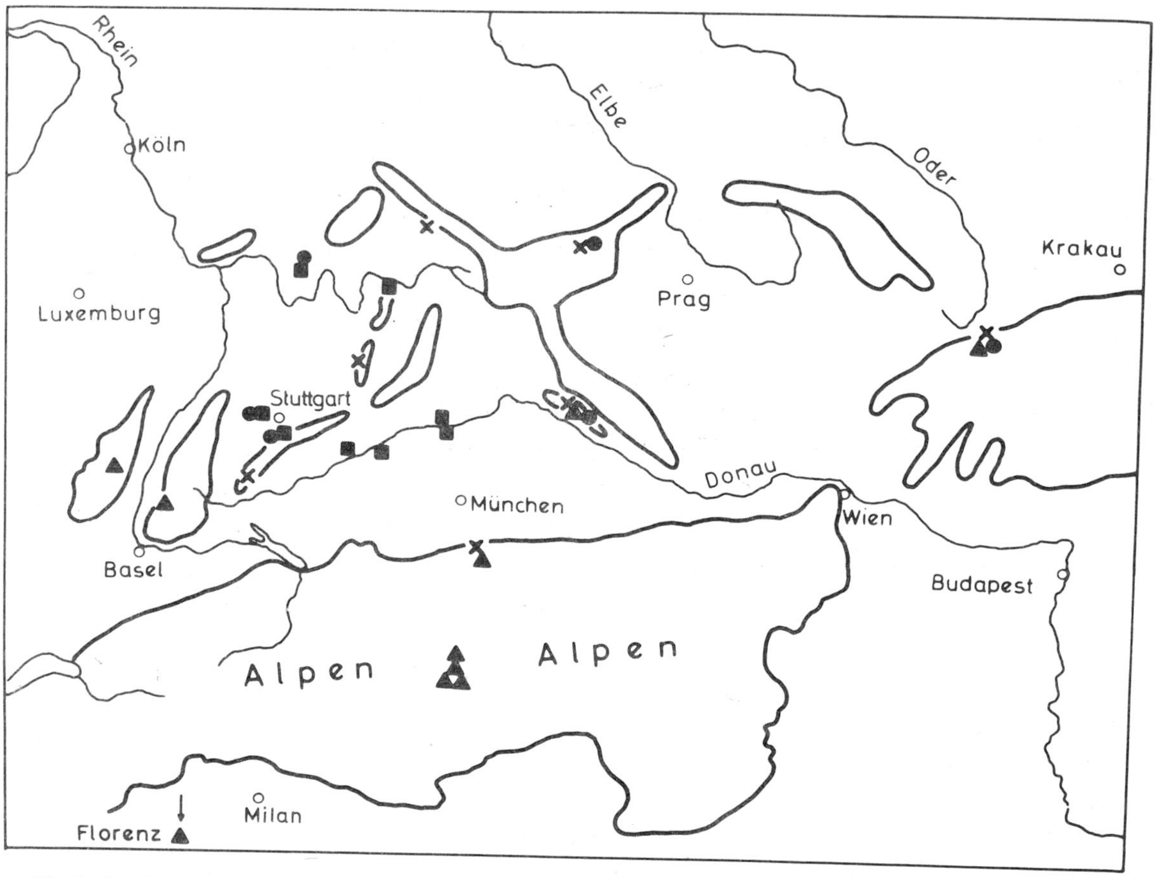

Fig. 1. Central Europe: Localities of chronologies for living trees sampled by co-workers of the late B. Huber and by the author. The thick lines enclose areas over 500-600 m high.
■ oak; ● beech; ▲ fir; × spruce.

low. Its climate in relation to oak growth is here called moderately continental, a description that applies to all the oak sites in the hilly areas of Central Europe.

The same tree-ring pattern prevails in Western Germany (Hollstein 1965), while indications were found by V. Kerner-Siebenlist that its pattern extended, for medieval samples, to western France and even to southern Britain (this volume, page 156). Recently Jahrig (1975) has found that the oak pattern in south-western GDR also matches this chronology.

Furthermore, the ring-width pattern of the Central European oaks of Neolithic, pre-Roman and Roman times extended some distance to the south and north. For example, the sub-fossil oaks deposited in fluvial gravels of the rivers Danube, Main, Rhine, Fulda and Werra have a common pattern with that of contemporary oak piles used in N.W. and S.W. Switzerland (Becker and Frenzel 1977; Becker, in press; Delorme 1976); with that of the oaks that form the pre-Roman and Roman chronology of the Rhein-Mosel area (Hollstein 1967); and with that of the oaks of the Weserbergland (Delorme 1973a).

The pattern ends abruptly at the transition in altitude to lowlands (below 200 m in altitude). This occurs to the north at the North German plain and to the northwest at the coastal areas adjacent to the North Sea. The correlation between different parts of this coastal area—in which the maritime climate is characterised by moderate summers and winters and by high humidity—has been studied at Hamburg. Different regional chronologies have been established for the Netherlands (Bauch _et al_. 1972 and this volume, page 133); for the northern part of Lower Saxony, and for the Hamburg area (Eckstein _et al_. 1970, 1972 and this volume,).

In contrast to this maritime coastal region, another lowland one with a uniform oak pattern exists around the western part of the Baltic Sea to the north-east of Central Europe. Eckstein's chronology for Schleswig, and Bartholin's one for Denmark and Southern Sweden (Bartholin 1975a and b, and this volume, page 125) relate to this zone, in which good synchronisation of oak extends to a long distance. Its climate may also be said to be moderately continental. However the winter temperatures and the mean annual temperatures are lower and its oak chronology does not match the one of Central Europe.

Fir

Fir provides the most instructive European example of a macroclimate producing a uniform tree-ring pattern. Its 1100 year Central European chronology (Becker and Giertz-Siebenlist, 1970) applies to fir growing in mountain forests in a very large region. For living trees, the same pattern covers the Beskid Mountains in Czechoslovakia, the Bavarian Forest, the Bavarian Alps, the Black Forest, as well as the Vosges to the west; and the Southern Alps (South Tirol) and Apennines to the south (Fig. 2). The recent analysis of living firs, more than 400 years old, from the Apennines (elevation, above 1000 m) to the east of Florence demonstrates that fir has the Central European pattern even in the Mediterranean region (Becker, unpublished).

This suggests that there exist similarities in the response of fir tree-ring patterns in mountain regions from the north to the south of the Alps and into

Southern Europe, though the differences in the precipitation regimes need to be taken into consideration.

The uniformity of the fir pattern is based on its occurrence in a subatlantic, mountain macroclimate (Rubner 1953), characterised by moderate summer and winter temperatures and a good water supply during the vegetation period. In the more continental climate of the higher elevations of the Central Alps, fir fails to produce a forest community.

Beech

The zones of oak and fir in Central Europe differ in their altitude. The two are linked by beech both in its native distribution and in its tree-ring pattern. The latter was found by a comparison of the signature years in oak, beech and fir of living trees (Becker and Giertz, 1970).

Beech forests extend from lowlands up to 1450 m or even higher in the fir belt. Their tree-rings match to a significant degree and they have a remarkably high mean sensitivity, but there is a tendency for rings to be missing. Long-distance uniformity in the tree-ring pattern is indicated by the similarity of samples from the Beskid Mountains in the CSSR (living trees, Müller-Stoll 1951); of chronologies from the Bavarian Forest and from the Spessart area (living trees, v. Jazewitsch 1953); and of living trees and medieval samples from the Stuttgart area. The chronology has been extended to the 14th century A.D. by medieval posts from Frankfurt (Hollstein 1973).

In comparison to the native area for Quercus robur (which covers the Russian plains as well as much of Southern Europe), the more sensitive beech is restricted to the atlantic and subatlantic-type climates in Central Europe, such as occur in parts of the oak distribution within the hilly areas and in parts of the mountain zone where firs flourish.

Spruce

This conifer favours the subatlantic-subalpine belt at the higher elevations in the mountains of Central Europe. It is the only one of the four species discussed which reaches both the treeline in the Central Alps as well as the forest border in North-western Scandinavia. Müller-Stoll (1951) found that spruce tree-ring patterns in mixed forest within the mountain fir zone present unusual difficulties compared to fir and beech. Our reinvestigation at these altitudes confirms the limited correlation between spruce trees. A comparison of Fig. 3 (for spruce) with Fig. 2 (for fir) shows that there are fewer coincident signatures for spruce.

On the other hand the dendroclimatological studies of Feliksik (1972), Ermich et al. (1976) and Schweingruber (this volume, page 89) indicate that in high-elevation forests, the growth of spruce is correlated significantly with climate. Therefore there should be a significant correlation of the tree-ring patterns over long distances. This holds at lower altitudes, where the spruce patterns become increasingly related to climate (see next section).

RESULTS AND THEIR SIGNIFICANCE FOR THE HILLY AREAS OF SOUTHERN CENTRAL EUROPE

Climatic and Site Effects on Growth

In areas of S-Central Europe where oak, but especially beech, are the dominating forest trees, conifers only occur on the top of the higher ridges. Recently, we have worked on samples from the region known as Mittelfranken, a zone located in the centre of the area between Stuttgart, Nürnberg and Würzburg. Here fir and spruce reach their lowest native distribution limit in Central Europe in a northwesterly direction. The limiting factors within this area seem to be the relatively warm winter temperatures, through their influence, especially for spruce, on insects and fungi; and low summer precipitation, in some areas less than 600 mm.

Spruce: The controlling effect of summer precipitation applies to spruce in the eastern part of the Black Forest. We studied two spruce forests near Villingen (Becker and Benckieser 1976), where the elevation is 700 m and the climatic environment (annual mean temperature 6.5°C, precipitation 880 mm) is relatively cool and continental. Frost temperatures can occur during the entire vegetation period; the water supply should be balanced because of the relatively low summer temperatures. Nevertheless, the spruce pattern in this mountain foothill region is correlated with summer precipitation, in particular with that of the months of May and July (Fig. 4). The temperature of the growing season (the limiting factor for spruce at high elevations) is not significantly correlated with the tree-ring patterns for the Villingen sites. Summer precipitation is not only important for the growth of spruce in the lower mountain region, but in general this factor is dominant for the spruce forests in the hilly areas of southern central Europe. This is seen in Fig. 5 where the chronology for living trees (sampled from the Frankenwald ridges at an elevation of 400 m) is plotted against the summer precipitation at Freudenstadt (Black Forest), one of the longest available climatic records. The correlation between years of low precipitation and the signature years for spruce (indicated on the chart by wider connecting lines) is apparent, though the mean values for precipitation are about three times higher at Freudenstadt.

Beech and Oak: A comparative study was made in 1977 of beech and oak forests at two sites in the vicinity of Stuttgart.

(a) At Pfullingen (Schwabian Alb Mountains). The oak forests of this area cover the steep, southfacing slopes. A shallow soil, 5 to 15 cm deep, overlies limestock bedrock. The site conditions—stress in water supply during summer—is indicated by the xerothermic (dry-warm) type of herbaceous flora. The beech forest covers the plateau and slopes in all other directions, where loamy brown-earth soils are present. The beech trees investigated were situated on a north-facing slope nearby the oaks.

(b) At Magstadt (Stuttgart). This forest is situated on hills 10 km to the west of Stuttgart. The soil is a deep loam overlaying sandstone and the water supply is balanced. The oaks grow on sites with a high water table that is reached by their roots. The macroclimate is similar to that at Pfullingen, the records used being those from the meteorological station at Hohenheim, located between the two sites.

The result of the dendroecological comparison between the four areas is plotted in Fig. 6. For the oak (Q. petraea) there is significant correlation between the two patterns and they are related to the precipitation in the period March to June. Their similarity, in spite of the fact that in one area water stress is occurring during summertimes and the other has a balanced water supply, suggests that the macroclimate is dominant and local conditions are less important for tree-ring formation in these oak forests.

The growth of beech at both sites is correlated with the summer precipitation, but their tree-ring patterns in addition show an inverse correlation with the temperature curve for the period March-June (Fig. 6). This is attributed to high temperatures causing a high transpiration rate and reduced growth in the more sensitive beech, whereas oak reacts more directly to precipitation.

To summarise, in S-Central Europe the limitation of water supply in the vegetative period has a controlling effect for beech, oak and spruce forests. This correlation is increasing from the mountain foothill zone (example Villingen) to the hilly areas (Stuttgart and Mittelfranken) and does superimpose remarkable site conditions, as the oak and beech forests nearby Stuttgart demonstrate.

Similarities in the Tree-Ring Formation of Oak, Beech, Fir and Spruce at Lower Elevations in S-Central Europe

The ecological interrelations between macroclimate and tree-growths becomes evident especially in those landscapes in which the lower mountain forests (fir, beech and spruce) gradually are replaced by the beech-oak forests of the hilly areas. The tree-ring formation along this so-called "transition zone" has been studied in SW-Germany. For example, the medieval chronology of spruce, derived from framework buildings of the Mittelfranken area (Wieser and Becker, 1975) could be established because spruce, the crossdating of which is remarkably reduced in the mountain fir zone, produces there under lowland climatic conditions a well-defined tree-ring pattern. Moreover, in this region similarities between tree-ring curves of different tree species becomes evident too. For example, an analysis of late medieval timbers used in a roof of the church at Murrhardt, near Stuttgart, including oak, fir and spruce, shows that the trees of these three species have nearly the same tree-ring pattern (Fig. 7).

From the study on living trees as well as on historical timbers of southwest Germany we find that the Southern German oak pattern (shown in part in Fig. 7) is strongly dominant for all species in the transition zone between the mountains and the hillsides. This pattern (typical for the entire southern region of Central Europe) can be regarded as a response of tree-growth to moderately continental climatic conditions in spite of marked differences in site conditions and of the differing adaptation of mountain conifers and of the deciduous trees of the hilly areas to the environment. In all the sites studied, the water supply, controlled during the vegetative period by low summer precipitation and high temperatures, is the factor that limits growth.

DISCUSSION AND CONCLUSIONS

The master chronologies of oak and fir and their uniformity over a considerable geographical range indicate the dominating influence of the macroclimate on tree-growth in Central Europe. The variety of site conditions within forests seems to exercise less influence and therefore is less important for tree-ring dating, particularly when based on signature years.

The following division of Central European forests into communities or types is based on dendroecology, i.e. the long-distance uniformity in tree-ring patterns:

A. The Stonepine/Larch Type: continental, mountain climates

Studies at high elevations show that one of the most limiting factors for tree-growth is the length of the vegetation period and the sum of radiation and temperature during the growing season. Response to these conditions has been described for Swiss stonepine (*Pinus cembra*) in the Alps (Artmann 1948) and in the Tatra Mountains (Bednarz 1973). Recent studies by Serre (this volume, page 88), by Schweingruber (this volume, page 89) and by Munaut (pers. comm.), as well as from Lappert, Innsbruck (pers. comm.), are evaluating the long-distance correlations within a species and between different species. This type is also shown by larch (*Larix decidua*), for the tree-ring behaviour of which Brehme (1951) and Rotlisberger (1976) have published details.

B. The Fir Type: moderately atlantic, mountain climates

This type exists in all the mountain forests of Central Europe and extends to the Apennines, its distribution being based on a moderate mountain climate in contrast to the high-elevation climate of Type A. As the fir and oak chronologies of the past 1000 years (Becker and Giertz-Siebenlist, 1970) showed certain similarities, the signature years of this mountain type must be dependent to some extent on summer precipitation, at least during those centuries in which a correlation with oak occurs. The variability of the spruce pattern is attributed to the effect of the moderately mountain climate on a species adapted to a continental mountain climate (compare the tree-ring behaviour of oak in maritime regions).

C. Oak Type: moderate continental hilly climates

The ecological interpretation of this community, formed at lower altitudes than Types A and B is that inadequate water supply and high summer temperatures limit growth. The former in particular is responsible for uniformity in the tree-ring patterns of different species. The type covers all the hilly areas of Central Europe from the Swiss lakes in the southwest to the Northern German plain on the north. Its range in the eastern-western direction is known only partly, but seems remarkably long.

In the maritime climate of the coastal areas along the North Sea, from the Netherlands to Northwest Germany, the oak being more xerothermic, cross-dating of its tree-ring pattern tends to occur within restricted areas only. An increasing influence of site conditions on oak-growth due to the maritime climate is also to be expected for Southern England.

The macroclimate of Central Europe is modified not only by altitudinal differences but by geographical differences which create a more continental or a more atlantic climatic environment for tree growth. Further tree-ring studies in Europe to find the limits of dendroecological units should lead to a better understanding of the tree-ring response to the circulation system and its variations in the past.

References

Artmann, A., 1948: Dissertation Universität München.

Bauch, J., D. Eckstein and M. Meier-Siem, 1972: Nederlands Kunsthistorisch Jaarboek, 23, 485-496.

Bartholin, T. S. and B. E. Berlund, 1975a: Fornvännen, Arg. 70, 201-208.

Bartholin, T. S., 1975b: Tree-Ring Bulletin, 35, 25-29.

Becker, B. and B. Frenzel, 1977: Paläoökologische Befunde zur Geschichte postglazialer Flussauen im südlichen Mitteleuropa. In: Dendrochronologie und postglaziale Klimaschwankungen in Europa, Erdwiss. Forschung, XIII, Steiner Verlag, ed. B. Frenzel. 43-61.

Becker, B., in press: Holocenen tree-ring series (oak) from Southern Central Europe for archeologic dating, radiocarbon calibration and stable isotope analysis. Proc. 9th Intern. Radiocarbon Conf. Los Angeles and La Jolla (1976).

Becker, B. and V. Giertz-Siebenlist, 1970: Flora, 159, 310-346.

Becker, B. and G. Benckiser, 1976: Jahrring- und Wachstumsanalysen zur Klärung des Wuchsstockungsproblems in zwei Fichtenbeständen des Stadtwaldes Villingen. AFJZ, 147/5, 99-106.

Becker, B., 1977: Jahrringdaten eines romanischen Tannen-Dachstuhls in der Martinskirche Sindelfingen. Forsch. u. Ber. der Archäol. in Baden-Württemberg, 4, 129-133.

Bednarz, Z., 1973: Badania dendroklimatologiczne nad limba (Pinus cembra L.) oraz innymi gatunkami drzew tatrzańskich. Dysertacja doktorska, Akademia Rolnicza, Kraków.

Brehme, K., 1951: Jahrringchronologische und -klimatologische Untersuchungen. Zeitschr. f. Weltforstwirtschaft, 14, 65-80.

Delorme, A., 1973a: Forstarchiv, 44, 205-209.

Delorme, A., 1973b: Uber die geographische Reichweite von Jahrringchronologien unter besonderer Berücksichtigung mitteleuropäischer Eichenchronologien. Prähistorische Zeitschrift, 48, 133-143.

Delorme, A., 1976: Neue Ausgrab. u. Forsch. in Niedersachsen, 10, 225-237.

Eckstein, D., J. Bauch and W. Liese, 1970: Holz-Zentralblatt, 96, 674-676.

Eckstein, D., K. Mathieu and J. Bauch, 1972: Abh. naturwiss. Verein Hamburg, NF 16, 73-100.

Ermich, K., B. Rutkowski, Z. Bednarz and E. Feliksik, 1976: Tree-Ring Bulletin, 36, 1-8.

Feliksik, E., 1972: Studia dendroklimatologiczne nad świerkiem (Picea excelsa L.). Acta Agr. et Silv., Ser Silvestris, 12, 39-83.

Hollstein, E., 1965: Bonner Jahrb., 165, 12-27.

Hollstein, E., 1967: Germania, 45, 60-84.

Hollstein, E., 1973: Forstwiss. Centralblatt, 92, 47-50.

Hollstein, E., 1974: Die Jahresringe vom Magdalenenberg. Guidebook of the Stadtmuseum, Stadtarchiv Villingen, 32 pages.

Huber, B. and V. Giertz-Siebenlist, 1969: Sitz. ber. Osterr. Akad. Wiss., Math. -Naturwiss. Kl., Abt. I, 178/1-4, 37-42.

Jährig, M. and K. Baumgarten, 1975: Dendrochronologische Untersuchungen an Bauernhäusern des mecklenburgischen Südwestens. Jahrb. f. Völkerkunde u. Kulturgeschichte, 18, 165-184.

Jazewitsch, W. von, 1953: Jahrringchronologie der Spessart-Buchen. Forstwiss. Centralblatt, 72, 234-247.

Müller-Stoll, H., 1951: Bibliotheca Botanica, 122, 1-93.

Röthlisberger, F., 1976: Die Alpen, 3/4, 59-152.

Rubner, K., 1953: Die pflanzengeographischen Grundlagen des Waldbaues. Neumann Verlag, Radebeul.

Wieser, E. and B. Becker, 1975: Die Entwicklung des spätmittelalterlichen Säulenbaues in Bad Windsheim und Uffenheim, Teil B: Jahrringanalyse der Bauhölzer, Jahrb. d. Bayer, Denkmalpfl., 29, 72-78.

Fig. 2. Masterchronologies of fir (Abies alba Mill.) from mountain sites in Central and Southern Europe
Thick lines indicate signature years (coincidence in direction in >75% of the samples)

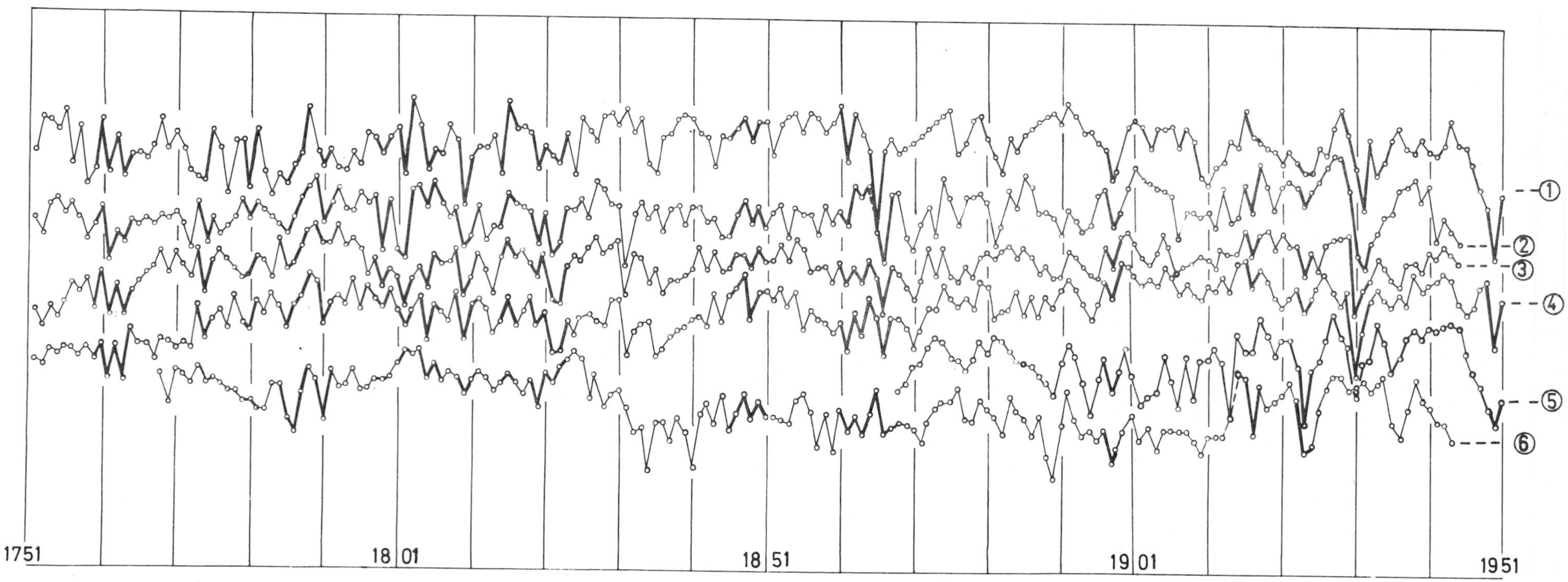

①Apenin Mts. 7 Trees
②Beskid Mts. 22 "
③Bayer. Wald Mts. 25 Trees
④Bayer. Alps 10 "
⑤Schwarzwald Mts. 30 Trees
⑥Vosges Mts. 5 "

Fig. 3. Masterchronologies of spruce (Picea abies (L.) Karst) from the mountain fir-zone of Central Europe

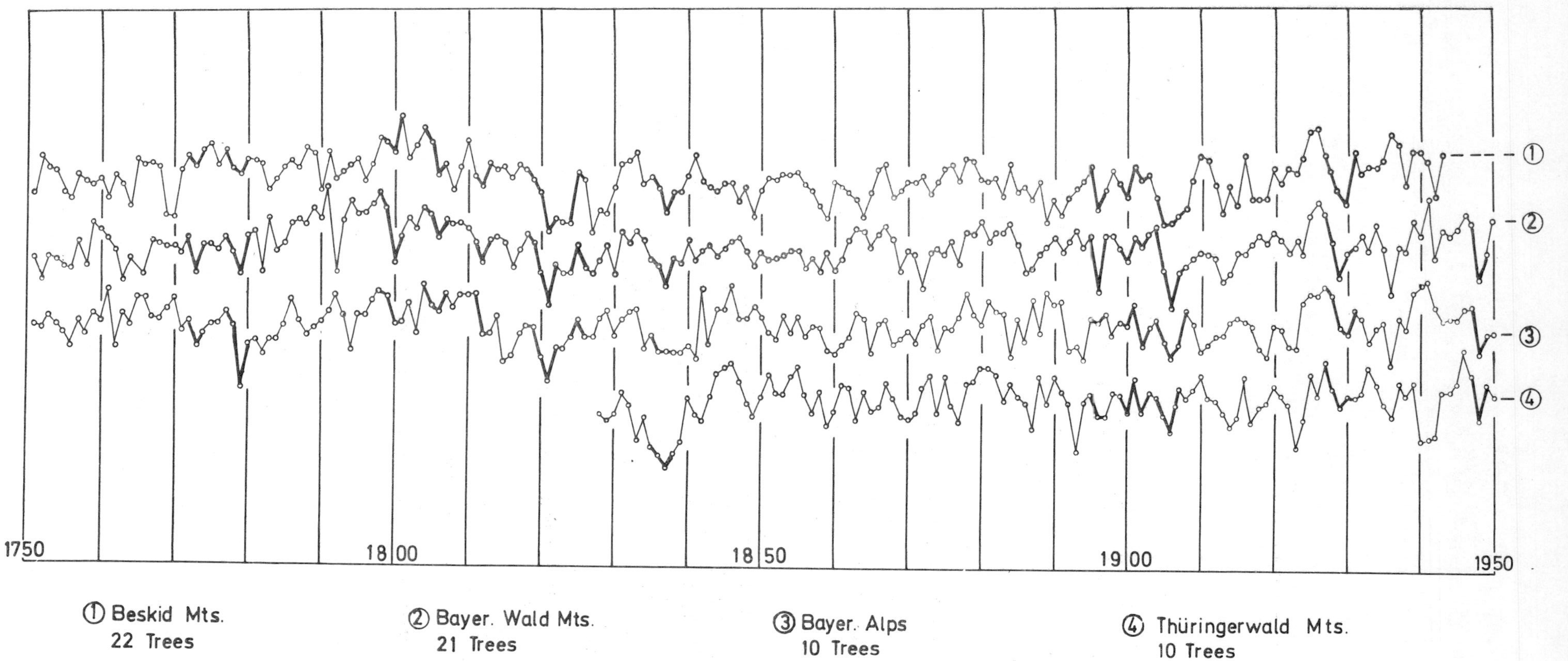

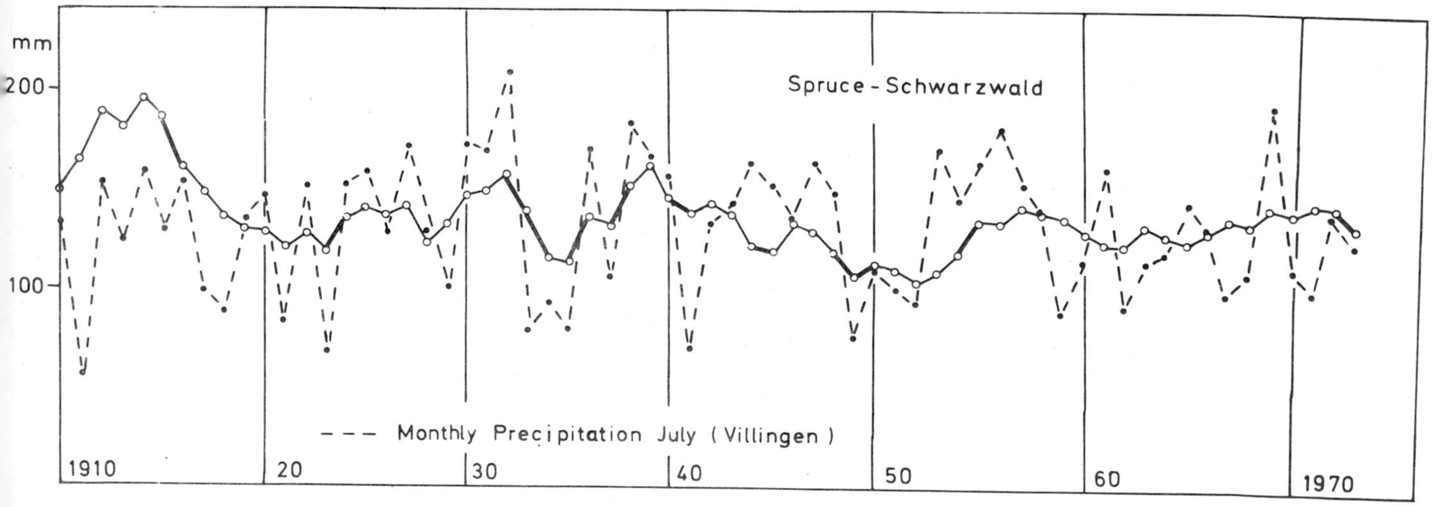

Fig. 4. Tree-ring curve with signature years. Spruce at Villingen.

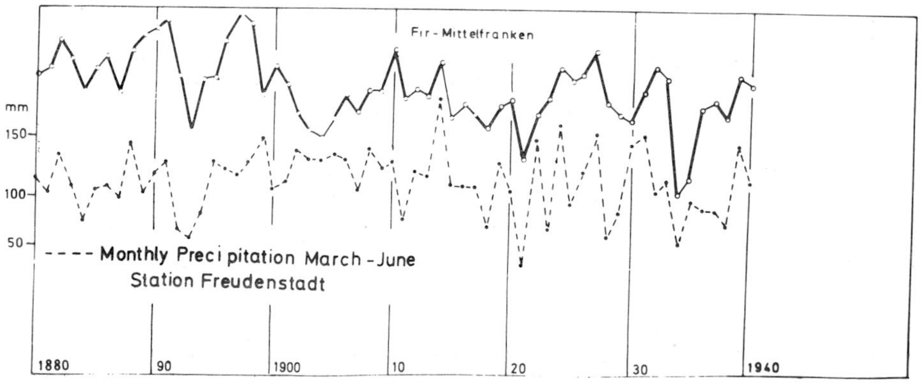

Fig. 5. Tree-ring chronology with signature years. Fir.

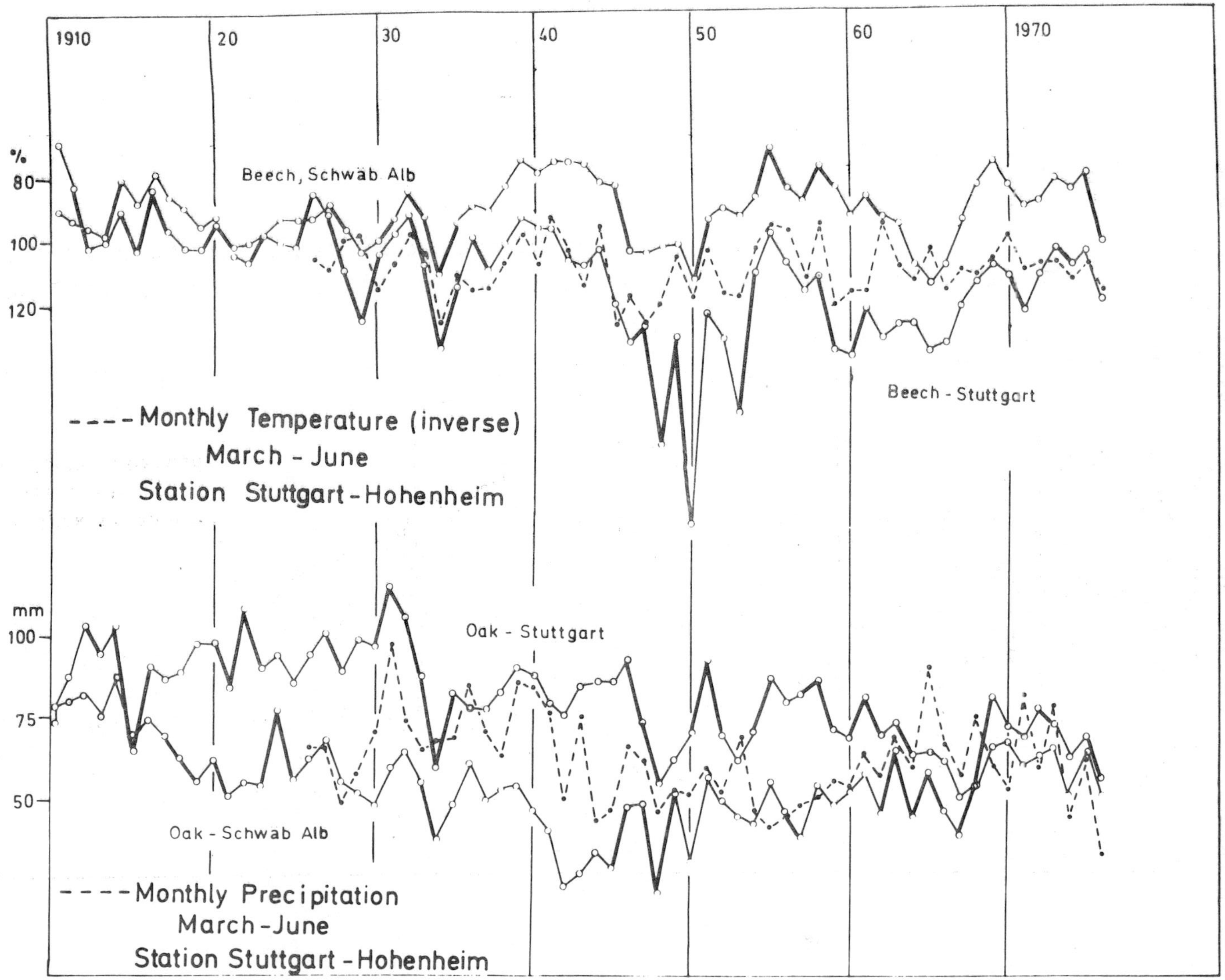

Fig. 6. Comparison of ring-width patterns with precipitation.

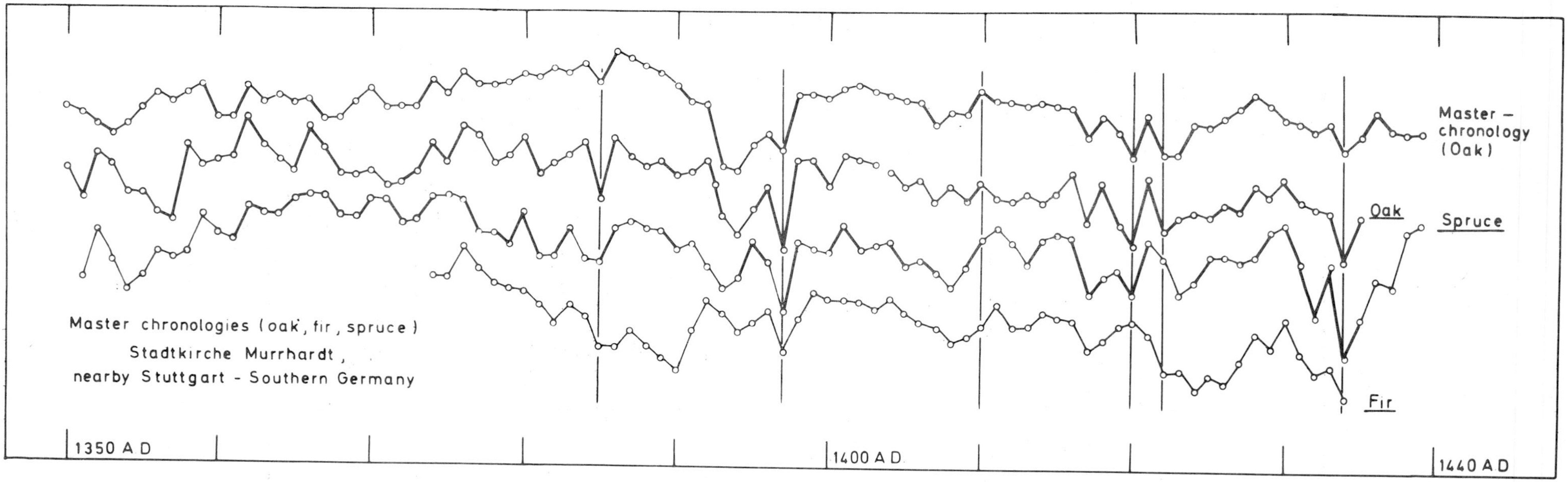

Fig. 7. Cross-matching of the ring-width patterns for late medieval samples of oak, fir, and spruce. Above, the South German master chronology for oak.

2.7 LONG CHRONOLOGIES BY X-RAY DENSITOMETRY FOR SUBALPINE CONIFERS AND THEIR VALUE FOR INDICATING HOLOCENE TEMPERATURES

F. H. Schweingruber and F. Röthlisberger
Swiss Federal Institute of Forestry Research

Chronologies based on the maximum annual density, started initially from living trees in the subalpine zone, have been extended to earlier times.

The good preservation of timber under alpine conditions has made it possible to obtain densitometric data from spruce used in historic buildings. Firstly, dated cores obtained in the Tyrol (kindly supplied by Giertz-Siebenlist, 1977) have been synchronised by Lamprecht (1978) in terms of their densitometric data and give a chronology back to 1330. Secondly, a 707 year series from Lauenen in the Bernese Oberland has been used (Schweingruber, Schär, Bräker, 1978). Both are well correlated (r = 0.6). The latter give a high correlation between the maximum annual density and the average temperature for the months of July to September (Schweingruber, Bräker and Schär, this volume, page 91) as shown in Fig. 1.

Further extension of the chronology is only possible with very good quality material from below moraines (Röthlisberger, 1976). Fortunately the wood of larch and of *Pinus cembra* from moraines and moors is suitable for densitometric work. In the cold conditions that prevail in the subalpine zone their resistance to decay is high and their attack by anaerobic organisms is very slow (Fig. 2). Radiocarbon dating of samples from many tree-trunks (Röthlisberger 1976) has established that throughout the last 9000 years many trees were uprooted either by advancing glaciers or carried onto them by avalanches and subsequently deposited in moraines or moors. Based on approximate C14 dates it appears that data for many larch, aged more than 400 years, can be synchronised to form long, floating chronologies.

We think that this material may therefore be able to provide yearly values of summer temperatures in the Alps over a long period of the Holocene.

References

Giertz-Siebenlist, V., 1977: Der gegenwärtige Stand der Jahrringchronologie aus dem obersten Oetztal. *Erdwiss. Forsch.* 13, 25-27.

Lamprecht, A., 1978: Radiodensitometrische Untersuchungen an Fichten aus der subalpinen Stufe des Tirols im Zeitraum zwischen 1360 und 1975. Unpublished work in the Geography Institute, University of Zurich.

Röthlisberger, F., 1976: Gletscher- und Klimaschwankungen im Raum Zermatt, Ferpècle und Arolla. *Die Alpen*, 52, 59-152.

Schweingruber, F. H., Schär, E. and Bräker, O. U., 1978: Jahrringe aus sieben Jahrhunderten. *Saaner Jahrbuch* für 1976, 1977. In course of publication

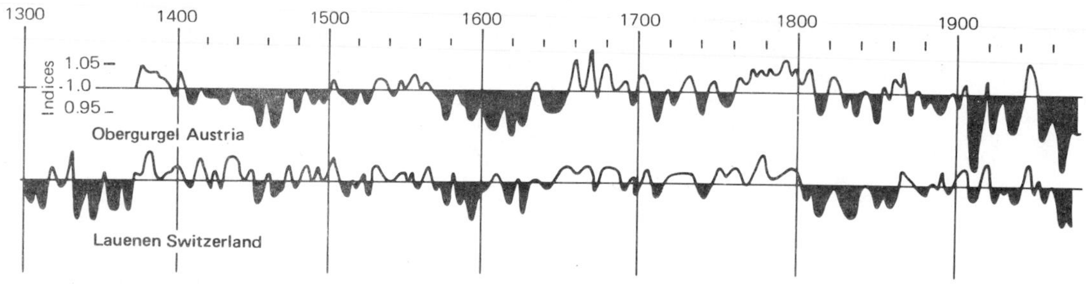

Fig. 1. Smoothed chronologies (based on 5 year running means) showing agreement between the deviations on the mean curves of the maximum annual density of spruce on sites 200 km apart in the Alps.

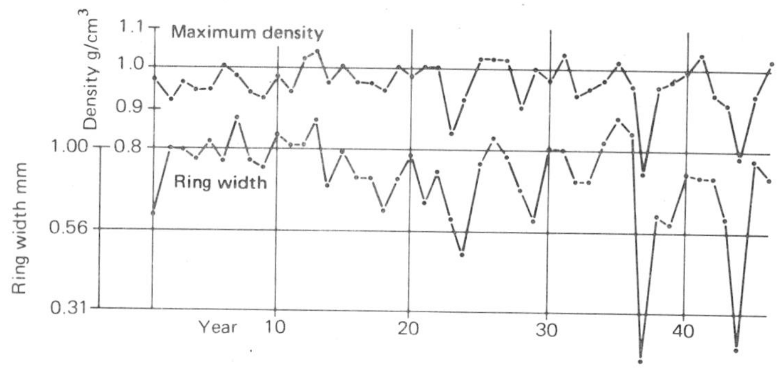

Fig. 2. Part of the annual ring sequence for a 500 year old larch from a moor near the tree-line in Canton Uri (Central Alps). The stem has lain subfossilised for about 3,400 years.

DISCUSSION—Section 2

Central European Chronologies and their Relation to Climate and Ecology

In a reference to Dr. Becker's remarks about applying European tree-ring chronologies to radiocarbon calibration, Baillie said they were in the same situation at Belfast. He considered there could be no meaningful comparison between the bristlecone pine calibration and floating European calibrations. The European chronologies must be dated to decide whether there is or is not any systematic difference in the concentration of radiocarbon between the New and the Old World. What did Becker consider his chances to be for completing a 6,000-8,000 year chronology? Becker said that all depended on luck; it might be during the next six months or within the next ten years.

- - - - -

Monsieur Polge, in answer to a question from Bräker about interaction between species, considered that a difference in age between the species used in his two experiments did not affect the result. He warned those about to undertake density measurements that maximum density is so sensitive that great attention must be paid to the soil and site when analysing results. In answer to a question from Prof. Chaloner, whose research on fossil wood makes it necessary for him to use light densitometry, Polge said he had not compared results derived from X-rays with those from light measurements.

Schove commented that the maximum density in the curves both of Polge and Schweingruber showed the zig-zag effect of a biennial oscillation.

SECTION 3 Oak Chronologies of Northern Europe for Dating

3 1 REGIONAL TREE-RING CHRONOLOGIES ALONG PARTS OF THE NORTH SEA COAST

D. Eckstein, Hamburg

When in the middle of the sixties the dendrochronological method was pursued extensively in Northern Germany it became obvious that for certain biological reasons the South-German Master Chronology for oak trees (Huber and Giertz, 1969, see this vol., p. 27) did not apply to the northern parts of Germany. In addition, it was impossible to obtain one single chronology for all the coastal regions along the North Sea (Bauch, Liese, Eckstein, 1967). Therefore the whole area from Denmark to the Netherlands has been divided empirically into several sub-regions and for each of these independent tree-ring series have been established.

The temporal and geographical range of the regional tree-ring chronologies

Six different tree-ring chronologies are being constructed at the Institute for Wood Biology of the University of Hamburg (Eckstein, Bauch 1977). The areas concerned, shown in the map in Fig. 1, are Denmark (1), Schleswig-Holstein (2), the region of Hamburg (3), the Lower Saxony coastal area (4), the northern, the more inland part of Lower Saxony (5), and the Netherlands (6). Their present temporal range is shown in Fig. 2.

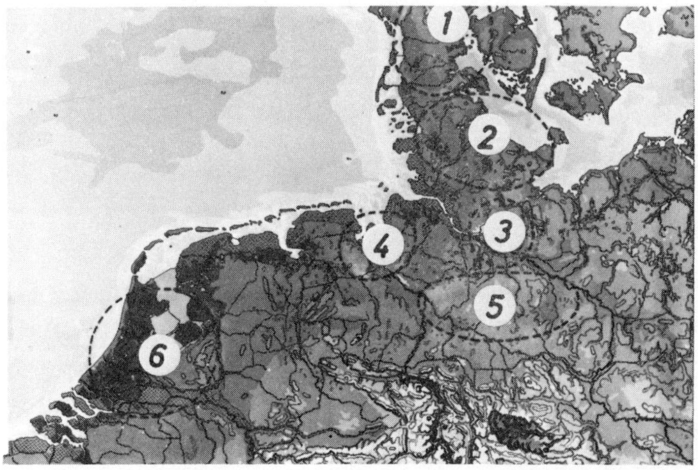

Fig. 1 Dendrochronological regions 1-6 along the coast of the North Sea.

The chronology for Denmark (1) started with modern oak trees provided by Bartholin (1973). In the meantime it has been possible continuously to prolong this back to 1310 A.D. (Eckstein, T. E. Christiansen, unpubl.). A further comprehensive chronology derived from the timber of late medieval buildings and from some Viking excavations covers the time between 1191 and 614 A.D. Additionally there is a 400 years long chronology from the Bronze Age built with the ring series of oak coffins (Eckstein, Schmidt, unpubl.). that are mainly in Danish museums.

The regional chronology for Schleswig-Holstein (2) is a coherent one going back to 436 A.D., in which the early medieval time has been verified several hundred times with oak samples from five prehistoric excavations (Eckstein 1976). The tree-ring series of a few subfossil oak stems cover some shorter periods between 4000 and 1000 B.C.

Further to the south there are now sub-regional chronologies for the area of Hamburg (3) back to 1338 (Eckstein, Mathieu, Bauch 1972), and for the coastal part of Lower Saxony (4) back to 1344 A.D. (Schwab 1975). For the northern, inland part of Lower Saxony (5) there is a tree-ring sequence from the 16th to the end of the 10th century, which due to its length has been synchronized with the chronology from the hilly area of the upper Weser and Leine rivers (Delorme 1973).

For the Netherlands (6) it has been necessary to establish two different chronologies (Eckstein, Brongers, Bauch 1975). The problems concerning these two chronologies are described by Bauch (this vol. p. 133). The timbers excavated from the centre of the city of Dordrecht (13th/14th century) as well as from the early medieval settlement of Dorestad at the river Rhine (8th/9th century), have been dated absolutely by the West-German Master Chronology (Eckstein, van Es, Hollstein 1977).

Each of the afore-mentioned chronologies exhibits to a large extent its own individual character. However, between those for directly adjacent regions there are obvious similarities.

Applications

The main applications of our tree-ring chronologies are the dating of archaeological, architectural, and art historical objects as well as the extraction of climatological and ecological informations.

Our largest archaeological dating project is the excavation of Haithabu (Hedeby), some 150 km northwards from Hamburg. This Viking settlement was one of the most important trade centres of Northern Europe during the early middle ages. The archaeological situation of Haithabu and the application of dendrochronological dating are described elsewhere (Eckstein, this vol. p. 267).

The applications to architectural studies have been focused on the investigation of the rural and urban architecture in Hamburg during the 15th to the 18th centuries (Eckstein, Grote, Mathieu 1977). The buildings analyzed were ones which it was not possible to date from records in archives or by architectural style.

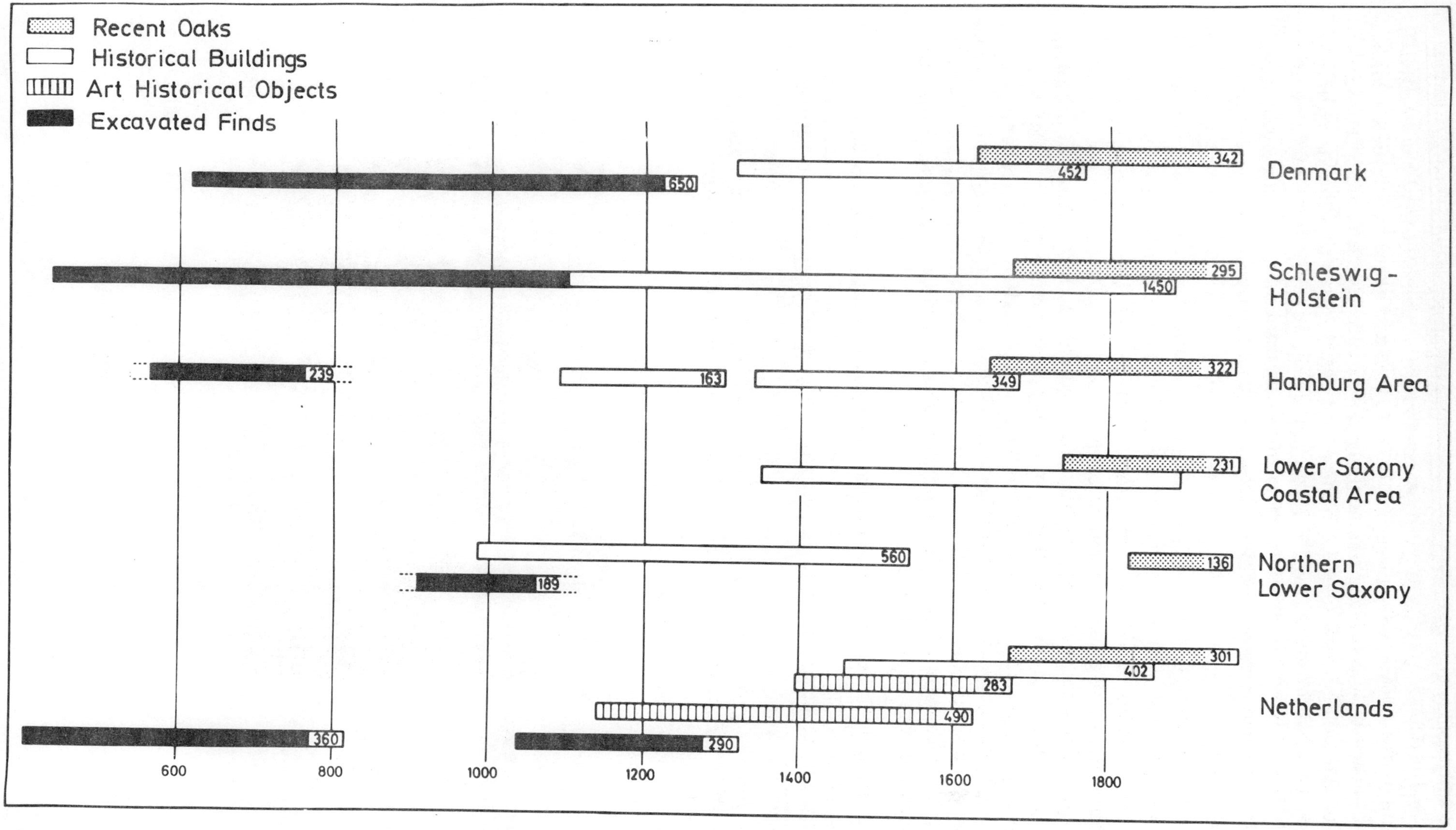

Fig. 2 The temporal range of the regional tree-ring chronologies 1-6 (Eckstein, Bauch 1977).

The rural buildings are farm houses of the type "Niederdeutsches Hallenhaus" (Fig. 3) as well as barns and store-houses. The oldest of the dendrochronologically dated farm-house, of 1473 A.D., had already attained a high cultural standard, possessing presumably a separate living room. The carpentry of the 16th century used both the older scarf joints and the younger tenon joints simultaneously over a long period; however, the tenon joint ultimately gained more popularity. The imposing dimensions of the buildings as well as luxurious facades can be interpreted in terms of social economical status.

The town buildings belong to the "Landhaus" (Fig. 4) and "Bürgerhaus" type of the 16/17th centuries. The "Landhaus" is a country house of town people, sometimes in combination with a farm-house; its existence is an expression of the increasing settlement of the town folk of Hamburg outside the town-wall. The dendrochronologically dated wings and alterations of buildings reveal the changes in the coexistence of town people and farmers. For the "Bürgerhaus" type (Fig. 5) some of the business dwellings of merchants and simple domestic houses were investigated. The dating of the diverse constructional periods within the buildings is interpreted in changes of function as well as in the arrangement of rooms. This provides an indication about the economical development of the town.

To summarize, dendrochronological dating has contributed to a better understanding of the records and to more precise knowledge about the living standards and customs of the town and rural folk of Hamburg in the 15th to 18th centuries.

The use of dendrochronology to date the oak panels of paintings and of sculptures is discussed elsewhere in this volume by Bauch (p. 307).

Fig. 3 Farm house in Schleswig-Holstein, 1569 A.D.

Fig. 4 "Landhaus" in Hamburg, 1598 A.D. (photo: Staatl. Landesbildstelle Hamburg).

Fig. 5 Business premises in Hamburg, 1663 to 1686 A.D. (photo: Staatl. Landesbildstelle Hamburg).

Our regional chronologies are also contributing to dendroclimatology. The amount of climatic information within a tree-ring chronology can be obtained by an analysis of variance which divides the total variance of the ring widths into components (Fritts 1976). Our chronology of Schleswig-Holstein for the recent past consists, for example, of rings (of average width 1.94 mm) from 15 oak trees which span a period of 158 years. Of the total variance of their ring-widths, 29% is caused by differences within the trees, 33% by differences between the trees and only 37% is preserved in the mean curve, due it is supposed largely to climatic factors. The coefficient for the serial correlation of the tree-ring sequences equals 0.36 in the case of the oaks of Schleswig-Holstein and thus indicates that the ring-widths depend substantially on the growth of previous years. In order to quantify the complex relationships between ring-width and climate a so-called "response function" was calculated using multivariate statistics of Fritts (1971). It is presented in Fig. 6 (Eckstein, Schmidt 1974): the ordinate axis indicates the relative influence of temperature and precipitation as well as of the three years prior to the current growth; the vertical bars include the 95% statistical significance. To summarize, 26% of the growth-ring variation is determined by the climate and 36% by previous growth.

A wide annual ring of the oak trees of Schleswig-Holstein is produced by above-average temperature in July and October of the previous season and from January to August in the current vegetation period as well as by below-average temperature in foregoing August and November. In addition growth is favoured by above-average precipitation in December and May and by above-average increments in the first and second season prior to the current vegetative year. With contrary climatic conditions a narrow tree-ring is produced.

On the basis of this model the climate for Schleswig-Holstein was reconstructed — according to Fritts (1976) — by means of tree-ring sequences to times for which no climatic records are available. This has been described in detail by Schmidt (1977). In Fig. 7, reconstruction for both temperature and precipitation of the season of spring are shown back to 1820 A.D. Around 1880 there was a short period of moist and cool springs, whereas about ten years earlier the springs had combined below-average dryness with an average temperature. From 1860 to 1864 and from 1827 to 1831 springtime was again moist and cool; around 1840 it was, however, warm with average precipitation.

If these preliminary results stand up to further verification the reconstruction of climatic conditions can be projected back in time.

Conclusions

At the start of our dendrochronological research in Northern Germany and adjacent areas the necessity of establishing several independent regional tree-ring chronologies seemed to be a handicap as it was time-consuming work and dating applications came late. For dendroclimatology this situation turned out to be advantageous because differences between regional chronologies has also provided climatic information. During our future dendrochronological work the afore-mentioned six regional chronologies will be extended in order

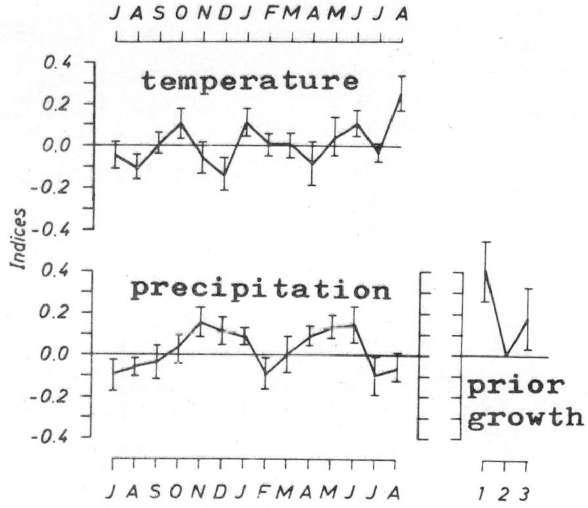

Fig. 6 The influence of temperature and precipitation from previous July to current August, and of the prior growth, on the ring width of oak trees near Schleswig. The vertical lines give the 95% confidence limits.

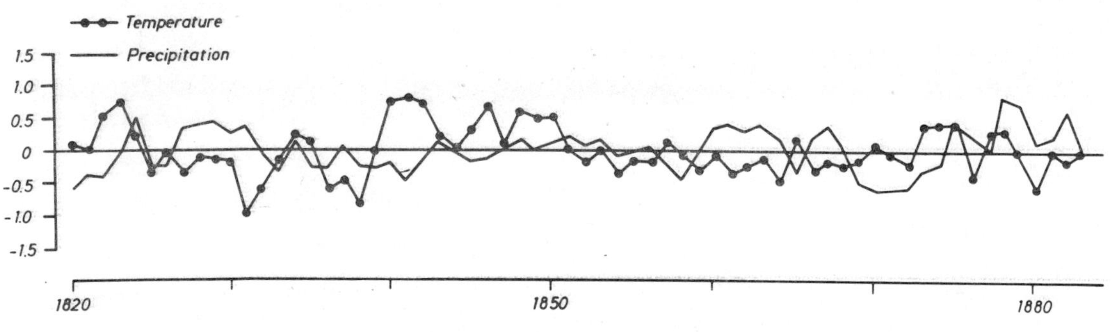

Fig. 7 The reconstructed changes of spring temperature and precipitation from 1883 to 1820 A.D. (Schmidt 1977).

to evaluate them climatologically and to use them for dating purposes in archaeology, architectural history, and art history.

References

Bartholin, T. S., 1973, Det forstlige Forsgsvaesen i Danmark 33, 217-241.

Bauch, J., Liese, W., and Eckstein, D., 1967, Holz als Roh- und Werkstoff 25, 285-291.

Delorme, A., 1973, Forstarchiv 44, 205-209.

Eckstein, D., 1976, Naturwiss. Rundschau 29, 81-84.

Eckstein, D. and Schmidt, B., 1974, Dendroklimatologische Untersuchungen an Stieleichen aus dem maritimen Klimagebiet Schleswig-Holsteins. Angew. Botanik 48, 371-383.

Eckstein, D. and Bauch, J., 1977, Aufbau von Regionalchronologien entlang der Nordseeküste von Dänemark bis zu den Niederlanden. Erdwiss. Forschung 11, 1-8.

Eckstein, D., Mathieu, K., and Bauch, J., 1972, Jahrringanalyse und Baugeschichtsforschung — Aufbau einer Jahrringchronologie für die Vier- und Marschlande bei Hamburg. Abh. Verh. naturwiss. Ver. Hamburg 16, 73-100.

Eckstein, D., Brongers, J.A., and Bauch, J., 1975, Tree-Ring Bull. 35, 1-13.

Eckstein, D., Grote, J.G., and Mathieu, K., 1977, Dendrochronologische Untersuchungen zur ländlichen und städtischen Architektur Hamburgs im 15.-18. Jahrhundert. Deutsche Kunst- und Denkmalpflege 35, 33-74.

Eckstein, D., van Es, W. A., and Hollstein, E., 1977, Beitrag zur Datierung der frühmittelalterlichen Siedlung Dorestad/Holland. Berichten van de Rijksdienst voor het Oudheid kundig Bodemonderzoek 27, in print.

Fritts, H. C., 1971, Dendroclimatology and dendroecology. Quatern. Res. 1, 419-449.

Fritts, H. C., 1976, Tree rings and climate. Acad. Press London.

Schmidt, B., 1977, Dendroklimatologische Untersuchungen an Eichen nordwestdeutscher Standorte. Diss. Univ. Hamburg.

Schwab, F., 1975, Aufbau einer Jahrringchronologie für den niedersächsischen Küstenraum. Dipl.-Arbeit Univ. Hamburg.

3.2. DENDROCHRONOLOGY, WOOD ANATOMY AND LANDSCAPE DEVELOPMENT IN SOUTH SWEDEN

T. S. Bartholin, Lund

Summary

A master chronology for oak for the last 1400 years and covering the southernmost part of Sweden has been constructed.

Very narrow tree-rings in the oak boards from coffins at Lund show that the trees came from a natural forest, untouched by axe for at least 500 years. Wide tree-rings in the stave timbers from the church at Lund show that the trees came from an open woodland pasture that had been developed before 800 A.D. Difficulties with transportation indicate that this pasture forest must have surrounded the town.

Wood-anatomy examination provides details of the species in the natural forest and in the open woodland pasture and their changes in the period 1000-1200. The natural forest disappears during the period and becomes an open woodland pasture. The oaks in buildings later than 1150 have, not only in Lund but in the whole southwest Scanian landscape, few but broad rings.

Thanks to the dates obtained and the anatomical evidence of changes in wood usage, we can now give a more precise reconstruction of the early medieval landscape and its changes during a relatively short period, than can be obtained by, for example, pollen-analysis.

Introduction

A Swedish research project on dendrochronology based on oak has been running for the last four years. Its aim is to establish the applicability of dendrochronology using oak in South Sweden and to collect samples for such a chronology.

Parallel with this, a wood-anatomical service is being given to archaeologists and Quaternary geologists, principally in order to give detailed information on the wood consumption of man to support the reconstruction of the changes in vegetation caused by man and/or nature, which is more usually interpreted from pollen analysis.

In the following an account is given of the present stage of this project. An example shows the possibility of combining wood-anatomical analysis with dendrochronological dating and from this a model of the landscape and its changes during the period 1000-1200 A.D. is derived.

It was the renewed interest in dendrochronology in northern Germany (Bauch, Liese and Eckstein 1967), and especially the results from the Viking town of Haithabu (Eckstein 1969), which seemed to offer the prospect of using dendrochronology based on oak in southern Scandinavia. The method first proved applicable in Denmark (Bartholin 1973), and it was therefore natural for investigations to continue in southern Sweden.

Methods

The tree rings are measured on an ADDO electronic measuring machine. Visual synchronization of the curves also seems suitable with Swedish oaks; the only undated samples often show atypical growth, have few tree rings, or are derived from objects of which only very few samples are available.

As a support for the visual synchronization, the Huber method (Huber 1943) for calculating the agreement is used in the form suitable for computer analysis (Eckstein and Bauch 1969). The tree-ring curves are drawn on semi-logarithmic paper as suggested by Huber (1941). Mean curves are calculated from the basic value of the width of the tree rings.

An oak chronology for Scania and Blekinge, southern Sweden

Investigations of modern oaks in Scania and Blekinge, the two southernmost provinces in Sweden, show that conditions for dendrochronology in this area are good (Bartholin 1975, Bartholin and Berglund 1975).

Within the area, there seems to be little variation going eastwards, particularly in comparison with chronologies from northern Germany and southern Denmark. Samples from SW Scania (Fig. 1 no. 3, 5, 13, 14) show much better conformity with those chronologies than samples from NE Scania and Blekinge (Fig. 1 no. 17, 18, 21). The reason for this may be due to climatic differences. The conformity between SW Scania and southern Denmark/ northern Germany is so good that chronologies for the latter areas are to a large extent applicable for dating in Scania. This fact has been of great help in the dating of samples from SW Scania.

The difference between Scania and Blekinge, is sufficiently small to allow the inclusion of results from within the whole area when compiling a chronology. This is especially important for the 13th and 14th centuries, centuries for which good samples are scarce in SW Scania. Taking the long-term view, it is expedient, however, when sufficient samples have been collected, to have two chronologies, one covering SW Scania and the other one NE Scania and Blekinge.

To the work along the west and east coasts, (thus including the islands of Öland and Gotland) we expect the south Swedish chronology to be applicable in the same way as the German-Danish one has been in Scania.

The chronology for southern Sweden today goes back to 578 A.D. Samples have come from churches, castles and other buildings, as well as from archaeological excavations. The localities are plotted on the map, Fig. 1.

Anatomical identification of the wood in the coffins from Lund

A reconstruction of the quantitative and qualitative uses of trees in the past communities can provide information about the way of life and of the immediate surroundings. A condition for such interpretation is the identification of the trees and shrubs in the widest sense.

An extensive investigation of this type has not yet been made in Lund, but from a knowledge of the species used for the coffins from the two cemeteries

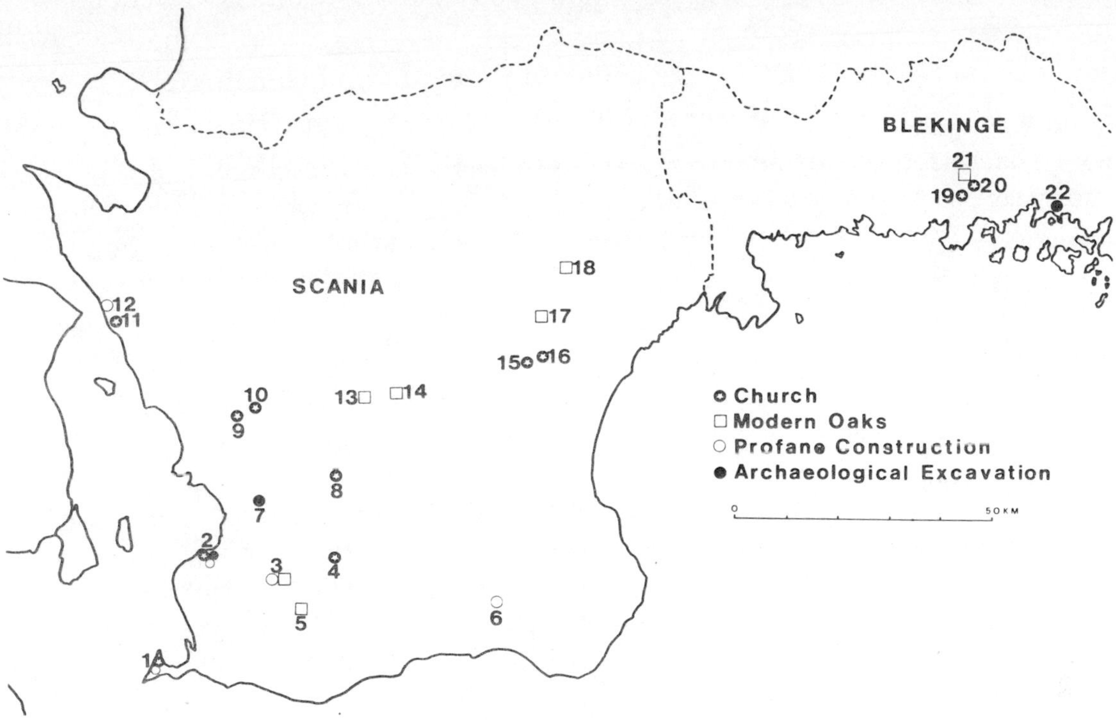

Fig. 1 Localities in Scania and Blekinge

A, Archaeological excavation; C, Church; F. Forest.

1. Skanör, C and windmill
2. Malmö: St. Petri, C
   Malmö: Kock's House
   Malmö: Kv. von Conow
3. Torup, Castle and F
4. Genarp, C
5. Borringekloster, F
6. Orup Castle
7. Lund, A
8. Hammarlunda, C
9. Norryidinge, C
10. Reslöv, C
11. Raus, C
12. Kärnan Castle
13. Bosjökloster, F
14. Fulltofta, F
15. O. Vram, C
16. Lyngsjö, C
17. Uddarp, F
18. Torsebro, F
19. Edestad, C
20. Hjortsberga, C
21. Johannishus, F
22. Lyckeby, A

TABLE 1

Tree species of the coffins

Tree species	cemetery 1 Number	cemetery 1 %	cemetery 2 Number	cemetery 2 %
Quercus sp.	396	53.01	275	73.92
Fraxinus excelsior	167	22.35	20	5.37
Tilia sp.	93	12.44	18	4.83
Tilia sp., bark	2	0.26		
Acer sp.	5	0.66	6	1.61
Ulmus sp.	1	0.13	1	0.26
Fagus silvatica	32	4.28	26	6.98
Alnus sp.	17	2.27	1	0.26
Pinus sp.	16	2.14	22	5.91
Populus sp.	14	1.87		
Corylus avellana	2	0.26		
Salix sp.	1	0.13		
Picea sp.	1	0.13	3	0.80
Total	747	99.93	372	99.94

(Bartholin, this volume, page 275) it is possible to draw some conclusions particularly as they can be related to information from the tree rings.

The trees used in the coffins do not of course represent the variety of the surrounding vegetation. Species which are easy to work, or which were used by tradition, were favoured in the first place. But a strikingly large number of species were used in the coffins (Table 1) and this may indicate the utilization of trees found accidentally.

Coffins made entirely of one species consist of Quercus, Fraxinus, Tilia or Pinus. The remaining species only occur together with others. The dendrochronological examination showed that many of the oak coffins were made of planks from only one tree. It is conceivable that coffins made of only one species were used for well-planned burials, whereas coffins made of more than one species, often children's coffins, were used in the case of sudden, unexpected burials, when coffins would be made out of any planks that happened to be available. One example is the use of Picea, an imported species, with the nearest natural stands of Picea more than 100 km from Lund. Reused Fraxinus, mostly from old troughs, was also employed.

The smaller number of species in the cemetery No. 2 may be due to more orthodox burial customs, also to be seen from the better-planned organization of the graves in the cemetery. But, as shown later, natural causes may also be of significance.

A model of the Early Medieval landscape around Lund, illustrated by observations of the wood

The establishment of a new town demands a big consumption of wood which has great effect on the landscape.

Difficulties of transportation of the wood result, in the first place, in the utilization of the nearest forest. This exploitation may have been so thorough as to convert the ground nearest to the town to arable land. At a further distance from the town the forest becomes pasture-land, and beyond this grazing-ground with single trees is the real forest. The borders between these elements of the landscape are not fixed but follow a decreasing and increasing demand for arable land, pastures and wood.

From the timber excavated in Lund we can see that this picture of the development of the landscape applied around this town. The oaks used in the coffins and for the stave church have an average diameter of at least 60 cm. Although the trees in the church were no bigger than those used for coffins, they were younger and have broader tree rings. For the period 921-1020 the tree rings in the staves were 1.57 mm on an average, and even before this period they were broad (see fig. in this volume, page 275). The modern oaks in the Scanian standard curve have nearly all been growing as solitary trees or in very open clumps. Their tree rings were 1.59 mm on an average for the period 1875-1974. This seems to indicate that the oaks used for the stave church had been growing as solitary trees in an open landscape.

The oaks used for coffins planks, on the other hand, have very small tree rings, such as will only be seen in oaks growing in very dense forests. For the period 921-1020 their tree rings were only 1.08 mm on an average. Also the Fraxinus used for coffins have extremely small tree-rings. The oaks used for coffins were, furthermore, very old—more than 200 years old, on average, many of them being more than 500 years old when cut down. This, together with the fact that the greater part of the trees used in the coffins are Quercus, Fraxinus and Tilia (87.8%, see Table 1), shows that in the neighbourhood of Lund a forest untouched by the axe had also grown for a long time.

The choice of broad-ring oaks for the stave church can only imply that near the town, at the time of its foundation, there must have been an open landscape with single oaks easy to transport to the nearby building site. The trees for coffin planks were easier to transport over longer distances, as they could be cut down to smaller dimensions. Thus in the neighbourhood of Lund there must have been pasture-land. At a further distance from the town a natural forest with Quercus, Fraxinus and Tilia.

On the pasture-land may have also grown the Fagus, Alnus and Populus used for coffins. The Pinus must have grown at least 10 km from the town since the soils in the neighbourhood of the town are too clayey for Pinus and provided optimal conditions for broad-leaved trees.

The changes in the composition of tree species used for coffins in the period of the younger cemetery (No. 2, see Table 1) may be interpreted as a continuous utilization of the forest, but, as mentioned before, changes in burial traditions may be another reason. The oak played a more dominating role as the tree preferred for coffins, and it is still the oaks of the natural forest that are used for this purpose. But shortly after this period we find only young oaks with broad and irregular tree rings: the natural forest has been cut down within a hundred years.

There is a pronounced fall in the use of Fraxinus and Tilia for coffins in cemetery, No. 2. These trees are wanted not only for the wood, but also because their twigs and leaves are used for fodder. Furthermore, Fraxinus is found on ground that would, in the first place, be used for cultivation of grass, and the bark of Tilia is also good for rope. These two species, therefore, have small chance of regeneration while the utilization of the forest is increasing. The increasing use of Pinus seems to indicate the necessity of transporting trees from distant places.

We can see that the exploitation of the forests continues. Not only from Lund but from the whole of SW Scania it has been impossible so far to date samples of oak from the period 1150-1350. All samples examined up to now from this period have had too few and too irregular tree rings for dating. Timber from oak branches as well as other species are used more and more. This, in other words, is the period when the open, woodless landscape of Scania was shaped.

From pollen analysis we know that human activities in the Scanian landscape increase around 700-1100 A.D. (Berglund 1969). Fauna-historical investigations of animal bones from Medieval Lund show the occurrence of both dense forests and pasture-land (Ekman 1973). However, due to the mixed character of the material it is not possible to trace any faunal succession.

Possibly the fauna of pastures and arable land becomes more common in younger culture layers. Our research has made it possible to reconstruct the Early Medieval landscape and its changes more precisely than by pollen analysis.

References

Bartholin, T.S., 1973: An investigation of the possibility of dendrochronological dating of oak in Denmark, particularly in North Slesvig. Forstl. Forsøgsv. Danm., 33, 215-241.

Bartholin, T.S., 1975: Dendrochronology of oak in Southern Sweden. Tree-Ring Bulletin, 35, 25-29.

Bartholin, T.S. and Berglund, B.E., 1975: Dendrochronological dating on oak in Skåne and Blekinge, South Sweden. Fornvännen, 70, 201-208.

Bauch, J., Liese, W. and Eckstein, D., 1967: On the dating of oakwood in northern Germany by the dendrochronological method. Holz als Roh- und Werkstoff, 25, 285-291.

Berglund, B.E., 1969: Vegetation and human influence in South Scandinavia during Prehistoric time. Oikos Suppl., 12, 9-28.

Eckstein, D., 1969: Diss. Univ. Hamburg, 1-113.

Eckstein, D. and Bauch, J., 1969: Forstwiss. Centralblatt, 88, 230-250.

Ekman, J., 1973: Early Medieval Lund - the fauna and the landscape, Archaeologica Lundensia, 5, 1-110.

Huber, B., 1941: Aufbau einer mitteleuropäischen Jahrring-Chronologie. Mitt. Akad. Dtsch. Forstwiss. 1, 110-125.

Huber, B., 1943: Uber die Sicherheit jahrringchronologischer Datierung. Holz als Roh- und Werkstoff, 6, 263-268.

3.3. A TREE-RING CHRONOLOGY FOR OAK FROM THE GOTHA RIVER AREA, WESTERN SWEDEN

A. Bråthen, Trollhättan, Sweden

A tree-ring chronology has been prepared from oak (Quercus robur), grown within a range of 100 km from the medieval town Lödöse on the river Gotha. It covers the period 831 to 1975 A.D. and more than 1100 samples have been used in building it with excavations from Lödöse contributing about 3000 logs, covering the years 831 to 1350 A.D. A number of churches from the 12th century have been dated by it.

In a paper in course of publication, I have attempted to trace the summer temperature from 1400 to 1860 A.D. For this purpose I have used only those rings formed 150 years after the trees started to grow. Between 10 and 18 trees, which have not been exposed to violent environmental factors and which started to grow at the same time, cover each period of 100 years, and the periods overlap each other by 50 years. It is feasible to get reliable estimates of summer temperature in the past particularly if more samples are collected and added to those already used. The present results show a cold period from the beginning of the 15th century to the end of the 17th century (see Fig.)

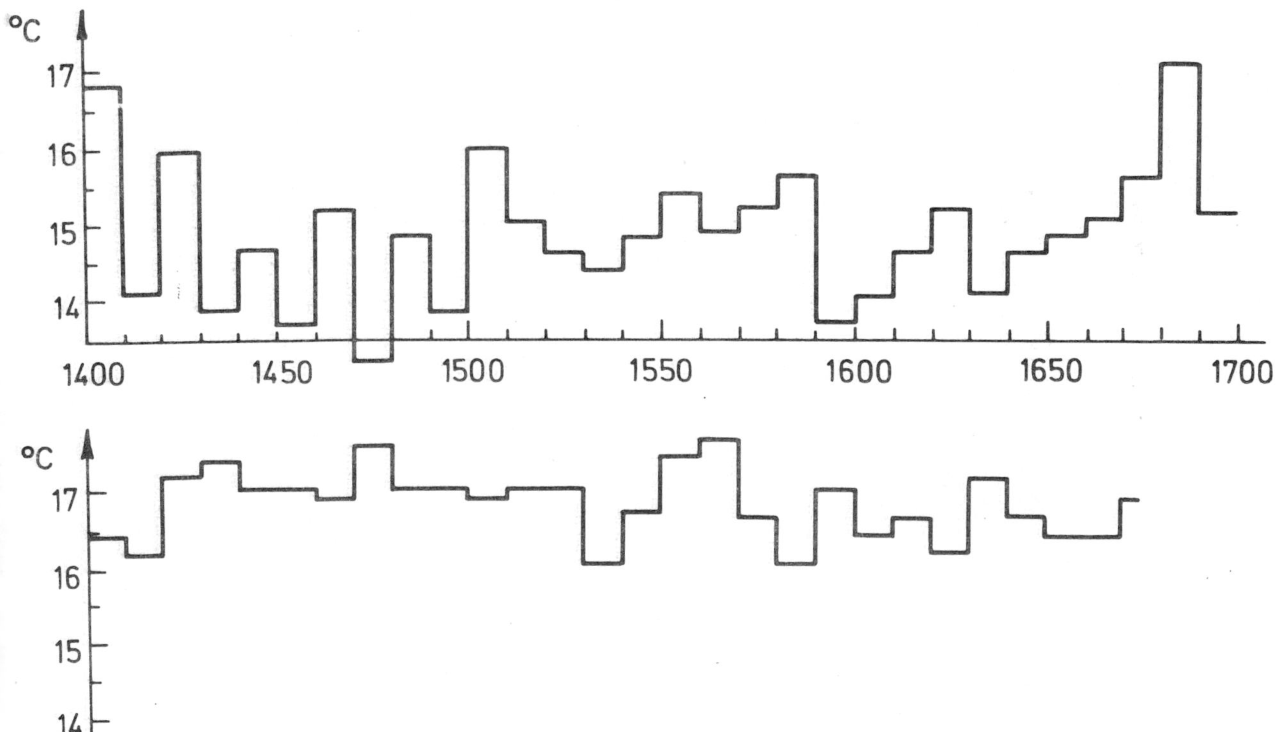

Fig. Summer Temperatures, with estimates for 1400 to 1860 A.D. from annual rings of an oak chronology.

DENDROCHRONOLOGY IN POLAND

M. J. Dabrowski, Warsaw

Under the influence of Huber, dendrochronology was started in Poland by the late Prof. K. Ehrmich. His pupils in Cracow have continued by studying Abies alba, Picea abies and Pinus cembra at the timberline of the Tatra Mountains. In the Polish literature there are about a dozen papers, all published after the second world war.

The construction of a chronological scale began recently for the northern part of Poland for oak (Quercus robur) and Scots pine (Pinus silvestris), the species found most frequently in excavations. Altogether nearly a hundred pine and more than 350 oak samples have been analysed. In our data-bank more than 300 tables from Denmark, Germany, USSR and Poland are stored.

The majority of the material is from the early medieval period but about 40 samples may be of Late Bronze Age origin. There is a gap in the 15th/16th century A.D.

Ring widths are tabulated as averages and for each sample the ratio of the width of a certain year to that of the previous one is calculated and normalised statistically. The history of a tree is in the form of the series of ratios and two samples are correlated in time by finding the smallest difference of these ratios. The same is done with the normalised series. Methods of correlation are discussed.

Editorial Addendum

The following papers appear in English in Bioecological Fundamentals of Dendrochronology, ed. T. Bitvinskas, Vilnius/Leningrad, 1975 as part of the Proceedings of the XII International Botanical Conference at Leningrad 1975: (cf. this volume, page 56)

E. Feliksik, Present State of Dendrochronological Investigations in Poland. Pages 21-25 (19 refs).

Z. Bernarz, Geographical Range of Similarities of Annual Growth Curves of Stone Pine (Pinus cembra L.) in Europe. Pages 75-83 (25 refs).

3.5 TREE-RING CHRONOLOGIES FOR THE NETHERLANDS

J. Bauch, Hamburg

Summary

Two independent tree-ring chronologies of oak have been established in the Netherlands. One starts at the present time and goes back continuously to 1385 A.D.; another part of it, derived from archaeological objects, covers the period from 1036 to 1325: this chronology is synchronised with those for Southern and Western Germany. The second chronology covers the period from 1109 to 1637 and matches that developed recently for the lowlands of England. It appears that oak of this origin was no longer available after about 1650.

Dendrochronology has been developed systematically in the Netherlands only since 1965. At that time the Institute of Wood Biology at the University of Hamburg began to date the oak-wood panels used by Dutch and Flemish artists (Bauch, 1968). Brongers (1973) a few years later attempted to estimate the geographical range of the modern chronology for the Netherlands by the examination of recent oak trees from different locations. Cooperation between the two groups made it possible to establish master chronologies for the Netherlands and Belgium (Eckstein, Brongers, Bauch, 1975). Results obtained during the last two years which extend these chronologies are given in this paper. Also an interpretation is offered for the two separate growth ring patterns which have been found for the Netherlands.

Materials

The oak samples used for the recent Dutch chronology consist of discs from 27 trees that were up to about 300 years old when felled. They came from six sites. To extend the chronology, increment cores and sections of oak beams were taken from 13 mills, the sites of which are shown in the map of Fig. 1 by the larger square symbols. The Netherlands still has about 900 mills, of which 40% are in working condition.

A third source for growth-ring analysis was provided by oil paintings by Dutch and Flemish masters on oak panels (Fig. 2). For chronology I, 40 panels, characterised by light bars in Figs. 1 and 2, were used; for chronology II, 182 panels, shown by a dark bar. In addition measurements were made on a few wooden sculptures.

For a possible extension of the reference charts back to earlier centuries, wood from excavations at Dordrecht and Dorestad was measured; their symbols on the map are the smaller squares.

The Master-chronologies

All dendrochronologically investigated objects are included in the summary bar-chart (Fig. 2). Each of the horizontal bars for the oak trees and for the mills represents material from a different site, the number of samples constituting a given bar being indicated on the right side.

Chronology I: The overlapping bars from AD 1972 to 1385 represent a regional tree-ring chronology of 588 years. All the sequences in it possess the charac-

Fig. 1. Sources in the Netherlands of the Oak Wood for Chronologies I and II (Key see Fig.2).

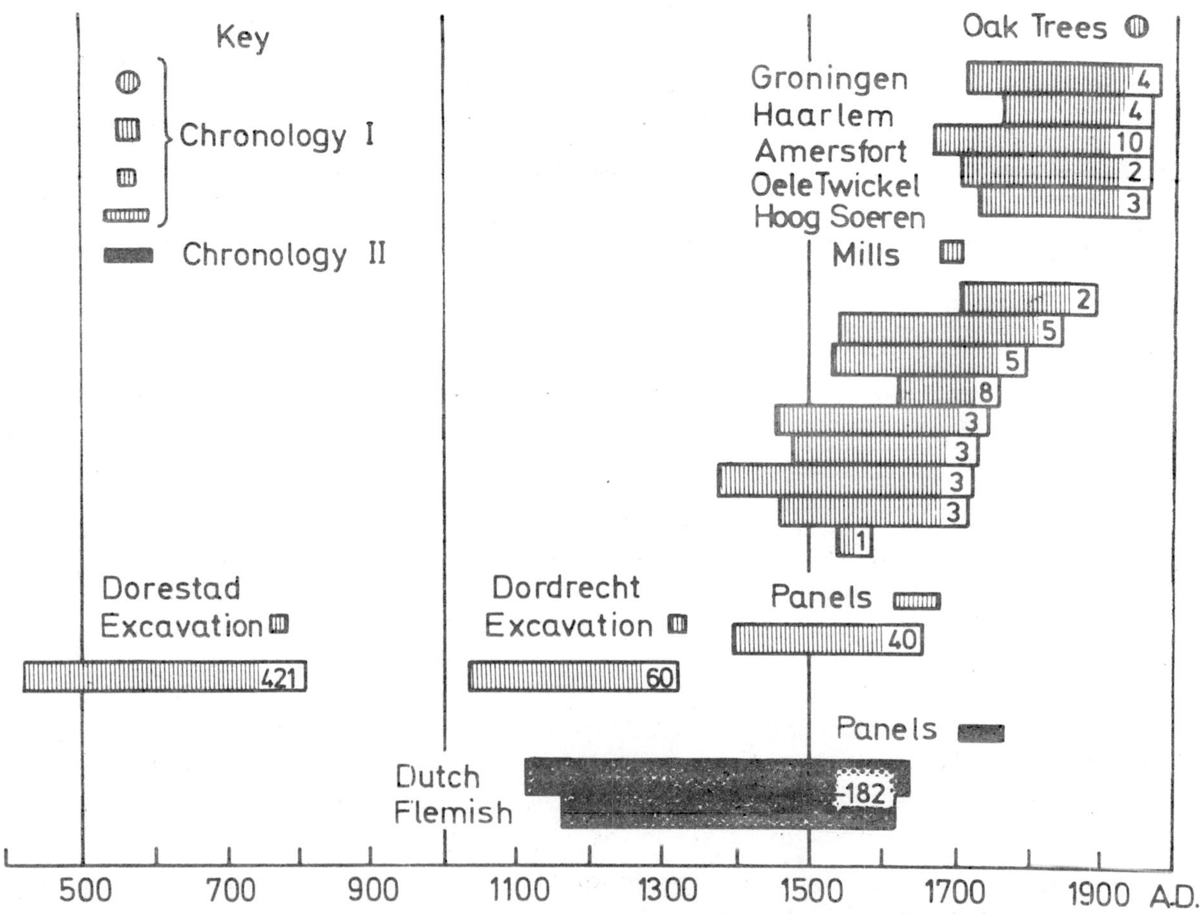

Fig. 2. Component Tree-Ring Sequencies for the Netherland Chronologies I and II

teristics of those for oaks in the southwest German chronology of Hollstein (1965). The earlier part is largely based on 40 panels, mainly executed by Rembrandt (1606-1669), Wouwerman (1619-1668) and Van der Werff (1659-1722) and covers the period from 1400 to 1687.

We have also derived a tree-ring sequence from AD 1036-1325 that is based on oak construction timbers from excavations in Dordrecht, and from 420 to 817 A.D. for oak samples from the Dorestad excavation. However, as the latter, according to Eckstein, van Es and Hollstein (1977), was imported from Germany it is not incorporated in chronology I.

Chronology II: A quite different growth ring pattern was obtained from 182 panels used for Dutch or Flemish paintings. The mean curve from the Dutch paintings extends from 1109-1637 and is derived from works of art mainly painted in Amsterdam or Leiden. The Flemish component (from 1167-1613) is very similar and is based on paintings, mainly those of Rubens, created in Antwerp. Because at present the master-chart can only be proved with a precision of ± 1 year, the curves for this chronology have not yet been published.

Origin of the Wood

For the recent trees a striking similarity exists with the German chronologies of Hollstein (1965) and of Huber and Giertz-Siebenlist (1969). Even trees from Harleem and Gronigen—typical coastal sites—could be synchronised with the trees from hilly and more continental areas that apply to the two German chronologies. This occurs in spite of the coastal trees being *Quercus robur* and the ones from hilly terrain being mostly *Q. sessiliflora*.

The timbers used for the mills also have growth-ring patterns which fit the German chronologies as do the panels that were made later than about 1650 and the wood from the settlement at Dordrecht. For the latter and especially for the timber at Dorestad the wood could have been derived from German forests and transported down the Rhine.

What, it must be asked, was the origin of the large amount of wood that was used for over two hundred years (from about 1400 to 1650) to make panels for paintings and blocks for sculptures in centres such as Amsterdam, Leiden, Haarlem and Antwerp? With one exception, the growth-ring pattern is very different to that of chronology I and forms the self-consistent pattern that we call chronology II. As a result of our close cooperation with the tree-ring laboratory at Oxford, Fletcher (1974) found some years ago that our chronology II has the same growth-ring pattern as that for panels for paintings executed in the London area. This happened, by chance and as a surprise, when measurements on the National Gallery painting of *The Judgement of Venus*, attributed to Rubens when at Antwerp about 1600, were made in London on our behalf. The sequence obtained was readily matched by the existing English chronology and the match confirmed by our own. The subsequent extensive study of English and Flemish paintings (Fletcher, 1977) has confirmed the similarity (W, 68.6%) between the reference chronology (REF 1) which applies to the south-east and lowland part of England, and our chronology II in the period 1354-1588.

One hypothesis is based on the wood for Dutch and Flemish panels prior to 1650 being derived from coastal and local forests in the Netherlands or

Fig. 3. Dendrochronological dating of the wood of paintings of Ph. Wouwerman (1619-1668)

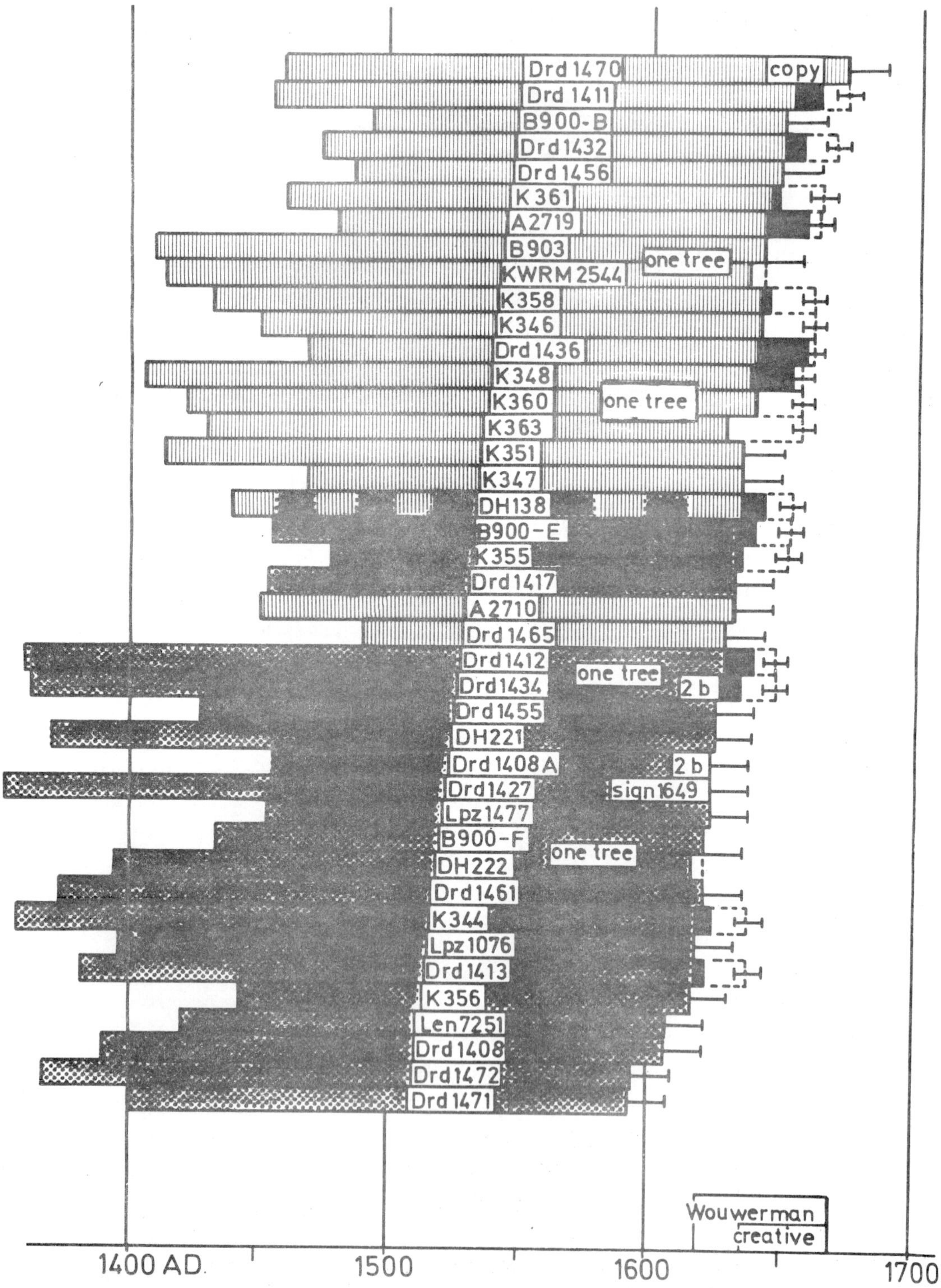

Galleries: A Amsterdam, B Berlin, DH The Hague, Drd Dresden, K Cassel, KWRM Cologne, Len Leningrad, Lpz Leipzig.

felling date of the tree, earliest possible; felling date, sapwood rings, chronology I, chronology II

Flanders that were exhausted completely by 1650 due to the heavy demands for shipbuilding etc. Consequently, the oaks used subsequently for panels and mills in the late 17th and 18th century had to be obtained from elsewhere and were transported by river either from woodlands in the eastern part of the Netherlands or from the adjacent hilly parts of Germany and had a different growth ring pattern (chronology I) in their inner parts. Another hypothesis, that the wood for the Dutch and Flemish panels in the period 1400 to 1650 came from another country, e.g. from England, is scarcely tenable as it would imply that all the timber for the panels was imported.

Fig. 3 demonstrates how the two chronologies apply to the panel paintings of Ph. Wouwerman (1619-1668). This artist lived exclusively in Amsterdam and it can be assumed that the panels for his paintings (several hundred in all) were made there. From 41 dated panels it can be proved that up to about 1650, the growth-ring pattern belongs to chronology II, but for later panels exclusively to chronology I. The hope of finding examples which would link the two patterns has only been realised to a very limited extent, indeed only two panels, one from each, have been tentatively synchronised.

Attempts will continue to be made to confirm the origin of the wood of the Dutch and Flemish paintings and sculptures that form Chronology II. Although up till the present there has been no success it may be possible to find, in the Netherlands or in Belgium, building timbers, e.g. from old mills or cathedrals, which came from local trees cut between 1400 and 1650.

Acknowledgements

The investigations were carried out in cooperation with Prof. Dr. D. Eckstein, Hamburg and Dr. J. A. Brongers of Amersfoort, the Netherlands. We would like to thank Mrs. G. Brauner for her assistance.

References

Bauch, J., 1968, Kunstchronik 21, 144.

Brongers, J. A., 1973, Berichten van de Rijksdienst voor het Oudheidkundig Bodemonderzoek 23, 456.

Eckstein, D., Brongers, J. A. and Bauch, J., 1975, Tree-Ring Bull. 35, 1.

Eckstein, D., van Es, W. A. and Hollstein, E., 1977, Berichten van de Rijksdienst voor het Oudheidkundig Bodemonderzoek 27, in print.

Fletcher, J. M., 1974, Archaeometry 16, 31.

Fletcher, J. M., 1976, Studies in Conservation 21, 171.

Hollstein, E., 1965, Bonn. Jb. 165, 12.

Huber, B. and Giertz-Siebenlist, V., 1969, Sitzungsbericht Akad. Wiss. 178, 39.

Prof. Dr. Joe Bauch about to use an eyepiece to measure the ring-widths on the edge of a panel painting.

Dr. John Fletcher with photographic reproductions of portraits of Jane Seymour. The larger shows Holbein's original, now in Vienna; the smaller a copy at Woburn Abbey, loaned for display in the Dendrochronology Exhibition at the National Portrait Gallery, July–September 1977. (Westminster Press Ltd.)

3.6 OAK CHRONOLOGIES FOR EASTERN AND SOUTHERN ENGLAND: PRINCIPLES FOR THEIR CONSTRUCTION AND APPLICATION: THEIR COMPARISON WITH OTHERS IN NORTH-WEST EUROPE

J. Fletcher,
Oxford University

Summary

Reference chronologies have been developed for trees of slow to moderate growth for the medieval and Tudor periods in eastern and southern England. Their pattern varies with the mean ring-width. They belong to a type which extends around the southern basin of the North Sea and appears to be one of two shown by oak in central and north-west Europe. A model is proposed.

I became aware of the potentialities of dendrochronology in 1935 as a result of a visit to the 11th century cliff dwellings at Mesa Verde in Colorado that Douglass had tree-ring dated. So dendrochronology seemed in the 1960s to be a possible method for dating the large number of early timber-framed buildings that we were discovering in the villages around my home in North Berkshire (Fletcher, 1965/6 and 1968). Microfilms of relatively obscure local publications in Germany describing the success of Huber and v. Jazewitsch at Ziegenheim in Hesse were obtained and kindly translated for me by Dr. John Sykes, then of the staff of the library at the Atomic Research Establishment, Harwell. With computers available to assess the reliability of comparisons in every position between two curves, the outlook seemed at least as promising as for many technical problems in nuclear energy that I had seen successfully solved in the post-war era. A niche in the Wood Structure section of the Dept. of Forestry at Oxford to set up X-ray densitometry under J. F. Hughes provided the opportunity to make ring-width measurements on discs from oaks felled on identifiable sites in local woodlands as well as much-needed contacts with persons engaged in forestry practice and research.

The function of European tree-ring chronologies is to provide information of economic or scientific value, or to match and date the annual rings of wood of unknown age. For the latter, the usefulness of a particular chronology may be judged by the number and variety of samples that can be reliably dated by it.

Chronologies are derived from values for several samples that are meaned in an appropriate manner to show common features useful for matching. Successive widths of annual growth-rings have been extensively used in Europe because such rings are easy to identify and their widths easy to measure for some species. For others, identification of the annual rings is difficult and for this and other reasons (e.g. missing rings) no European chronologies have been constructed for diffuse-porous species such as poplar and walnut.

Ring-width chronologies were successfully constructed early this century for conifers of recent growth in Scandinavia and Eastern Europe; and rather later, for conifers and certain deciduous trees, oak and beech, in the hilly areas and sub-alpine zone of central Europe. Ring-widths are, however, a crude reflection of the many environmental factors that affect the growth of

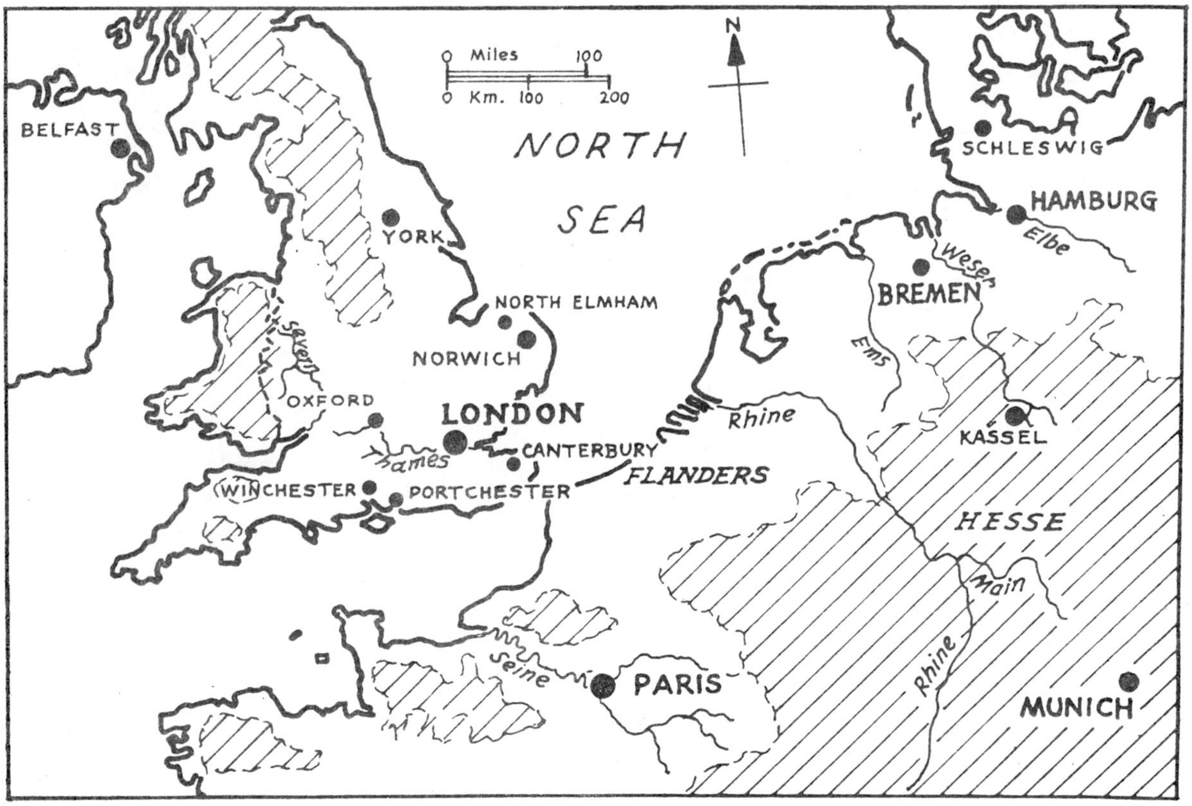

Fig. 1. North-western Europe. Shaded Areas are mainly over 200 m in altitude.

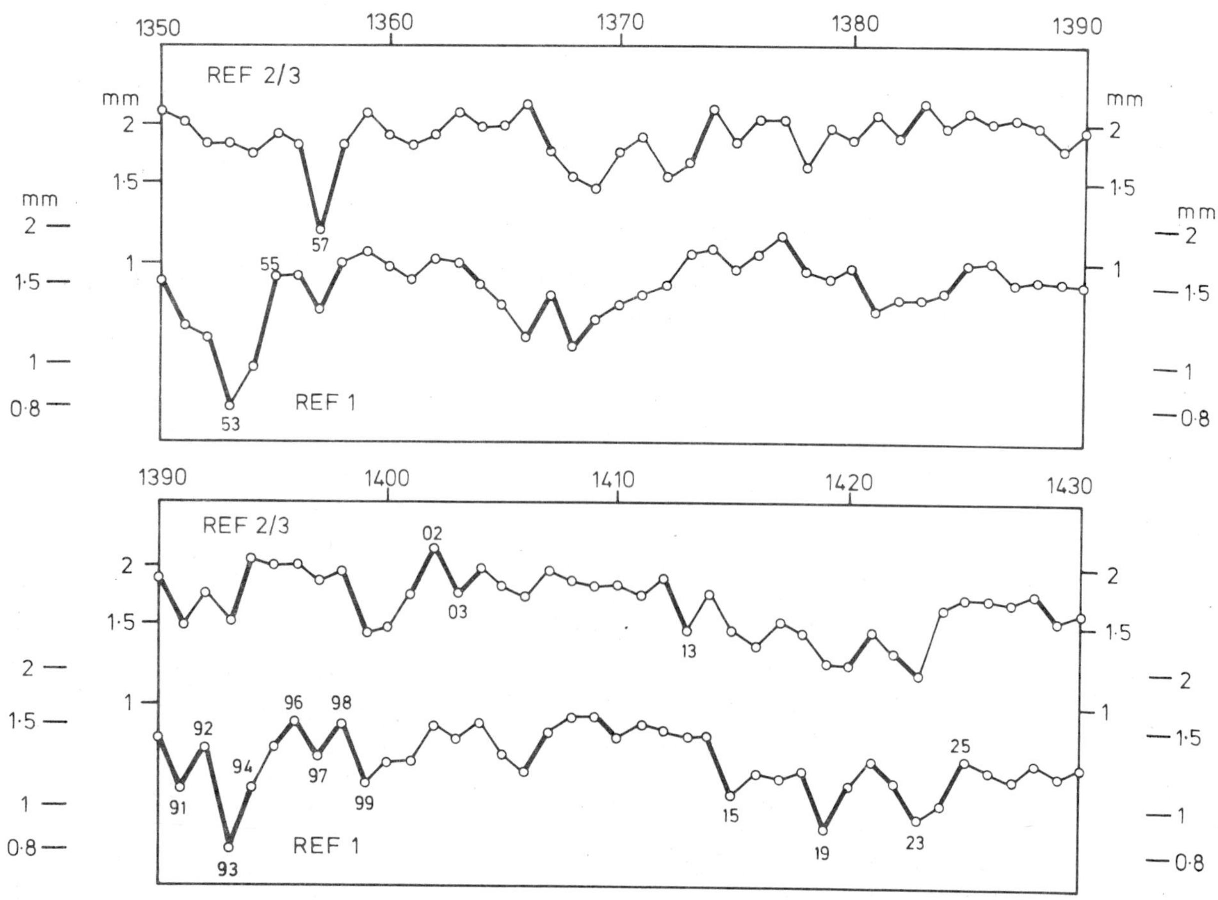

Fig. 2. Parts of the REF 2/3 and REF 1 Chronologies showing Indicators (thick lines) and Signatures (years numbered).

trees in temperate regions. For maritime areas in particular, as research on oaks in the North German plain indicated in the early 1960s, the construction of a tree-ring chronology based on ring-widths was difficult.

Those familiar with the British countryside and woodlands are aware not only of the influence of terrain and micro-climate, but of the effect of damage by insects and fungi on the growth of trees - particularly on oak and elm. The multiplicity of factors that affect growth was believed to make the existence of a common growth pattern, even in trees of the same species, a remote possibility. In the 1940s their experience had convinced many British foresters and forest-scientists that ring-widths were an almost hopeless basis for dendrochronology in southern Britain. Such scepticism had scarcely weakened even twenty-five years after Huber and his co-workers had demonstrated the potentialities of dendrochronology in Germany, partly because his research and results were virtually unknown here.

The success that has attended our research, begun systematically in 1971, has therefore caused some surprise. Ring-width chronologies, comparable in reliability, but not in length, to that of Huber, have been constructed for medieval oaks growing in southern and eastern England by adapting principles evolved in Germany to the peculiarities of the growth of oaks in this part of Britain. The obvious source of material, construction timber from barns and buildings, was, however, unsuitable. Though easy to sample and convenient for laboratory measurements, its fast growth made it virtually impossible to use for that most difficult stage, the start of a chronology. This point had already emerged by 1970 from the results by Siebenlist-Giertz at Huber's laboratory in Munich on samples collected at the instigation of Horn and Charles from English barns. As the starting material for historic times, we used, on the advice of Prof. Bauch in 1970, the radially-divided, oak boards of panel paintings. Such boards are usually derived from the trunks of slow grown trees. In the first place measurements were made on paintings of known date as they provided ring-width sequences ending at a time, known approximately.

CHRONOLOGY BUILDING

We have aimed to construct oak chronologies, suitable for matching and dating purposes, with a reliability and quality comparable to the 1000 year chronology of Huber and Giertz for oaks from hilly areas (Fig. 1) in central and southern Germany (this volume, page 29). The narrowness of the annual rings (on average 1.25 mm) of that chronology implies that the trees were subject in their growth to considerable stress; the high correlation (in the order of 70%) between the individual curves that formed the chronology, that the trees were subject to similar stresses. In fact, Huber deliberately excluded the faster-grown pedunculate oaks from valley sites and the chronology was largely formed from slow-grown sessile oaks from altitudes of over 200 m. It is to be noted that the precipitation in the Spessart Forest, for example, is of the order of 1000 mm per annum.

In the lowlands of southern and eastern England (Fig. 1), oak woodlands are situated almost entirely at altitudes below 180 metres (600 feet). The annual precipitation from the Pilgrim's Way, 40-50 km inland from the south coast, to the northern extremity of the plain of York, a distance of 400 km, is low—in the range 500 to 750 mm. Westerly air currents from the Atlantic are depleted of much of their moisture by the time they reach this eastern zone and other pockets of land,further north and near the east coast of the British Isles,that lie in rain shadows. These areas are also affected by cool air from the continent, modified by the North Sea. The oaks are mainly pedunculate.

About 70,000 ring-widths have been measured on 350 'historic' samples almost all of radial section and mainly from trees likely to have grown around London.

Panels: 230 samples (as boards) from 160 panel paintings, about 18 are Flemish and 5 French, have provided the backbone for our evidence about the patterns shown by oaks of a slow or moderate rate of growth. The number of rings per board has averaged over 170 and many of the trees must have been over 250 years old, in exceptional cases at least 400 years old, when felled.

We find that wide rings (up to about 30 in number and over 3 mm wide) at the beginning of such sequences do not have a typical pattern. On the other hand, a long series of very narrow rings (<0.9 mm on average) at the end of a sequence have such slight variation in width that they add little of value to a mean curve. Consequently —

(i) Even from sequences with 250 or more years, the number of rings suitable for use in a chronology does not always exceed 200;

(ii) Shorter sequences with about 150 rings may have only 120 suitable.

A number of these long oak sequences, unlike those of conifers, maintain almost the same mean width throughout their length.

Bauch, Liese and Eckstein (p. 24) pointed out for the North German plain that a slight change in altitude, or difference in the soil, can decide whether a warm dry summer checks or advances tree growth. We appear to have been the first to recognise an implication of this effect in a lowland area, namely that the ring-width pattern is likely to change with the rate of growth. We find indeed that the patterns of narrow-ringed and wide-ringed sequences differ. This applies not only to different trees but within the stem of individual oaks as results on discs, taken at a height of about 12 feet, from trees felled in Bagley Wood, near Oxford, show. When the pith is highly eccentric, the ring-width patterns along the longest and shortest radii can be sufficiently different to lower the agreement value between them considerably:

		Average Ring Width		Longest/Shortest Radius	
Tree	Pith	Longest radius	Shortest radius	Ring Width ratio	Agreement value (W)
1.	Relatively central	2.6 mm	2.4 mm	1.1	79%
2.	Relatively central	2.1 mm	1.4 mm	1.5	80%
3.	Eccentric	3.1 mm	1.5 mm	2.0	67%

It follows that sequences, or parts of sequences, in which the ring widths are substantially different should not be meaned together, either in deriving a 'single' curve from samples from the same tree, or in joining single curves to make mean curves. Furthermore, removal of age trend, though possibly beneficial with other species, is not only unnecessary for well-replicated mean curves of oaks (cf. Delorme, this volume, page 53) but, in temperate areas, such as southern and eastern England, is likely to be positively misleading in chronologies for dating purposes.

Methods

For objects such as chests or panel paintings measured in situ, ring-widths have been recorded directly on transparent semi-logarithmic paper of standard size. In sequences from some boards the presence of rings with abnormal early vessels (see page 339) together with prominent signatures (vide infra) indicated a likely date even at the plotting stage. Sequences are compared by computer (using the Hamburg and CROS programmes) with more than one reference curve and possible matching positions examined visually for rejection or acceptance.

The ease and reliability of visual matching has been greatly helped recently by marking Huber-type indicators and signatures (see pages 22 and 48) on the reference curves by thick lines. Their presence draws attention to consistent year-to-year changes in the same direction (Fig. 2). We mark indicators, not in terms of a fixed percentage, but in relation to the number of component single curves and at a level of random probability, P, chosen as $<10^{-2}$. This requirement gives the following relationship:

		7	10	15	20	25	30	35	40
No. of single curves in mean curve		7	10	15	20	25	30	35	40
Minimum percentage of common direction for	(i) $P<10^{-2}$	100	90	86.5	80	76	73	71	70%
	(ii) $P<10^{-4}$	-	-	100	95	88	86	82	80%

When P is $<10^{-4}$, we regard the change as a signature and add the date of the year concerned (Fig. 2). A few changes, 1392 to 1393, and 1456 to 1457, are so consistent that all, 31 and 30 respectively, of the single curves of REF 1 move in the same direction. These conditions give rise to the following percentages for different parts of the REF 1 English and REF 2/3 curves:

No. of single curves in mean curve	Year to Year Changes: No. of Indicators to Total No.	Year to Year Changes: No. of Signatures to Total No.
23 to 34	31%	12%
10 to 22	16	6

The values suggest, as recommended by Delorme in Germany (this volume, page 47), that 30 single curves provide good coverage for an oak mean curve.

Table 1. Comparison of Type A1 and A2 Reference Curves, Late Medieval and Tudor Periods

	Type A1 REF 1 English	A2 REF 2/3
Number of single-curve components (trees)	36	33
Number of samples (boards)	57	49
Mean no. of annual rings per component	184	118
Rings used to total measured and dated	89%	78%
Years spanned, overall	1210-1602	1224-1618
Years spanned, with < 7 single curves	1266-1535	1341-1561
No. of indicators in above years	59 (22%)	42 (19%)
No. of signatures in above years	21	13
Mean ring-width	1.34	1.90 mm
Mean ring-width, range for components	1.0 to 1.9	1.3 to 2.9 mm
Mean fluctuation, range for components	12 to 30	14 to 26%
Mean fluctuation	12.4	11.3
Coefficient of variation	16.2	14.3%

t between components of REF 1 with REF 1: mean 9.1, range 8.0 to 15.4
t between components of REF 2/3 with REF 2/3: mean 8.0, range 4.6 to 11.3
t between components of REF 2/3 with REF 1: mean 3.8, range 1.6 to 4.9

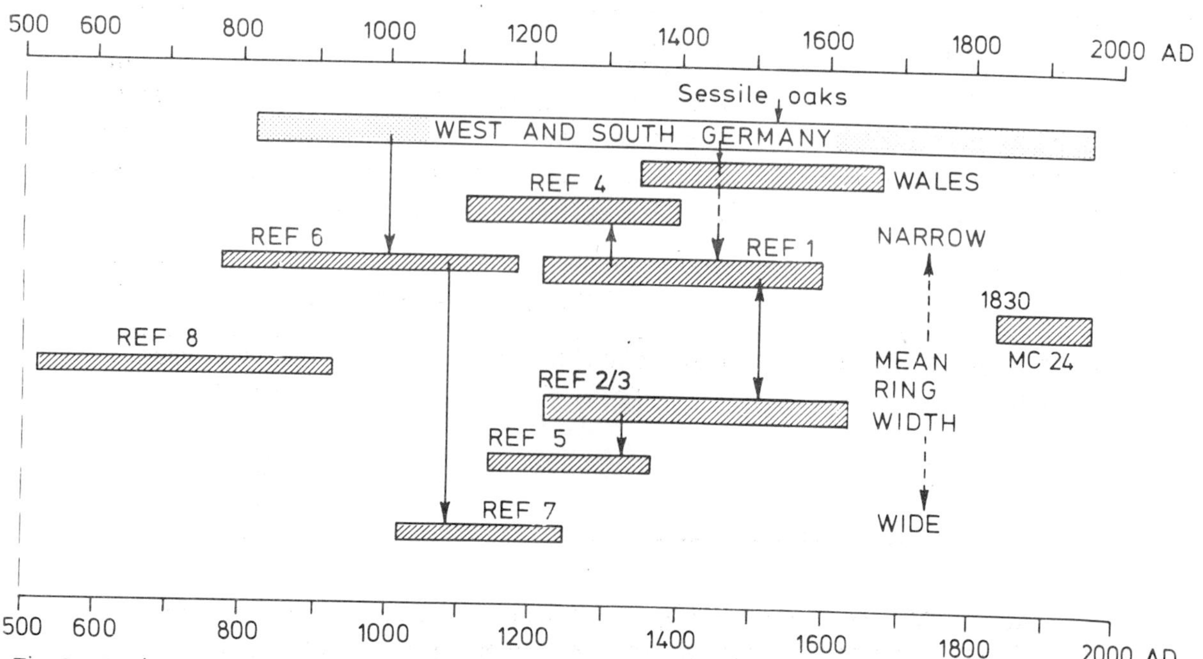

Fig. 3. Periods spanned by Type A Reference Chronologies. Vertical arrows show cross dating.

THE CHRONOLOGIES CONSTRUCTED AND A MODEL FOR THEIR INTERPRETATION

TYPE A

About 85% of the historic samples mentioned above have been matched and from them the chronologies named as this type and shown in Fig. 3 have been formed. REF 8 is 'floating', REF 6 and REF 7 are dated directly by the Huber-Giertz and Hollstein chronologies.

<u>Slow Growth, Type A1: REF 1</u>: The agreement between the sequences from the panels of paintings is not only greater when the two sets of width are comparable, but also when the rings are relatively narrow, i.e. in the range 1 to 1.6 mm. A mean curve of this narrowness was therefore that most readily constructed in the first place (Fletcher, Tapper and Walker, 1974). As more and more samples were matched, the number of components combined into mean curves was increased stepwise from 4 to 8 to 18. A few components were from Flemish rather than English panels.

The components of REF 1 in its latest form, shown as a bar-chart in Fig. 4, consist of 36 single curves that are almost certainly of English origin, hence its name REF 1 English. Each single curve is believed to represent a separate tree and is often the mean of measurements on two or more boards, or of both edges of a board.

The sections of single curves excluded in REF 1 are unshaded in Fig. 4. They consist either of wide rings in the early part of sequences excluded due to poor matching; or very narrow rings at the end of other sequences excluded as being unrepresentative and therefore misleading. The parts included have mean ring-widths that are similar to one another and the agreement of each with REF 1 is very high (Table 1). New sequences enter the chronology continuously, the average interval between them being about 6 years.

<u>Moderate Rate of Growth, Type A2; REF 2/3</u>: Suitably matched single-curves in the two wider-ringed sequences REF 2 and REF 3 previously published (Fletcher 1977) have been combined, and, with others of similar width, used to form REF 2/3 included in Fig. 3. Details are given in Table 1 and it is also shown as a bar-chart in Fig. 4. Four single curves, itemised in Fig. 4, have sections both in REF 1 and in REF 2/3.

<u>Comparison of REF 1 English and REF 2/3</u>: Although the average widths of REF 1 English and REF 2/3 vary by no more than 40%, the corresponding difference in growth on a weight basis is about 100%. For an overlap of 300 years the agreements are W = 68% and t = 6.9 between the two chronologies. In certain decades, however, there are low W values while overall the indicator years and signatures do not coincide. Thus for the period 1341 to 1535, in which both curves are based on more than 6 single curves, the disagreement with regard to indicators is as follows:-

In REF 1	53 instances;	41 different to those in REF 2/3.
In REF 2/3	40 instances;	28 different to those in REF 1.

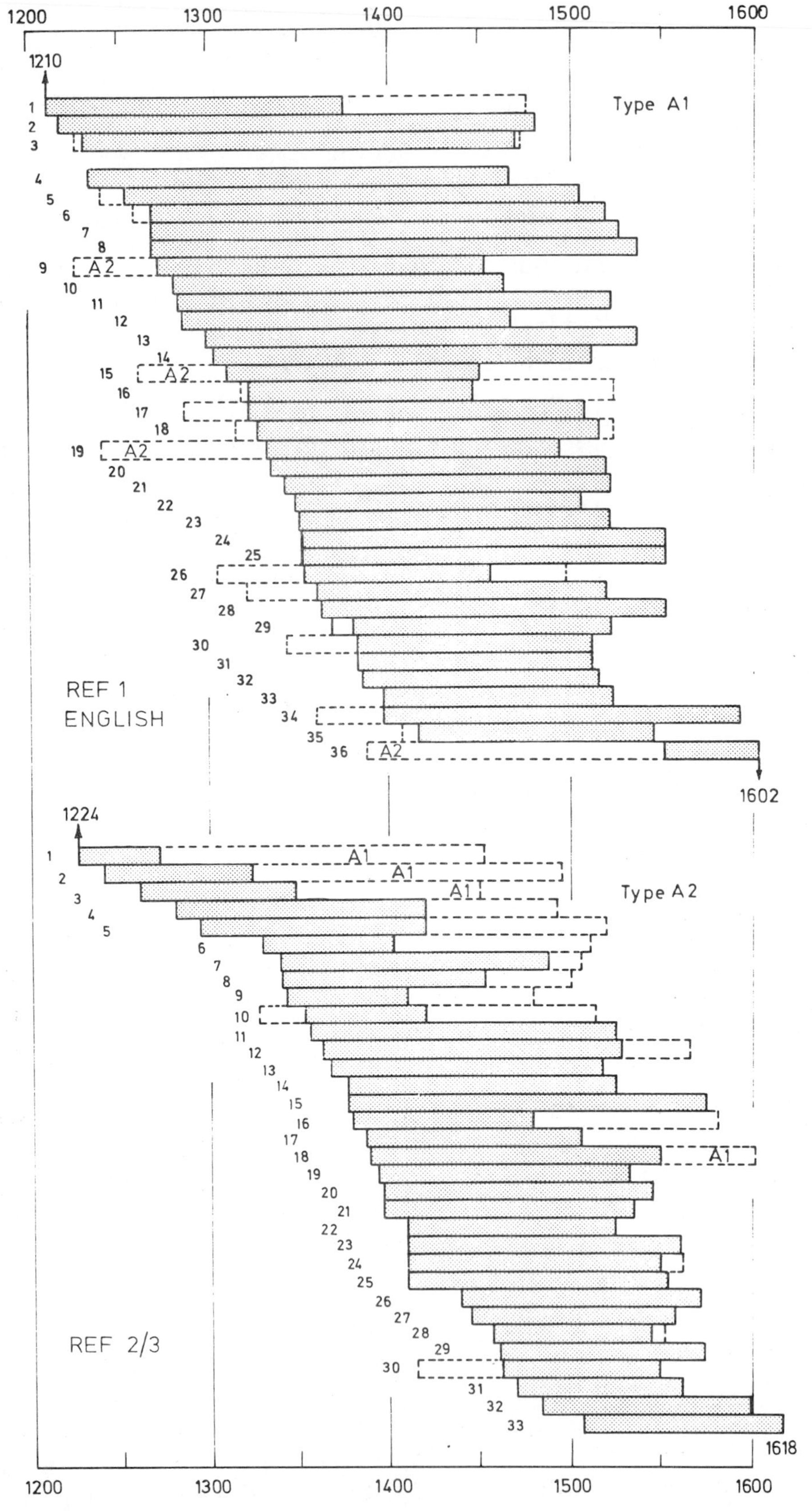

Fig.4. Components (69 single-tree curves, 103 boards) for two Reference Chronologies for London area and Southern and Eastern England.

Upper: Type A1 for rings of average width 1·3 mm

Lower: Type A2 for rings of average width 1·9 mm

Four of the curves have their earlier ring-widths included in the Type A2 Chronology and their, later, narrower ones in the Type A1 Chronology.

Components (except for three from Oxford chests) are derived from panels likely to be from English trees and used for paintings.

18 other suitable components from panels of Flemish or French origin are excluded.

Marked differences between the signatures occur in the 1350s and 1390s (see Fig. 2).

Agreement with Continental Chronologies: The good agreement with others of Type A (Part I, Table 2) extends at least as far as the Netherlands. With Type H ones (Part II, Table 2) there is little or no agreement in this period, 1250-1550. In the preceeding period, 1000-1250, the agreements, however, between the relevant Type A1 chronology (REF 6) for southern Britain on the one hand, and the Huber-Giertz and Hollstein chronologies for Germany on the other hand, are sufficiently high (67% for an overlap of 233 years) for REF 6, and, in turn, REF 7 to be dated by them (Fig. 3).

REF 4 to 8: Each of these, depicted graphically in Fig. 3 and already decribed elsewhere (Fletcher, 1977), is based at present on few (5 to 13) trees derived in all from about 100 samples that include chests, cupboards, covers to manuscripts, a ship, Viking houses and Saxon wells and a mill. I am grateful to Ruth Morgan for making available, for REF 6, ring-widths both for timbers from Seal House on the London waterfront (this volume, page 223) to supplement our own results on a few timbers from that site; and from Viking houses in York.

REF 4 is in course of integration with REF 1 English to form REF 1/4 back to 1100 A.D. Similar amalgamation of REF 2/3 with REF 5 is in hand to extend the wider type back to about 1150.

TYPE H (Fig. 5)

This nomenclature is used for the key pattern first recognised in the oak chronologies of Huber, Hollstein and Delorme for the hilly areas of central and west Germany. It is also the pattern in the chronology (MK 8) of Siebenlist for hilly areas of Wales /W. Midlands (this volume, page 156), and in her chronology for hilly parts of Normandy (unpublished); it occurs as well in the mean curve of Ruth Morgan (1977) for the timber of a house in Sheffield. Further to the west, Baillie's chronology (1977) for the Belfast region is a variant of it.

In Hesse and Northern Ireland, Type H is the only pattern so far identified for oak. In other regions both Type H and Type A occur. Yorkshire is one where the two types are found together. Another is the West Midlands where there are woodlands at altitudes that vary from 60 to 200 metres (200 to 700 feet). Overlapping of the types in these two areas was suggested by the results obtained by Siebenlist-Giertz on the samples she collected in England in 1972 (see page 156); of 77 samples from 9 sites in Yorkshire, Shropshire and the Midlands, she was able to match only 4 (from 2 sites) by the Munich reference chronology of Type H. We now find that some of the other sequences, e.g. from Coventry are Type A.

REF 2/3.

It has perplexed some dendrochronologists that the Type H pattern should prevail in forests as far apart as Wales and Hesse and yet be absent in most of the intervening ones. It is significant that there is almost no land over 120 metres in altitude in the direct line from the western fringe of Hesse to the

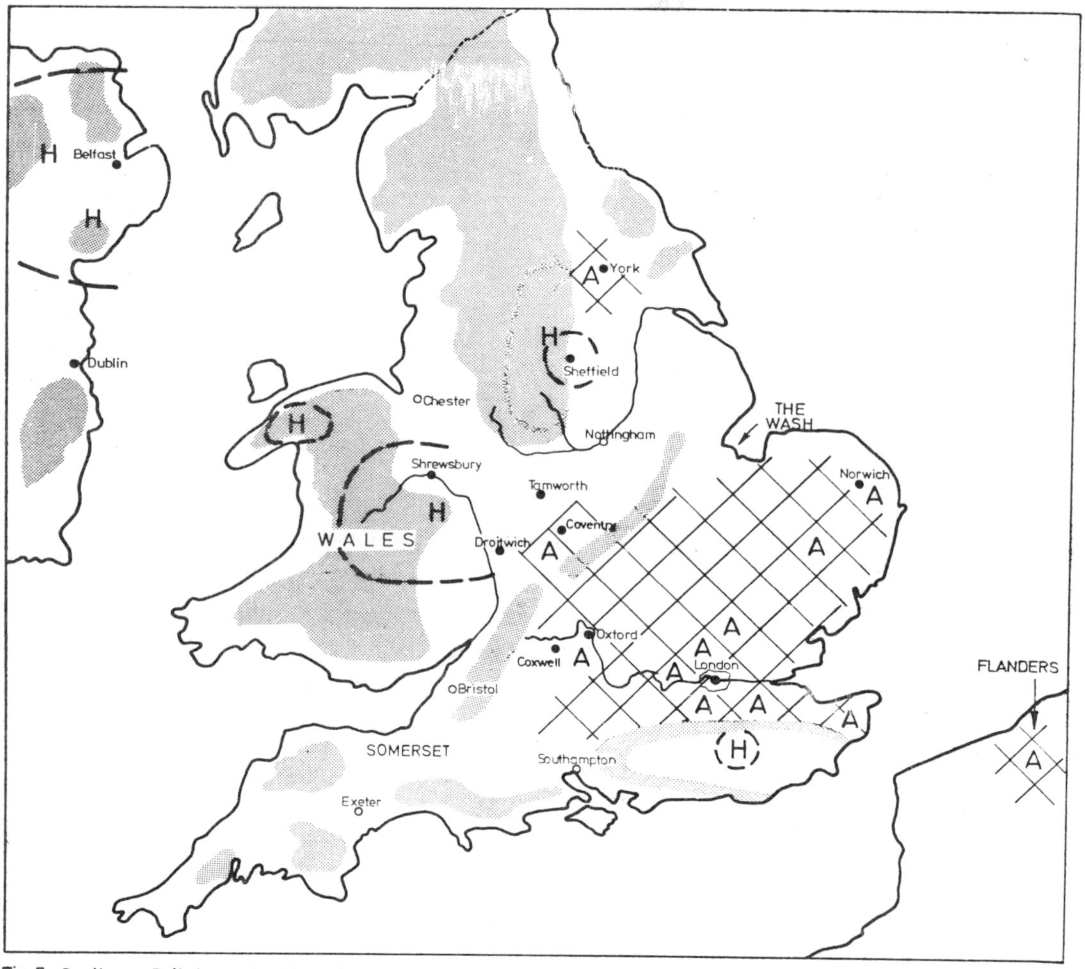

Fig.5. Southern Britain and adjacent areas showing land over 200 metres as shaded.
Type A chronologies apply, 1250 – 1550
Type H " " " "

Plate 1. York Minster from the south-east (National Monuments Record)

west Midlands and central Wales (Fig. 1). Slightly to the south of that line, however, Type H occurs in the hilly areas of the Eifel and Ardennes; also we find Type H not far south of London on the southern (Weald) side of the North Downs which form a ridge at altitudes of over 200 metres (Fig. 5).

The location can account for sporadic examples (2 out of about 125) in the ring-width sequences for panel paintings likely to have been used in the London area. The following shows their agreement with the Siebenlist curve for Wales:-

	No. of Rings	Mean Width	Versus Siebenlist W	Versus Siebenlist Overlap	Years Spanned
Henry IV, 2 boards (Private Collection)	218	1.4 mm	70%	180years	1353-1517
Lady Margaret Butts, 1 board (Isabella Gardner, Boston)	132	2.2 mm	68%	131years	1381-1512

The W and t values between these two sequences are 66.5% and 4.5 respectively. The W values for the Lady Butts sequence with continental Type H chronologies are comparable to those given by the Siebenlist curve for Wales:-

	With Southern Germany	Normandy
Siebenlist, Wales	67.5%	67%
Lady Butts curve	64%	67%

The impeccable origin and period—used by Hans Holbein the Younger within a year of 1542—implies that the Lady Butts panel came from a tree grown in south-east England not far from London.

Proposed Model

Comparison of the single curves for Henry IV and Lady Butts with REF 1 is of interest. Each has a greater mean width than the mean of those in REF 1. In the period 1414-23, known from historical records to be a series of dry years (Titow, 1970), the rings of the trees of REF 1 became progressively narrower (Fig. 2). The trees of Type H, however, have reduced widths from 1415 to 1417 only; 1418, 1420 and 1423 appear as peaks on the curve.

These differences suggest a model as follows:-

(i) Type H occurs either in hilly and windy areas where precipitation is high, or at lower altitudes, e.g. parts of northern Ireland, where evaporation is low; high stress by lack of moisture is not usually a dominant factor in determining its ring-width pattern; in warm, dry summers, when the growth of Type A is restricted, that of Type H may be enhanced.

(ii) Type A occurs in lowland areas or in rain shadows where precipitation is relatively low or soil moisture becomes depleted to levels which play an important role, through the stress so created, on the growth pattern.

Table 2. Agreement Values between Type A English Chronologies and Other Reference ones; Period 1250-1600

Chronology	Author	Years or centuries	No. of components	Mean width (mm)	REF 1, 1250-1550 W	REF 1, 1250-1550 t	REF 2/3, 1356-1536 W	REF 2/3, 1356-1536 t
I. Type A								
Flanders	Fletcher	1204-1504	14	1.35	83%	17.8	69%	9.7
Netherlands II	Bauch	1354-1624	25	1.3	69	7.9	62	6.2

II. Type H					Position of First Ring: W 1250 (Late)	W 1249 (Middle)	W 1248 (Early)	W 1356 (Late)	W 1355 (Middle)	W 1354 (Early)
Netherlands I	Bauch	1400-1600	13	1.6	58	Abs	61	Abs	Abs	Abs
West Germany	Hollstein }	9th to	Many	1.15	"	"	Abs	"	"	"
Central Germany	Huber Giertz }	20th	"	1.3	60.5	"	58	"	57	"
Normandy	Siebenlist	1287-1456	11	Variable	Abs	"	Abs	"	64	"
Wales/Midlands	"	1348-1636	13	1.25	"	"	"	"	57.5	"

Chronology	Author	Versus REF 2/3 1344-1567 in middle position*		
Netherlands I	As above		t = 2.7, overlap	220 years
West Germany	" "		t = 1.7, "	224 "
Normandy	" "		t = 1.3, "	113 "
Wales/Midlands	" "		t = 3.2, "	220 "

Note: Abs means any agreement is of less than 95% reliability

* t = 0.0 in the early and late positions

(iii) Climatic swings between atlantic and continental air streams over southern Britain modify the growth pattern in certain decades and in certain centuries to bring areas predominantly Type A into greater or less agreement with Type H.

This model does not exclude biological factors, such as insect attack, and other climatic ones, such as late frosts, from playing additional roles in influencing the growth patterns.

Dating of REFs 1 to 5: 1200-1600

This conerns Bauch's Chronology II for the Netherlands (page 133) through the correlations (W, 69%; t, 7.9) between it and REF 1, and those (W, 62%; t, 6.2) between it and REF 2/3. Since 1972, when the research teams at Hamburg and Oxford first became aware of the similarity of chronologies for the Netherlands and for SE England, both laboratories have encountered difficulties in obtaining evidence to identify precisely to the year their respective chronologies. From independant evidence, such as the date of panel paintings, the area of doubt extends only over two years: the Oxford laboratory has favoured and used (Fletcher, 1977) dates two years earlier than those favoured by the Hamburg laboratory.

Although Type H chronologies, well dated by the Hollstein and Huber-Giertz masters, exist in this period both in the Netherlands and in Southern Britain, no correlation between them and the Type A mean curves has been consistently convincing. Comparisons, in the three adjacent positions, between the recently developed and well-s bstantiated REF 1 English and REF 2/3 chronologies with the Type H ones are, however, shedding new light on the problem in two ways:-

(a) The conventional W values given by REF 2/3 (Part II of Table 2) show a consistent but low measure of agreement for the 'middle' of the three positions. This may be significant as, on the model proposed above, the agreements with Type H should be greater for the wider REF 2/3 than for the REF1 (stressed by moisture): this view is the reverse of what we accepted as likely prior to the formulation of the model.

(b) The number of signatures common to pairs of chronologies can now be derived for the three positions; with the Siebenlist chronology for Wales and West Midlands, that is the Type H one nearest to the London region, the numbers common to those in REF 1 and REF 2/3 are:-

		No. of Signatures	Position of First Ring	No. in Agreement	No. in Opposition	No. Neutral
REF 1	1305-1517		1305	5	8	4
	212 changes	17	1304	6	8	3
			1303	5	7	5
REF 2/3	1388-1546		1388	3	8	2
	158 changes	13	1387	8	2	3
			1386	4	4	5

The significance of high agreement by signatures between REF 2/3 and the Siebenlist chronology in the 'middle', rather than in adjacent positions is supported by t values (Table 2). The 'middle' position is also favoured over the other two for the following reasons:

(i) The most consistent ($P < 10^{-6}$) of all the directional changes in REF 2/3, (21 constituents all showing an increase in width) corresponds in that position to a peak, for the year 1481, on the Siebenlist curve; and

(ii) The ring most frequently found to have abnormal early vessels (page 339) then applies to the year 1432, for which the spring and previous winter were of abnormal severity.

The dating given in our 1974 and 1977 publications is retained here but it seems likely that the middle position may prove to be correct; if so, the dates given for the rings of REFs 1 to 5 would all be one year earlier.

This progress with a difficult dating problem emphasises the value in a temperate region of well-substantiated reference chronologies: it also draws attention to a weakness of mean curves of ring widths, namely that many of the year-to-year changes are of very little significance and it is beneficial to disregard them.

To provide continuous coverage by at least 10 trees between the end of REF 1 English and the beginning of our mean curve for recent trees, about 20 single curves, 200 years long, would be required. This assumes that on average 90% of their rings would be sufficiently well-matched to be usable. Samples from at least 50 trees would be needed to allow for some sequences being of inappropriate width and others not having a suitable spread of years. Although many 17th to 19th century panels exist in churches and country houses, their edges are not accessible and the provenance of the wood is not necessarily English. We believe it is unnecessary to bridge a gap in time by local samples few in number if matching can be achieved with well-replicated chronologies from other areas.

Applications of Type A Reference Chronologies

Flemish Panels: We find that the ring-width patterns derived from oak panels used for paintings of the 14th to 16th centuries in Flanders and south-east England are very similar. Differences in the quality of the panels and the unlikelihood of trade in panels on a large scale between the two countries, indicate that they were made locally. The similarity in ring-width patterns is therefore attributable to slow-grown oaks in the two areas producing virtually the same pattern. Our reference curve (MC 14 Flemish) made from 14 single curves, based on the boards of Flemish panels, spans the years 1236-1498, and its mean width is 1.3 mm. Agreement values are included in Table 2.

English Panels: The high percentage (85%) of the 135 English panels that have been matched and dated by Type A reference curves can be attributed to:-

(i) Few boards had less than 120 rings;
(ii) The benefit of finding abnormal earlywood vessels (Page 339) as a chronological indicator in the 15th century;

(iii) The high quality of the Type A1 and A2 ring-width reference curves that have progressively been built up; and

(iv) The availability of these two sub-types rather than the less distinctive chronology that would result from meaning them together.

The earliest panels from paintings that have been matched and dated in the course of this research were used in the late 14th century in Norwich; the latest in the 1640s. The 15% not yet dated consist mainly of sequences with 90-120 rings of average width well over 2 mm.

Chests and Cupboards: The planks and boards of those at Westminster Abbey are mainly too early for matching by REF 1 or REF 2/3, but two chests and a large cupboard at Corpus Christi College, Oxford have been dated to about 1520, i.e. shortly after the college was founded; W values with REF 1 are about 70%

Building Timbers: More than 100 samples from 15th and 16th century building were collected in southern England prior to 1970; most are from fast-grown timber. Reliable matching of such samples demands high quality reference curves and so we are only now beginning a systematic attempt to date those samples using the recently constructed REF 1 and REF 2/3 for Type A, and the Siebenlist Wales curve and the Lady Butts curve for Type H.

An example which shows the occurrence of Type A in the Midlands is the dating of the sequence of 145 rings, mean ring-width 1.5 mm, for six timbers from the medieval house at 169 Spon Street, Coventry. With REF 2/3, the agreement values are W = 64% and t = 3.5 for the position 1244-1388, with sapwood beginning at 1371.

On the other hand, an example that reveals the locality in which Type H can be expected is provided by the match (W = 75%) between the Siebenlist curve for Wales and that for a post of sessile oak taken from a Wealden-type house, reconstructed at the Weald and Downland Museum, Sussex. The farmhouse of which it formed part stood to the south of Winkhurst Green, which is a wooded area on lower greensand, an acid soil, about 3 km south of the North Downs in west Kent.

Applications to Archaeological Material of Norman, Saxon and Roman Times

These have been limited to excavations in which our assistance has been sought on an *ad hoc* basis. Samples, on the one hand, from Saxon timber-lined wells as far apart as Portchester and North Elmham (Fig. 1) and, on the other hand, from the Graveney Boat (Fletcher, Tapper and Walker, 1978) have been correlated to form site curves. The tree-ring work has been associated with radiocarbon analysis.

DISCUSSION AND CONCLUSIONS

This research in maritime and temperate lowlands adjacent to the North Sea has mainly concerned a region where oaks are subject to stresses that are different from those that apply at higher altitudes in central and western Europe, or at the higher latitudes of Northern Ireland. As the lowland oaks are not near the limit of their geographical habitat those of slow growth can be attributed to factors such as overcrowding.

Distribution of Type A and Type H in the British Isles

Fig. 5 shows the areas where Type A and Type H have been identified in the British Isles in the period 1250-1600, a period during which their patterns were remarkably different. Lack of agreement between them occurs in north Germany, in the Netherlands, and in areas of England as far apart as Kent, the Midlands and Yorkshire. It is of interest that no examples of intermediate type have been found.

These points suggest a possible relationship between the two types and the two species—sessile (Quercus petraea) and pedunculate (Q. robur). Though the latter is widespread in the British Isles (Gardiner, 1974), sessile oak is mainly limited to western areas. This may be due to its preference for acid soils as well as to its suitability for hilly terrain. On the limited evidence at present derived from our characterisation of the species (see page 329) in samples used for ring-width measurements, it appears that Type H is usually sessile and Type A pedunculate oak in the period under discussion. The growth pattern would be affected by several physical and biological factors. The species distribution in the British Isles and Europe indicates that the soils and sites favoured by sessile and pedunculate oaks are dissimilar; while, due to different times of flushing, the damage from defoliating insects varies.

In considering the likelihood of oaks having the pattern of Type A or Type H in the British Isles it must be remembered that annual rainfall tends to decrease from west to east, while the summer temperature decreases from south to north. On the basis of our model, with hilly terrain and annual rainfall of, say, over 40 ins (1000 mm) as the requirement for Type H, there should be many woodlands from Dartmoor and Exmoor in S.W. England up to Western Scotland where this type existed. If, however, there is additionally a need for the species to be Q. petraea, occurrence may be limited to sandstone strata and other areas with neutral or acid soil.

Type A is to be expected throughout the lowland belt on the eastern side of Britain with annual rainfall under 30 ins (760 mm). Also, intermingled with Type H, in areas of altitude from 100 to 200 metres with 30-40 ins rainfall, such as coastal parts of southern and western England, the Midlands and S.W. Scotland. In particular, Type A can be expected in woodlands situated where rain shadows are produced, such as East Lothian—an area near the coast to the east of Edinburgh—and even as far north as the oaks in the dry belt adjacent to the Moray Firth (Atlas of Britain, Clarendon Press, 1963: pages 34 and 43).

Types A and H may occur over distances of 1000 km in north-west Europe. Type H is already recognised in regions as far apart as Northern Ireland and Northern France to the north-west, Switzerland and Central Europe to the south-east. Type A may be as widespread, embracing lowland coastal areas round the whole southern basin of the North Sea from Denmark to Scotland and penetrating inland into northern France, the Midlands and certain western areas in the British Isles.

As a general principle, the higher the agreement between single curves and a chronology, the wider the geographical range of its applicability. For oaks of slow growth, we expect our Type A1 to apply at least to the eastern half of Britain: for Type H, two may cover the whole British Isles—the Belfast chronology together with a composite one made up from (i) the Siebenlist Wales/

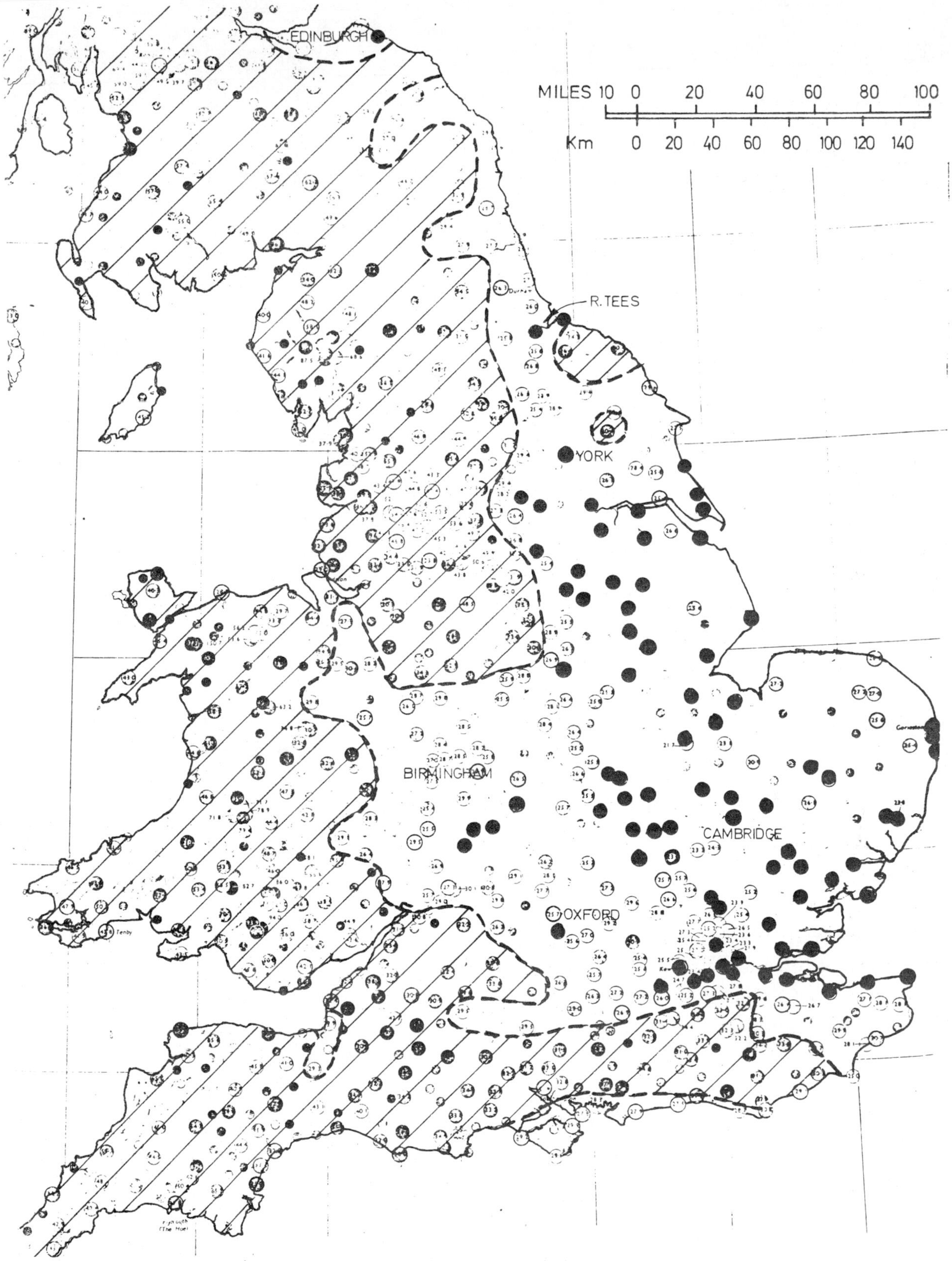

Fig. 6 Areas of low average annual rainfall in the eastern part of southern Britain.

— — — 30 ins per annum limit.

● Stations that recorded less than 25 ins.
(Based on Atlas of Britain, Oxford University Press, 1963.)

Midland one, (ii) that of Morgan for the Sheffield area and (iii) our own for parts of Kent/Surrey/Sussex. Oaks of faster growth match one another less well, so more chronologies may be needed, e.g. alternatives to REF 2/3 for north of The Wash or of the Trent, and for S.W. and N.W. England.

In York Minster (Plate 1) all the major roofs were constructed of timber rather than stone vaults. This reflected the proximity of the extensive Forest of Galtres which came to the very gates of York on its north side (Victoria County History, Yorks, Vol. 1, 1907, 501). Doors of the set of 16 medieval cupboards that line the walls of the Zouche Chapel in the Minster have produced results that are informative.

The original cupboard doors, made about 1400, are slow-grown and of Type A1 pattern as expected for woodlands in the Forest of Galtres which lay in the Vale of York. The agreement (W = 72%) with the pattern for chests and a cupboard in Westminster Abbey is high in view of the distance (300 km) involved. Some of the cupboard doors were, however, replaced. They are dated (by Ruth Morgan) to the last decades of the 17th century by the agreement with Hollstein's chronology—by trees at an average distance of about 600 km from York. Their Type H pattern is explained by the disappearance of the Forest of Galtres and the need to use timber from hillside woodlands on the Wolds.

Thus the timber in the Zouche Chapel shows the same change in type and thereby in origin as the Dutch panels examined by Bauch which changed from Chronology II (equivalent to lowland Type A) to Chronology I (equivalent to hilly Type H). Both changes can be attributed to exhaustion of a local source of timber. Fortunately, the terrain near York and the history of the forests around the city enable the York examples to be interpreted easily.

I am grateful to the Leverhulme Trust and to the Ancient Monument Branch of the D.o.E. for grants which have supported this research; and to Margaret Tapper for much help in preparing this paper for publication.

This paper follows the lines of that presented at the Symposium but has been modified to take account both of points raised in discussion and of the publication in December 1977 of a paper giving details of the chronologies constructed.

References

Baillie, M. G. L., 1977. In B. Frenzel, ed. Dendrochronologie und Postglaziale Klimaschwankungen in Europa. Steiner, Wiesbaden, 9-15.

Fletcher, J. M., 1965/6. Berks. Arch. J., 62, 45-58; 1968 Oxoniensia, 33, 71-88.

Fletcher, J. M., 1977. J. Archaeological Science, 4, 335-352.

Fletcher, J. M., Tapper, M. C. and Walker, F. S., 1974. Archaeometry, 16, 31-40.

Fletcher, J. M., Tapper, M. C. and Walker, F. S., 1978. In V. Fenwick, ed., The Graveney Boat. B.A.R. British Series 53, 111-124.

Gardiner, A. S., 1974. In M. G. Morris and F. H. Perring, eds. The British Oak. Faringdon, Classey, 13-26.

Morgan, R., 1977. Vernacular Architecture, 8, 809-814.

3.7 THE CHRONOLOGY, 1341-1636, FOR CERTAIN HILLSIDE OAKS FROM WESTERN ENGLAND AND WALES

V. Siebenlist-Kerner (formerly, Munich)

Earlier research (Berger, Giertz, Horn, 1971; Charles, 1971) showed that tree-ring curves of English oaks fit the master chronology for southern Germany (see this volume, page 27) only if the trees grew on hillsides and in large forests. The expedition to Southern Britain in 1972 therefore concentrated on certain hilly parts of England and of Wales in a search for medieval, or later, buildings constructed of close-ringed oak timbers from which samples with long tree-ring sequences could be obtained to establish an English/Welsh reference curve.

Method: We took samples, as cores, from every suitable building. After returning to Munich, I prepared each sample, measured the widths of the annual rings, made their ring width curves. I first tried by eye and by computer to synchronise the individual curves for each building. Then the mean curve for each building was compared with English reference curves and our southern German master curve.

Results (Table 1)

For Yorkshire, 4 sites, no positive results;
For Shropshire, 2 sites, one dated curve;
For Mid-Wales, 4 sites, 100% success;
For the Midlands, 3 sites, one dated curve.

Fig. 1 compares the ring-width sequences from three of the dated Welsh sites. The relative positions of the six dated mean curves are shown in bar-chart form in Fig. 2. It includes the dated sequences for three samples from the Bailiffs House, Bewdley, and four from the Priory House, Droitwich, as well as the sequences for five recent oaks from Brockhampton (Herefordshire) and four from Cumberland (Giertz, Horn and Charles, 1971). Table 2 gives the ring-widths of the mean curve derived from the 17 samples of ancient timber and those of the mean curve for recent oaks.

Fig. 3 compares our English/Welsh mean curve with our southern German one and with one for four market halls and barns in northern France. The signature 1417/20/21 in the German curve is one of our best in the middle ages: it appears in about 150 single curves. One can see that the English/Welsh curve has a different signature: 1416/19/22 and that the French curve has similarities to both: 1417/19/22.

References

Berger, R., Giertz, V. and Horn, W. 1971. Vernacular Architecture 2 (1971) 3-6.

Charles, F. W. B., 1971. Forestry and Home Grown Timber, 30 Oct., p. 14.

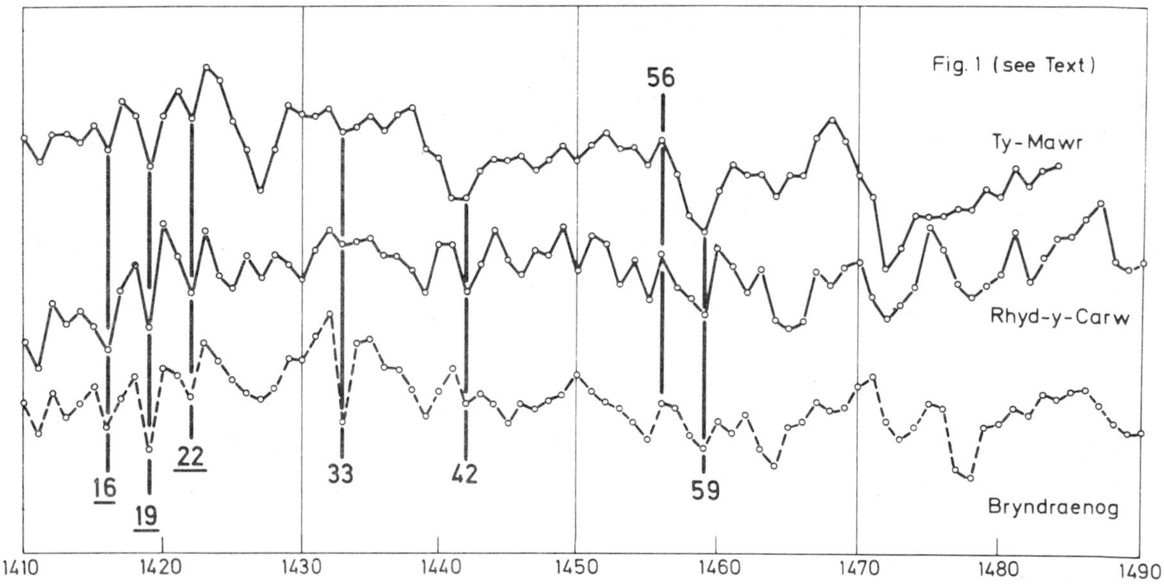
Fig. 1 (see Text)
56
Ty-Mawr
Rhyd-y-Carw
Bryndraenog
16
19
22
33
42
59
1410
1420
1430
1440
1450
1460
1470
1480
1490

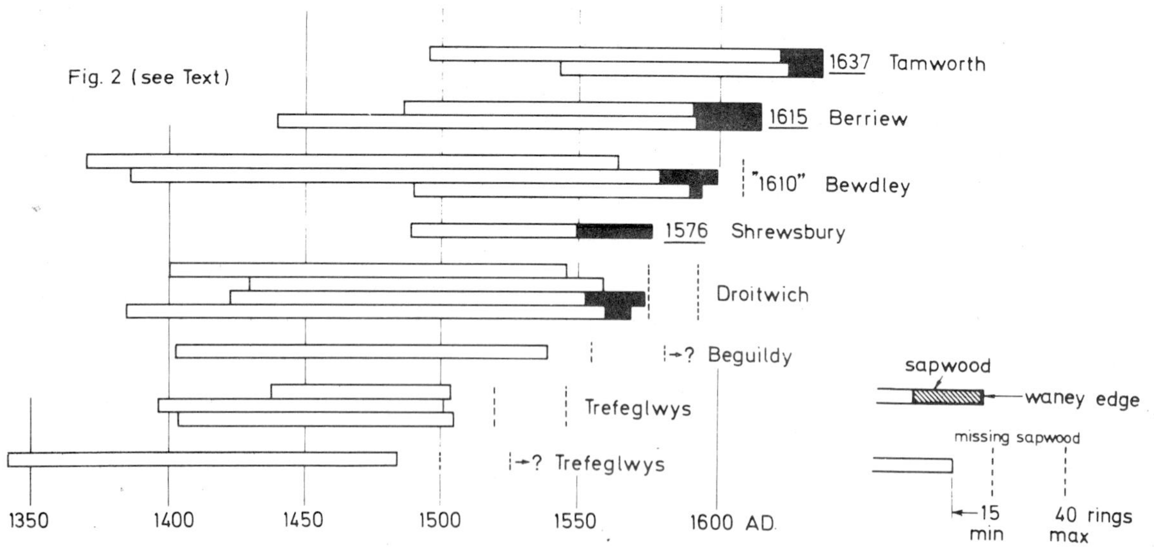
Fig. 2 (see Text)
1637 Tamworth
1615 Berriew
"1610" Bewdley
1576 Shrewsbury
Droitwich
→? Beguildy
Trefeglwys
→? Trefeglwys
sapwood
waney edge
missing sapwood
15 min
40 rings max
1350
1400
1450
1500
1550
1600 AD

Table 1. Results for Samples Taken in 1972 from English and Welsh Sites

	Site	Building	Number of Samples	Site Mean Curve	Agreement with other Mean Curves
	Yorkshire				
1.	Worsbrough	Houndhill	8	No	None
2.	Summerbridge	Banger House	12	Yes, for 4 samples	None
3.	Beningbrough	The Hall	4	Yes, good	None
4.	York	Minster, Central Tower	4	Yes, good	None
	Shropshire				
5.	Alberbury	Church of St.Michael & All Angels	8 from pew-ends	Yes, for 2 trees, 325 years long	None
			14 from Tower	Yes, 3 for 8 samples	None
6.	Shrewsbury	Bear Steps Restaurant	12	Yes, two	*One dated, 1489-1576
	Mid-Wales				
7.	Berriew	The Old Vicarage	4	Yes, for 2 samples	*Dated 1439-1615
8.	Trefelglwys	Ty-mawr	1	Yes	*Dated 1341-1484
9.	Trefelglwys	Rhyd-y-Carw	4	Yes, for 3 samples	*Dated 1396-1505
10	Beguildy	Bryndraenog	4	Yes, for 1 sample	*Dated 1402-1539
	Midlands				
11.	Nottingham	Severns	3	No	None
12.	Tamworth	Manor House	2	Yes	*Dated 1496-1637
13.	Coventry	8-10 Much Park Street	12	Yes	None

Note: *Outer ring of sapwood intact.

Table 2

Ring-widths for oak (1/100 mm). Above, for Middle Ages, 1341-1636 A.D. Below for recent trees; 4 trees from Cumberland, 5 from Herefordshire. n is the number of trees.

	1	2	3	4	5	6	7	8	9	0	n
1341	387	419	155	65	65	97	129	213	235	174	1
	245	245	371	306	435	268	323	258	226	119	1
	135	213	258	284	248	242	194	119	171	194	1
	200	190	194	184	181	197	152	139	187	165	2
	142	165	152	145	129	165	155	142	165	142	3
1391	181	145	135	103	77	94	97	139	135	155	4
	152	116	155	132	139	165	139	145	181	142	7
	119	152	135	135	135	110	126	139	87	155	8
	145	126	127	148	139	142	142	155	161	158	9
	174	184	148	165	165	145	139	126	103	135	10
1441	129	97	119	145	126	116	123	123	139	119	12
	155	142	123	152	123	158	135	129	110	145	12
	129	116	123	87	94	106	129	132	142	152	12
	129	126	123	126	184	152	135	129	145	132	12
	165	129	142	152	152	168	184	129	123	142	12
1491	123	97	113	119	113	152	119	106	142	129	13
	113	110	94	113	139	135	123	116	126	113	11
	116	116	123	106	100	100	90	97	106	94	11
	106	113	90	103	81	87	94	106	100	84	11
	116	90	97	90	123	106	100	100	113	116	11
1541	123	94	113	113	116	106	87	110	123	113	10
	123	106	106	106	129	97	81	90	90	100	10
	94	103	87	106	94	77	74	87	116	116	8
	119	97	106	110	106	84	97	74	100	123	6
	97	81	81	94	126	126	110	113	129	81	6
1591	110	100	142	168	184	155	142	123	123	113	5
	110	97	129	132	87	116	135	100	97	84	4
	87	77	77	52	74	142	139	126	84	116	3
	284	271	268	206	129	158	187	248	268	171	2
1631	187	294	219	148	158	184	-	-	-	-	2
				*	*	*	*				
									345	387	1
1731	161	281	145	190	155	203	184	281	177	123	1
	161	187	255	300	187	242	394	294	332	245	1
	294	290	245	248	210	229	216	229	194	165	2
	216	210	345	265	168	332	300	281	239	229	3

	1	2	3	4	5	6	7	8	9	0	n
1771	161	174	174	255	248	300	300	297	229	274	8
	255	297	345	335	187	219	287	319	429	355	9
	377	381	287	319	326	258	332	300	329	319	9
1801	303	258	194	223	277	252	287	294	277	261	9
	306	281	261	255	213	216	181	216	226	177	9
	155	194	261	213	197	142	181	210	252	274	9
	213	171	197	245	223	232	219	187	171	174	9
1841	184	219	203	168	184	181	226	252	223	190	9
	155	155	181	184	213	306	200	165	139	161	9
	148	129	145	161	177	165	197	145	152	174	9
	194	190	161	132	187	145	139	165	155	142	9
1881	110	119	110	139	148	187	129	123	119	158	9
	132	161	142	135	168	119	148	139	155	216	9
	187	168	197	174	152	155	158	171	155	158	9
	168	203	197	165	161	155	187	171	174	148	9
1921	123	139	126	148	155	165	145	161	152	145	9
	152	126	87	84	94	155	116	106	135	135	9
	135	139	148	113	187	171	200	145	113	152	9
	155	139	123	129	139	113	90	135	123	145	9
1961	135	158	139	145	123	142	123	113	123	-	9

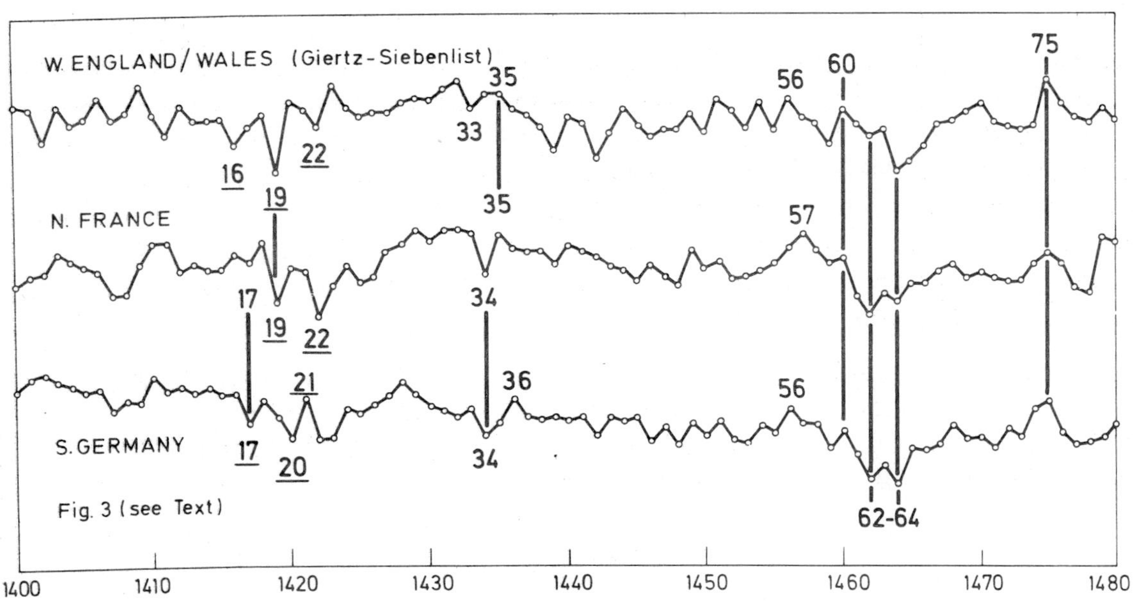

Fig. 3 (see Text)

3.8. DENDROCHRONOLOGY AND THE WINCHESTER EXCAVATION

A. C. Barefoot, W. L. Hafley, North Carolina State University, and J. F. Hughes, Oxford University.

With the uncovering of building timbers in the Brooks Street site of the Winchester excavations wood technologists began, in 1964, the dating of these oak specimens using dendrochronological techniques. The work at that time which showed the most promise of dating tree rings in temperate and cold zones was the work of Kolchin (1962) at Novgorod.

During the early searches for tree ring samples, A. W. G. Lowther made all of his results available to the authors. Also during this time of seeking out architecturally datable sequences, it became evident that a study of modern samples would be necessary for three reasons:

1. To learn the characteristics of growth of English oak in the Winchester Region.
2. To provide a known set for developing statistical and mathematical processes for building a standard chronology for Winchester.
3. To provide a precise year for anchoring a tree ring chronology for Winchester.

Thus, in 1967, 1970, and 1973 ring sequences from trees of known felling date were collected. The width of the rings on these and the architectural samples was measured to the nearest 0.001 mm. The plots of these soon indicated the necessity for standardizing and smoothing in order to make meaningful comparisons between ring sequences from different trees. A procedure for effecting such smoothing was introduced by Barefoot, et al. (1974) and Barefoot (1975). This procedure has been incorporated into a multistage process for the chronology development described in this paper.

Results have been obtained from a variety of ancient timber collected from many architectural sources. However, the extreme difficulty of cross matching many of these specimens using only architecturally established dates places great suspicion on the success of dating. Reuse of timbers, imports of timber, and imprecision in dating present uncertainties which suggest the elimination of certain specimens from use in the Winchester dendrochronology. We present here the development of the "best" chronology available at this time based on all the objective and subjective information we can bring to bear.

This paper is presented in two parts. Part I reviews the procedures used in developing the modern portion of the chronology based on forty one trees. Part II presents the decisions involved in developing Chronology A for Winchester and the uncertainties associated with it.

PART I

The Modified Tree Ring Series

Plottings of the original modern trees demonstrated considerable variation in width of growth ring from tree to tree among trees known to be growing at the same time and in the same location. These differences are attributable to a number of factors; the obvious ones being genetic characteristics of the individuals, modifications of stand structure either by nature or man resulting in changes in competition from neighboring trees, or the microsite variations in moisture and nutrient availability. We chose to group all of these factors into what we call "microclime" factors, features unique to the particular tree, as distinguished from "macroclime" factors, the general year to year weather and climatic changes.

The microclime factors are considered to provide long term, though not uniformly consistent, influences on growth. While the macroclime factors are considered to be short term factors which have the capacity to induce rapid, but relatively short lived, responses in the growth patterns of the trees. Given this bipartite nature of the growth pattern our interest centered on separating the two parts.

We chose to accomplish this separation by attempting to estimate the microclime long term effect and separating this from the total growth pattern. In a mathematical sense what we did was to smooth the growth pattern, accept this smoothed response as the estimate of the inherent growth character of the tree and subtract it from the measured growth pattern of the tree, i.e. the original ring width minus the one estimated by the smoothed response. The resulting series reflects the deviations from the inherent pattern induced by short lived weather factors and climate shifts.

There are a number of procedures extant for smoothing time series, probably the most well known being the moving average. However, we chose to use weighted exponential smoothing. Weighted exponential smoothing has been used primarily for forecasting time series (Brown, 1959, and Duffin and Schmidt, 1960). The feature of the technique that was most attractive to us is that the prediction of response at a given point in time is dependent solely on the information (growth influences) preceding that point, giving maximum weight to the most recent information and decreasing weight (decreasing to zero) to information farther back in time. A weighting factor, α, determines the amount of data which goes into the weighted average and, therefore, the amount of smoothing attained.

We must note, however, that our overall procedure is not dependent on weighted exponential smoothing. Any smoothing procedure may be substituted given one has confidence it is estimating the response of interest.

Since we are convinced that it is not feasible to leave the development of a chronology, or the fitting of patterns to that chronology, purely to mechanical techniques, a subjective, visual consideration of the matches must remain in the decision process. While the macroclime series consists of positive and negative values whose expected value is zero, the values still reflect scale differences that are related to age and tree size, and effect visual matching. Elimination of this scale variation is accomplished by normalizing the series through the division of each value by an estimate of the standard deviation. Resorting to the techniques of quality control we partition the series into subgroups of size four and estimate the standard deviation by subgroups.

Being concerned as to whether subgroups of size four might not be valid for our data we evaluated the lag correlations of the series. In terms of the raw data the lag correlations remained high, greater than .5, through 10 to 15 lag periods for most of the series. However, in terms of the 'macroclime' series the first lag correlations were rarely as large as .5 and by the fourth lag correlation were insignificant. This is an indication that our subsets are providing independent estimates of the standard deviation.

It was such normalized series that were then used in the development of our Winchester Dendrochronology.

Evaluation of the Matching Procedure

Quantitative estimates of comparison have been suggested by Eckstein, by Bauch and Eckstein (1970) and by Baillie and Pilcher. We have considered the use of Eckstein's 'W' and Baillie's 't'. Table I gives the results of using visual comparisons and Eckstein's 'W' to test match ten of the thirty nine modern trees which were especially shortened for test purposes. As a result of further comparisons we have chosen to emphasize the use of Baillie's 't' because it more closely conforms to the statistical literature of time series analysis. However, the statistic as computed by the computer programme of Baillie and Pilcher has a slight error and is not a true 't'; in our work we have used a modification of the Baillie and Pilcher programme which calculates a true 't' for each match.

The data of Table I demonstrate the result of smoothing and visual match using three different values of the weighting factor, α, and the match obtained from the best value of Eckstein's 'W'. The results tend to confirm our selection of a weighting of $\alpha = 0.2$. If we view the smoothing process as a filter, we are attempting to filter out all cycles in the series that are of duration of 7 or more years. The choice of 7 years is not altogether arbitrary. In discussions with tree physiologists and silviculturists it was estimated that the responses to the factors we classify as 'microclime' factors would be ex- years in duration, while responses to the 'macroclime' factors would be expected to be less than six years in duration. The choice of an α of 0.2 appears to provide such a filter.

Whether one is using visual or quantitative matching procedures there are often characteristics of the data that will produce best matches that are a result of pecularities of the data. A very good fit over a short span of the series may overly influence the overall statistical evaluation. We believe that relying solely on statistical matches may cause drastic errors. In fact, in the development of this chronology, we feel that combining visual and statistical information for the Winchester region is insufficient. Thus, we prefer to enter a third factor in the decision process, historical or C^{14} estimates. We have used the statistical evaluations to suggest alternative possible matchings, evaluated these visually and added the best archaeological, historical or C^{14} estimates of an appropriate time interval to arrive at a final decision.

TABLE I

Results of matching 10 modern tree ring series from five sites near Winchester, England. Correct dating is marked*

Number	Number of Rings	Known Year of Last Ring	Date Based on Visual Matching alone			Date Based on best Eckstein value
			$\alpha = .1$	.2	.3	
1	130	1961	1953	1961*	1956	1955
2	130	1970	1969	1951	1969	1969
3	80	1955	1964	1926	1969	1943
4	130	1965	1946	1965*	1967	1914
5	130	1964	1945	1964*	1966	1914
6	337	1964	1964*	1964*	1964*	1964*
7	130	1940	1940*	1940*	1940*	1940*
8	180	1939	1939*	1939*	1939*	1939*
9	80	1969	1961	1969*	1969*	1969*
10	74	1969	1969*	1969*	1969*	1969*

PART II

Chronology A and It's Limitations

In presenting Chronology A for the Winchester region, we have chosen ring series which have yielded pathways consistent with historical and statistical data. We have not, however, tested the curves against all sources of data in order to crosscheck proposed alternative acceptable dates.

Without doubt the 1635-1972 A.D. sequence is correct to the year. However, the scarcity of oak timbers cut and used in the mid-17th to early 18th century makes the bridging backward in time difficult. Most often the overlap with the modern curve has been too short; or questions about the historical perspective and re-use of timbers have clouded the acceptance of suggested tree ring dates.

Fortunately, planks from a staircase at 126 High Street, Winchester, said to be early to mid-18th century, were located and seem to yield a reasonable result. This chronology is based upon 17 individual curves from panels, stringers and treads and is 296 rings long. The house from which these timbers came was demolished in 1950 and the staircase was stored at that time in a museum. From the sole surviving photograph, Dr. Derek Keene (personal communication) believes that the house was probably mid-18th century rather than earlier. He reports knowledge of a similar staircase dated to 1726 A.D.

The matching of these 17 individual curves among themselves was very satisfactory. No sapwood was present on any of the timbers, therefore, there is no way to estimate how many rings were removed in machining or how many years of seasoning were employed. Also, the source of the timber is unknown. However, since sequences from panels, treads and stringers all match quite well one could argue that the preponderance of evidence supports the hypothesis that these members all came from the same source; most likely locally.

Table II. Possible Matches: A. For 127 High Street with Modern Curve
B. For Langton Chapel with 126 High Street*

	Year for End Ring	Overlap	t	W %	W VH	Visual	Felling Date‡
A.	1661	27	1.90			Can't say no	1701
	1673	39	2.24			Acceptable	1713
	1686	52	3.19			Good	1726
	1699	65		60.8	95	Not good	1738
	1713	79	2.36			Barely acceptable	1752
	1721	87		61.5	95	Not good	1760
	1734	100		59.5	95	Not good	1774
	1780	146	1.96	60.6	99	OK	1820
	1827	193	2.25	56.2	95		1867
	1828	194	2.61				1868
	1830	196	1.36	58.4	99		1870
B.	1421	31	2.47				1441
	1422	32	2.36				1441
	1433	43		65.1	95		
	1438	48	2.21				1458
	1460	70	1.58	63.6	95	Possible	1480
	1483	93		64.0	99	Not good	1503
	1493	103	1.89			Acceptable	1513
	1496	106		58.0	95	Not good	1516
	1503	113	2.00			Fair	1523
	1531	141	2.36				1551
	1577	187		64.7	99		1597
	1586	196	3.09				1606
	1636	246	2.41	55.7	95		1656

* Assuming 126 High Street begins 1391 and ends 1686 (see 3rd line of A)

‡ For A, felling date estimated by adding 30-40 years for rings removed by machining; for B, by adding 20 years as 5 rings of sapwood present on the specimen. For corresponding use date, add 5 years for seasoning and storage.

Unfortunately, the unequivocal selection of an over-lapping match of the end rings of 126 High Street to the modern curve is not possible. The feasible matches are given in Table II. Of these dates for the end ring only three, 1780, 1827, and 1830 have a coincidence in both 't' value and Eckstein's 'W'. Those dates, when adjusted for missing rings and seasoning, puts the years of use for all three at or past 1825, a date obviously too late according to the architectural observation. Therefore, if 1828 is too late, and 1661 is too early, there are six plausible dates remaining: 1673, 1686, 1699, 1713, 1721, and 1734. However, in Table IIA, the lack of any joint combination of 't' and 'W' for any of these remaining dates raises two conjectures:

(1) A correct match has not been located because in fact the architectural reasoning leads us astray; or,

(2) Mature wood of 126 High Street is being compared and dated with juvenile rings of the modern curve.

We reject (1) out of hand. In regard to (2) it is possible that growth of wood in these two zones does not always respond in exactly the same physiological manner to climatic stimuli. Certainly significant mast production is not likely in the juvenile tree. On the other hand, general weather trends favoring or inhibiting energy production in the tree should eventually be recorded in the overall growth trend, if not in a corresponding year by year pattern necessary for a significant 'W' test. The tests of our procedure for constructing the dendrochronology, summarized in Table I, were essentially of mature wood against mature wood. We have not tested rings of dated juvenile wood against dated mature wood. At this time we merely accept that the 't' correlation is the best statistic for comparing these two physiologically different zones of rings. On this basis and poor visual fits, we reject the 1699, 1721 and 1734 dates.

Therefore, of the six plausible dates three remain as acceptable for further analysis: 1673, 1686, and 1713. The date of the corresponding first ring of the 126 High Street chronology could be 1378, 1391, or 1418. These dates obviously suggest that the wood of the 126 High Street series began its growth in the latter part of the 14th century. Therefore, for extending our chronology, we needed a ring sequence beginning much earlier and ending preferably in the 15th and 16th century. The panels and wood carvings of the Langton Chapel, Winchester Cathedral, satisfy these criteria. Langton was Bishop of Winchester from 1493-1501 and was Archbishop-elect of Canterbury when he died. While there is no record as to the date of his chapel, the accepted assumption is that he ordered the materials and began its construction. In fact, it is possible that he did not quite acquire all the material for his chapel prior to his death; and, therefore, some material may have been supplied a few years immediately after his unexpected death.

There were twenty-one individual curves from Langton Chapel which were averaged to provide a chronology 306 rings in length. This chronology was then matched to the 126 High Street chronology. If we accept the 1391-1686 dates for the High Street data, since they have the highest 't' value and the best visual fit, Table IIB presents the feasible matches of the Langton Chapel chronology to that of 126 High Street. With the historical evidence in mind, there are five positions of fit which must be considered. These are 1460, 1483, 1493, 1496 and 1503. Of these 1503 provides the best visual fit, 1460 is the only date of coincidence for the 't' and 'W' evaluations, and 1483 and 1496 both give very poor visual fits in spite of their 'W' values.

The next overlapping series came from six panels of the Choir Stalls in Winchester Cathedral. The Choir Stalls were originally constructed in 1308 and the chronology from these panels is 229 rings long. Comparing it with that of the Langton Chapel gives a very satisfying position of fit (t = 3.65, W = 63.7%, visual good) at an overlap position of 92 rings. Based on the construction date of Choir Stall and the fit to the Langton Chapel curve the end of Langton curve should be 1493.

The final span of time in the Winchester chronology is covered by one ring series of A. W. G. Lowther (delivered personally by him to the authors). While known not to be specific to the Winchester region we are forced to use it for the time being since this is the only data we have for the ninth century. The data, consisting of 281 rings, come from the Lid of an Oak Chest, Westminister Abbey and were originally measured by Mr. John Harvey. The chest has been dated to the late 12th century; its last ring was assigned a date of c. 1138 by Shove and Lowther (1957).

Comparing the chest data with the Choir Stall curve results in the following two choices, one is at an overlap position of 71 years and one at 96 years;

Position of overlap	t	W	Reliability	Visual
71 years	2.69	63.4%	95.0%	Good
96 years	2.57	60.4%	95.0%	Good

One additional position occurs at 227 years of overlap with t = 3.37, 'W' — 56.2, 95.0. VH. and a good visual fit. Locating the match in this latter position would of course, place the fit, completely out of historical context. We note this possibility only to point up the pitfalls of blindly following the "statistical" data.

From the evidence cited for these chronologies there are, therefore, several possibilities for forming a working chronology. Choosing the more likely candidates — and ruling out in advance those clearly leading to spurious historical perspective — we chart in Table III the pathways of the several overlaps. The table begins with the three dates earlier noted for 126 High Street. Of these three choices, we prefer the 1686 date and the pathway leading to the 866 date in the final line. The normalized values for Chronology A for Winchester, based on the data for the matches for the years 866 to 1972, are available on request.

Conclusion

A method of combining the growth ring records of trees having disparate responses to major silvicultural effects of man or nature has been developed and is presented for constructing a dendrochronology for Winchester. Statistical aids for matching tree ring sequences are discussed and limitations noted on their use and interpretation.

TABLE III

Possible Choice of Overlapping Positions for Dendrochronology Curves, Winchester, England

Historical Period	Source	Choice of years for matching end rings of compared curves
Modern: 1635-1972 AD	Winchester Dendrochronology Curve: 41 modern trees plus five 18th C. trees from H.M. S. Victory. 338 rings in length	AD 1635 -- 1972
Early-Mid-18th Century	Staircase, 126 High Street Winchester: 17 curves, 296 rings, add 30, up to 45 yrs. for use date. Position of ring 296 . . .	1673 1686 1713
Very Early 16th Century	Langton Chapel, Winchester Cathedral: 21 curves, 306 rings, add 20-25 years for use date. Position of ring 306 . . .	1447 TE, 1470 1480 1490 / 1460 1483 1493 1503 / 1487 1510 NL 1520 TL 1530 TL
Late 13th or Early 14th Century	Choir Stall, Winchester Cathedral, constructed 1308 AD: 6 curves, 229 rings, add 30 (?) rings for use date. Position on ring 229 . . .	1256 TE 1266 TE 1276 / 1269 NL 1279 1289 TL / 1272 POS.
Mid-12th Century	Lid of Oak Chest, Westminister Abbey: one curve, 281 rings, no sap wood recorded. Position of ring 281... Position of ring 1...	1118 TE 1143 (863) / 1121 TE 1146 (866) / 1114 TE 1139

CODE

NL	Not Likely
TE	Too Early
TL	Too Late
Pos.	Possible

A tentative dendrochronology for Winchester has been developed. The chronology begins in the late 9th century and ends in 1972 A.D. This chronology is a first working base and may well be subject to modification as better data become available.

Post Script: After Symposium

During the course of the Symposium Mr. Martin Biddle raised a strong "consumer's concern" for the validity of the data base used in constructing chronologies. He suggested that whenever possible, radiocarbon dates be acquired for all non-modern specimens which are considered for incorporation into a chronology.

This concern comes directly from our experience and advice to Mr. Biddle in studying tree chronologies in the Winchester region. Our understanding of the biological variability and the lack of statistical certainty has led us to this most conservative advice. Mr. Biddle was reflecting our concern and we think he, as a consumer, has every right to request the most thorough validation of the data which he must rely upon for his archaeological statements.

Once the chronologies are adequately established, Mr. Biddle is not suggesting that each artifact submitted be dated with a C^{14} test.

We have stated our concerns less personally in the body of the paper. They are motivated by the fact that a large segment of our data collection are "unusable" because we are completely uncertain of the true growth period for the trees involved. Mr. Biddle was reflecting, therefore, our conservative concerns for the adequacies of the statistical and visual processes to identify the single proper matching point for tree ring chronologies.

Acknowledgments

Appreciation is expressed to the Winchester Research Unit; Richardson and Starling, Ltd. of Winchester; Ian Gourlay, Department of Forestry, Oxford; the architects and Dean of the Winchester Cathedral; the Bursar of Winchester College; the late A. W. G. Lowther; John Harvey; Dr. Derek Keene; to the Fulbright-Hays Commission and to many other unnamed individuals who have contributed to the work summarized in this paper.

References

Barefoot, A. C., L. B. Woodhouse, W. L. Hafley, and E. H. Wilson, 1974. Developing a dendrochronology for Winchester, England. Journal of the Institute of Wood Science, 6, 34-40.

Barefoot, A. C., 1975. A Winchester Dendrochronology for 1635-1972 A.D. Its Validity and Possible Extension. Journal of the Institute of Wood Science, 7, 25-32.

Bauch, J. and D. Eckstein, 1970. Improvement of a dendrochronological method for dating in Central Europe. In Tree Ring Analysis with Special Reference to Northwest America. Faculty of Forestry, The University of British Columbia. Bulletin No. 7, 37-40.

Brown, R. G., 1959. Less risk in inventory estimates. Harvard Business Review, 37, 104-116.

Duffin, R. J. and T. W. Schmidt., 1960. An extrapolator and scrutator. Journal of Mathematical Analysis and Applications, 1, 215-227.

Kolchin, B., 1962. Dendrochronological Method in Archaeology. International Congress of Prehistory and Protohistory Sciences. Academy of Sciences of the USSR, Institute of Archaeology. Moscow.

Schove, D. J. and Lowther, A. W. G., 1957. Tree-rings and medieval archaeology. Medieval Archaeology, 78-95.

DISCUSSION—Section 3

The Development in Northern Europe of Oak Chronologies for Dating

Prof. ECKSTEIN was asked how his chronology for the northern plains of Germany matched those for areas to the east and that for the Netherlands on the west. In reply, he pointed out that the Netherlands chronology only covers recent times. There are similarities between some of the mean curves but not between the single curves. There is no resemblance to Bauch's Chronology II for the Netherlands and little eastward to Jahrig's GDR one (see page 263). So it seemed that the North German plain and adjacent areas, such as Denmark and the Netherlands, needed to be handled separately.

Becker asked if this meant that there was no cross-correlation or whether over periods of 400-500 years there were some similarities especially in the signature years. Eckstein explained that their research originated for dating purposes and ten years or so ago it proved advantageous to build chronologies step by step for several regions. Afterwards it emerged that there were similarities especially between some adjacent regions, e.g. between Hamburg and Lower Saxony (but less so between Denmark and Lower Saxony).

Editorial Note: Questions to Eckstein on dendroclimatology are included in the Discussion at the end of Section 4.

— — — — — — — —

Becker asked whether there was any correlation between the results for Polish material and those for the German samples of Huber. In reply, Dr. DABROWSKI described his criteria for comparison of chronologies and gave the following general figures: less than 60%, correlation between samples from different sites within the same region; 60-70% for samples from the same site; and over 70% for samples from the same timber. The similarity with Huber's chronologies would be of the order of 50%.

— — — — — — — — —

Prof. Fritts asked in what way indices would not do the same job as ring-widths. He went on to say that people who fit a curve to remove the age trend are performing an extra operation that makes some chronologies much more straightforward and usable not only for dating but for environmental studies. He would like to see the question of standardisation discussed in as many meetings as possible. In reply Dr. FLETCHER referred to the reasons given by pioneers in European dendrochronology, Kolchin (Dendrochronological Method in Archaeology, page 7. English version of paper to VI International Congress of Prehistoric and Protohistoric Sciences, Section 1b, Moscow, 1962) and Huber (this volume, p. 16) for using ring-widths rather than indices in building chronologies for dating oak samples. Huber's plots of ring-widths on transparent semi-log paper of particular dimensions had been standard practice in central and other parts of Europe for many years. They enable mean curves made in different laboratories to be interchangeable for visual comparisons. He had always used Huber's method for dating and had found there were long term trends, sufficiently reproducible to act as signatures for dating, that covered three, five, ten or even longer periods of years; some of these would surely be less obvious if indices were used, particularly with smoothing as well. Also mean curves derived from wide and narrow sequences of oak had decidedly different patterns; indices do not indicate the widths of the rings concerned.

Dr. Laxton asked whether for dating short sequences of 50 to 60 rings smoothing and indices would be needed. Fletcher replied that such short sequences lay well outside the scope of his dating projects but agreed that on account of the prominent age trend shown by short sequences indices might be helpful for them.

Becker added that for dating purposes the semi-logarithmic scale in conjunction with visual matching by signatures forms the easiest and simplest method. It avoids the complications of using statistics to remove trends. With oak in central Europe, by working with a Master Chronology on which the signature years (in the Huber sense) are marked, visual matching or non-matching is definite. Computer comparisons are used to indicate possible positions of synchronisation. Fritts said that with indices the same kind of calculations could be made and that a variety of techniques are available for different purposes.

Participants expressed different opinions on the use of indices and of ring-widths. The general opinion was that ring-widths rather than indices should be made available.

Baillie asked whether the chronology for Winchester was one single continuous one, or is there still the possibility that it will need to be revised. Prof. BAREFOOT replied that as a forester he considered it is the best they can do with the data they have. He knows that they can operate with it, but if someone comes along and demonstrates to him that they are in error, he would be very glad to change it.

Mr. Drown asked whether the growth patterns of adjacent trees bore any direct relationship once they had been processed. Barefoot replied that sometimes they were exact duplicates, but you might not get exact duplication even from trees only 10 kilometres apart. However, when you look at average curves, they fit; but individual curves may differ.

Ms Gray said she had looked for correlation of Barefoot's ring-width data with climate and found 69% of the variation was accounted for by previous growth and local climate.

4.1. GROWTH RINGS IN THE SOUTH BALTIC AREA AND THEIR RELATION TO CLIMATIC FACTORS

L. Kairiukstis and A. Juodvalkis, Vilnius, Lithuania SSR

Summary

Biometric measurements have been made on well developed specimens of spruce, aspen, birch, ash growing in the South Baltic area. For spruce, the duration and extent of growth have been correlated with main indices of the phytoclimate. Duration is dependant on temperature in May and precipitation in August: the size of the annual increment is related to the temperature in May and June and to the precipitation in July and August. Climatic reconstruction for past centuries is feasible from ring-widths.

An understanding of the dependence of the magnitude of the annual increment on the climatic conditions of the particular year is desired by foresters and dendroclimatologists: by the former to form a model of the bioclimate and by the latter to explain the fluctuation in increment within climatic cycles and to reconstruct the climate in the past. Accordingly it is necessary to examine the increments at various stages of each growing season against climatic factors.

For over 15 years at the Birzhai Forest Hydrology Centre of the Lithuanian Forest Research Institute we have been studying the fluctuations in the seasonal formation of the annual rings and the height growth of trees. Biometric investigations with daily measurements have covered 20-25 well-developed trees of spruce, aspen, birch, ash and alder. Simultaneously we recorded the main indices of phytoclimate: light, temperature of air and soil, amount of precipitation, relative humidity, depth of water table, depth of snow and depth of soil freezing.

The individual species exhibit very different duration, rhythm and rate of growth. Well developed specimens of spruce enlarge their diameter on average for 87 days, birch and aspen for 91 days, grey alder for 93 days, black alder for 84 days and ash for 73 days. Formation of the annual ring starts earliest in spruce, on average on 17th May, and last of all in black alder, 27th May. Growth in height lasts for 109 ± 2 days in black alder, 103 ± 2 days in birch, 95 ± 1 day in grey alder, 75 ± 3 days in aspen, 62 ± 3 days in early-flushing spruce, 56 ± 3 days in late-flushing spruce, and 51 ± 5 days in ash. Furthurmore, the start, duration and finish of growth, together with the total increment both in diameter and in height in trees of a given species varies greatly.

The formation of the annual ring in spruce starts earlier or at the same time as the extension in height, but in broadleaved species it starts 5-15 days later. Height increment ends 5-25 days earlier than diameter increment.

In the formation of the annual increment, each species and (within species) each tree-class is characterized by a definite rhythm associated with the change in phytoclimate and the accumulation of growth regulators.

Table 1

Distribution of height and diameter increments by months

Species	Height increment %				Diameter increment %			
	May	June	July	Aug.	May	June	July	Aug.
Spruce	9	61	30	-	23	41	20	16
Birch	16	47	20	17	6	42	39	13
Aspen	7	74	19	-	3	36	48	13
Ash	16	52	22	-	4	45	46	5
Black alder	8	29	38	25	5	22	55	18

The variations in rhythm and rates of growth for different species causes significant differences in the formation of increments in individual months (Table 1). For spruce, the high fraction of the annual ring that is formed in spring and early summer (May-June 64%) indicates that the amount depends largely on stored substances laid down the previous year. For broadleaved species however, the increment is determined mainly by the assimilates and the climatic conditions of the current year. The height increment of spruce was found to be directly related to the size of the bud formed in the autumn, and this species survived and grew properly when all the needles were removed in spring: their increment due to stored substances was over 50% in the first year and 20% in the second year of that of normal spruce.

All this shows that one should not seek a direct and complete relationship between the amount of annual increment and the climatic factors of the current year. Other workers have found that it is difficult to establish a direct influence of climatic conditions in the current year on the annual increment. For accurate carbon dating, the spring and early summer fractions of the ring, especially in conifers, should be assigned to the previous year and the summer/autumn fraction to the current year.

As for the magnitude of the actual increment, there is a very close correlation with individual climatic indices in individual periods of the growing season. Mathematical analysis of the table of coefficients of correlation (r) shows (Table 2) that the decisive factor determining the formation of the annual ring and the amount of height increment in spruce in May and in June is the air temperature ($r = 0.918-0.978$).

The amount of precipitation in this period has hardly any effect on the increment because the soil is sufficiently moist in the first half of the growing season in the South Baltic region. However, in the second half of the growing season the soil moisture gradually decreases and the decisive environmental factor determining the amount of increment in July and August becomes the amount of precipitation ($r = 0.629-0.959$). Also, the more the precipitation the later the end of growth. Air temperature in July and August does not have a significant effect on the amount of increment.

Thus in normally developed spruce trees the duration of growth in the growing season is increased mainly by two environmental factors, the date

Table 2

Correlation (r) of spruce increments with climatic factors

Factor	Height increment			Diameter increment			
	May	June	July	May	June	July	August
Air temperature	0.965	0.978	-0.328	0.925	0.918	0.629	0.058
Amount of precipitation	-0.390	0.420	0.959	0.290	0.249	0.905	0.673

at which the temperature necessary for the start of growth is reached, and the amount of precipitation in July and August. Hence equations can be derived which express the height and perimeter increments in terms of the mean temperatures in May and June and of the precipitation in July and August. These equations enabled us to derive a general one showing the dependence of height and perimeter increments (in cms) on environmental factors in the conditions of Vaccinium myrtillus and V. myrtillus/Oxalis forest types in the moderately moist climate of the South Baltic region. They enable height increment during the growing season to be determined with a mean accuracy of $\pm$ 2 cm, and annual perimeter increments to be determined with an accuracy of $\pm$ 0.3 mm, from the mean monthly indices of climate.

For dendroclimatologists, the reverse expression of these equations is important as they would enable the main climatic factors in the past to be assessed from the size of the annual ring. Accordingly, we calculated the equations for the main climatic factors from the size of the annual ring. The mean air temperature (y) for May/June is related to the perimeter increment (x in cm) in these months as follows:

$$y = -13.212 + 43.104x \quad (r = 0.921)$$

Similarly the precipitation for July/August (y in mm) is related to perimeter increment (x in cm) for that period (on average, 0.36 of the annual total) as follows:

$$y = -79.435 + 484.559x \quad (r = 0.907)$$

These equations can give the accuracy indicated only when the appropriate climatic indices are determined in the forest itself. Our investigations have shown that on average over the growing season the air temperature in the stand is 0.5°C lower than in the open. Indeed, during intensive growth the difference is even greater. For more exact restoration of climatic conditions in accordance with the dendroscales by our suggested method it is necessary to use conversion tables. For example, to convert the temperatures at different times in the day, within a particular woodland site, to those that apply on an exposed site.

Editorial Note

Further details in English about this research, namely a translation of a paper by the same authors in the Proceedings of the Second All-Union Conference held at Kaunas, Lithuania in 1972, are given on pages 50-56 in Fletcher and Linnard, Russian Papers in Dendrochronology and Dendroclimatology, Research Laboratory for Archaeology, Oxford University, 1977.

References

I. Кайрюкштис Л.А., Юодвалькис А.И. I970. Особенности сезонного роста деревьев в свете дендрохронологических и дендроклиматических исследований. (The pecularities of seasonal growth of trees in the light of dendrochronological and dendroclimatic investigations.) Ж. „Лесоведение" № 3.

2. Кайрюкштис Л.А., Юодвалькис А.И. I969. Изменение климатических факторов в насаждениях в связи с постепенными рубками. (Climatic factor changes in stands in relation to gradual cuting.) Труды Литовского научно-исследовательского института лесного хозяйства, том XII. „Швеса", Каунас: I78-204.

3. Крамер П., Козловский Г. I963. Физиология древесных растений. (Physiology of woody plants.) Гослесбумиздат. Москва.

4. Молчанов А.А. I976. Дендроклиматические основы прогнозов погоды. (Dendroclimatic fundamentals of the weather forecast) Изд. „Наука" Москва.

5. Смирнов В.В. I964. Сезонный рост главнейших древесных пород. (Seasonal growth of the main species.) Изд. „Наука", Москва.

6. Kairiūkštis L. 1968. The growth of trees during vegetative period. Department of Forestry. Translation N 207. Canada.

7. Kairiūkštis L. 1973. Mišrių eglynų formavimas ir kirtimai. (The formation and cutting of mixed spruce stands.) Leidykla „Mintis" Vilnius, 1-358.

Lithuanian Oak Chronology

Editorial Note: Extracts from the paper 'Variations in the radial increment of oak stands in Lithuania, and their relationship with environmental conditions, climate and solar activity' by T. T. Bitvinskas and J. J. Kairaitis (Bioecological Fundamentals of Dendrochronology, Vilnius—Leningrad, 1975, 69-74) are given on the next two pages.

Decade	Year									
	0	I	2	3	4	5	6	7	8	9
I970	I0I	I09	I22	I06	62	–	–	–	–	–
I960	I03	I04	I09	II7	95	78	95	I0I	96	9I
I950	I07	I03	77	9I	88	75	89	II3	I20	II6
I940	70	97	96	I04	II0	I20	II7	I00	I06	II7
I930	95	96	9I	84	88	I06	I03	99	I08	98
I920	II0	I07	I22	96	II2	II3	I05	I07	75	I03
I9I0	I0I	I07	I09	I09	92	89	93	I04	90	II3
I900	9I	I09	99	II2	I03	I02	80	89	89	88
I890	I23	II4	I2I	I0I	I08	97	98	I00	II2	84
I880	I20	88	94	9I	93	93	II5	89	88	8I
I870	II0	99	9I	I03	85	90	I03	II0	I03	II6
I860	94	II3	II2	I03	II3	I03	I00	III	I02	I20
I850	II7	II8	I2I	II9	II3	88	83	86	72	8I
I840	II7	9I	I07	93	9I	85	93	80	84	I2I
I830	I04	II0	II6	I03	99	79	87	I04	87	I20
I820	I28	I0I	9I	II6	98	98	99	8I	II6	I00
I8I0	83	77	I04	96	93	I0I	I09	II5	96	I0I
I800	92	I45	I04	I02	II0	I08	76	96	77	98
I790	95	I02	I22	II4	84	87	I04	93	II2	II6
I780	I02	98	I07	II5	I08	93	96	97	84	9I
I770	I00	I00	I03	97	99	93	95	III	II2	I05
I760	88	99	72	90	90	8I	74	I04	96	98
I750	I09	I08	I04	I02	80	I2I	I32	II6	97	I05
I740	96	8I	85	84	76	90	88	92	I20	I06
I730	I27	I04	I20	I07	96	9I	92	95	I05	I20
I720	59	75	94	93	99	9I	I02	86	I23	I30
I7I0	–	–	–	–	–	I53	I40	78	60	62

Table of annual indices, calculated from cores of 2584 oaks (Quercus robur, L) in 43 sample plots of different forest types in Lithuanian SSR, 1974-1715.

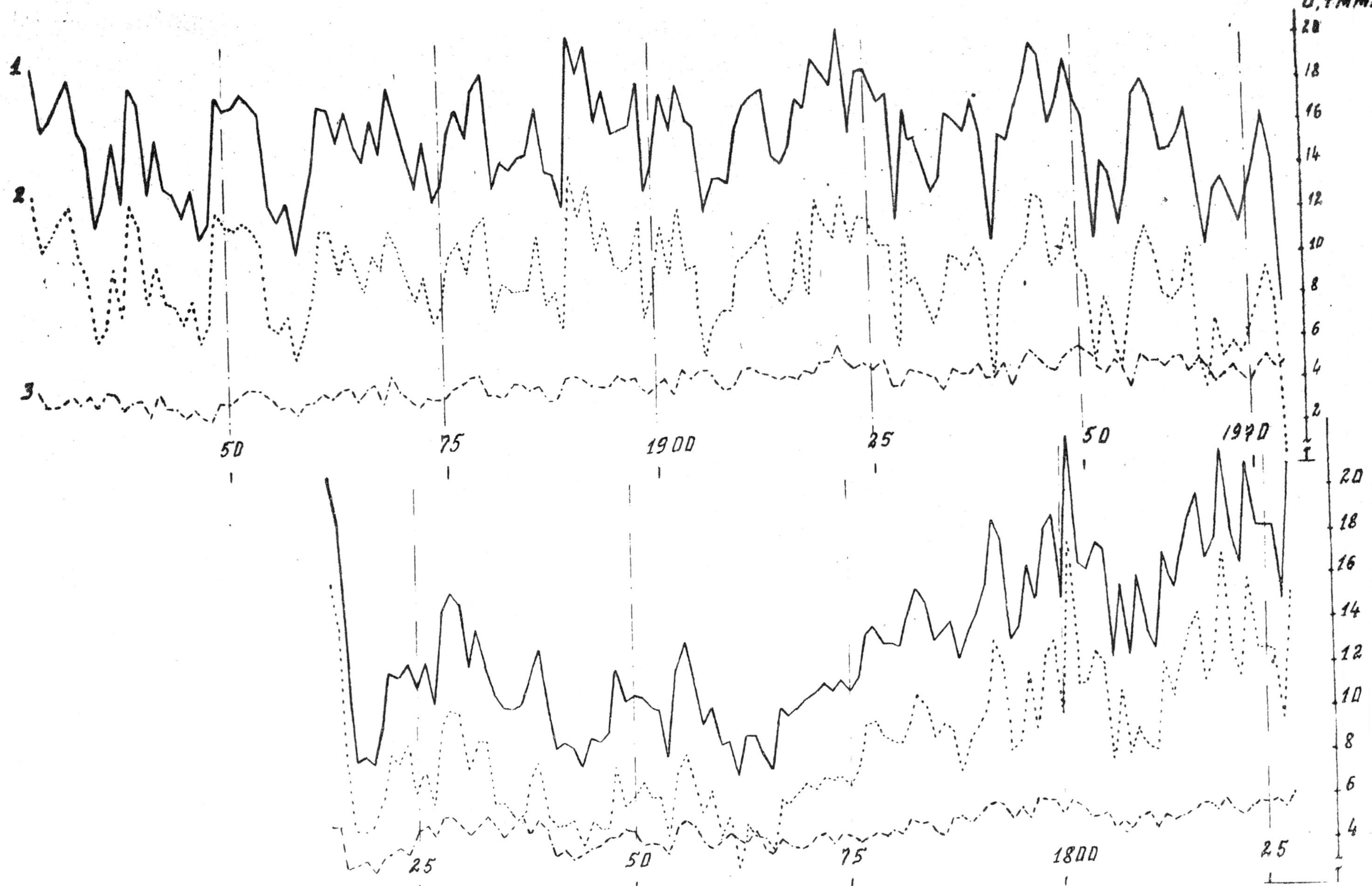

Curves showing: 1. The annual increments; 2. The latewood increments; 3. The earlywood increments for the same oaks (see Table on previous page).

4.2. THE EFFECTS OF INSECT DEFOLIATION ON THE GROWTH OF OAKS IN ENGLAND

G.C. Varley, Oxford University

Summary

The summer increment of oaks (Quercus robur) near Oxford is reduced by damage to the foliage in May caused by many species of caterpillars, amongst which the winter moth (Operophtera brumata) and the oak leaf roller (Tortrix viridana) are prominent. Caterpillar numbers were estimated on five trees from 1950-1969 and the damage to the foliage was assessed. Changes in summer increment correlated better with caterpillar numbers than with weather factors.

Some small oaks growing in full sunlight were defoliated by hand in 1960 and 1961. A tree defoliated in both years (which refoliated by July each time) was harvested late in 1964. It had put on no growth of summer wood in five years. Such enduring effects of damage would tend to obscure simple correlations which assume that growth in any year is determined only by the conditions of that year.

Using growth changes to build up a chronology over a wide area will be easy only if the insect damage is also widespread in particular years. The evidence for this is scanty.

Even if correlations between tree growth and weather factors can be established, they cannot be safely used to infer past weather changes if the effects of insect damage and of adverse weather are similar.

The oak, Quercus robur, has a bigger list of insect pests in Western Europe than any other native tree (Southwood 1961) and is more frequently defoliated than most other kinds of tree.

Authors have disagreed about the relative importance of the effects of insect damage and of weather factors on timber increment. Holmsgaard (1955) favoured the idea that the effects of weather were important, whereas Jüttner (1959) showed, by comparing the growth of undamaged oaks with that of other oaks which were completely defoliated by caterpillars in May each year, that the damaged trees put on between a third and a quarter of the growth of those which were undamaged.

We have assessed the populations of defoliating caterpillars on 5 oaks in a group of about 25 near the Northern boundary of Wytham Wood, 5km NW of Oxford. Winter moth (Operophtera brumata) and the Oak leaf roller (Tortrix viridana) were usually the most numerous species of defoliating caterpillars.

Timber increments were measured with a core sampler at breast height and the results were briefly reported by Varley and Gradwell (1962). The five trees differed in their average growth rate and, to our surprise, the tree with the greatest average radial increment also had the highest caterpillar population each year; this tree also opened its buds earlier than the other four trees. The caterpillars establish themselves in the oak buds as soon as they begin to open, and complete their growth as the leaves reach their full size. Defoliation has little effect on the Spring wood, which is laid down as the leaves grow; but a defoliated tree produces substantially less Summer wood than when undamaged.

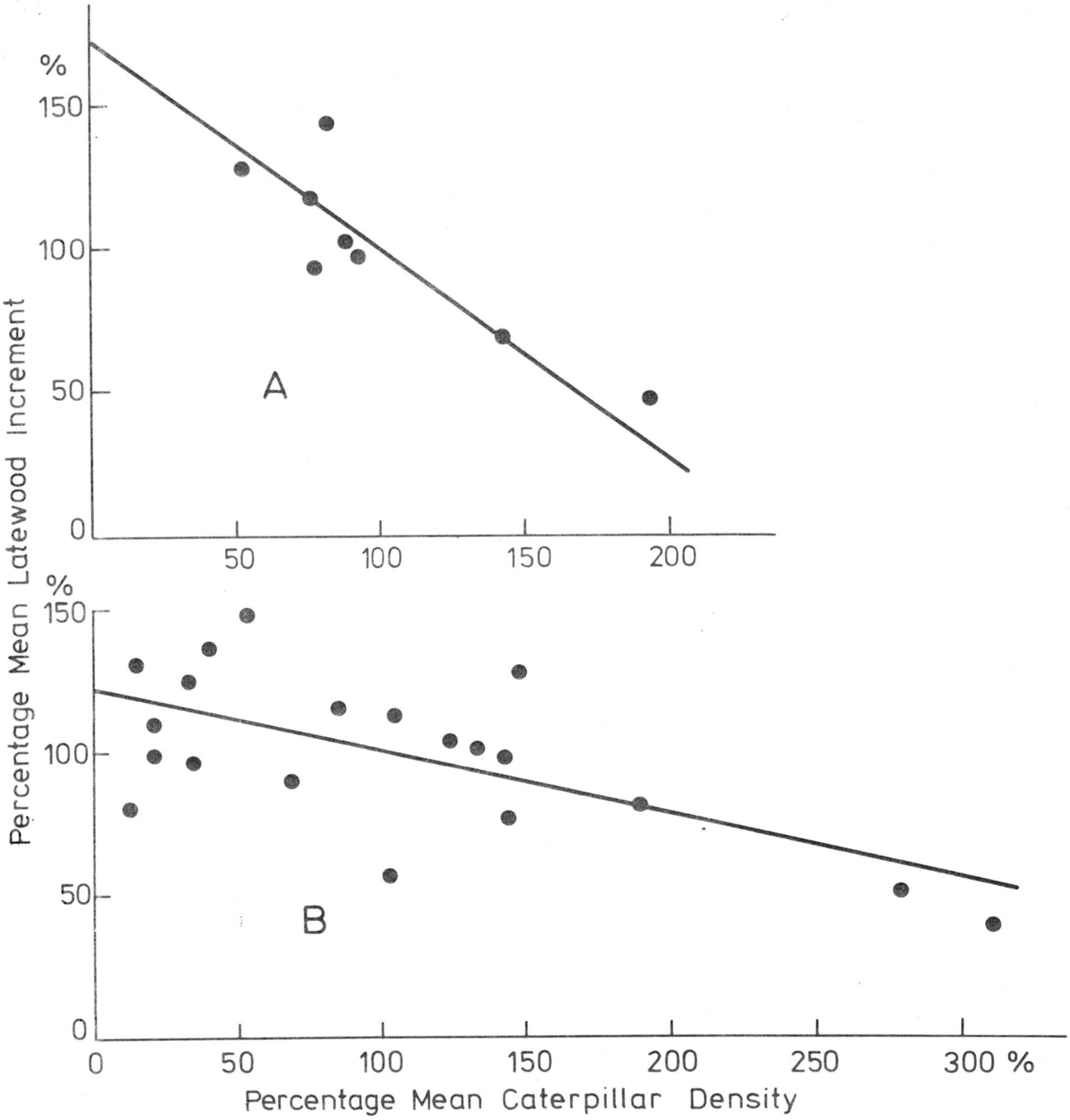

Fig. 1. Correlation between increment of Summer wood and caterpillar density for five oaks (Q.robur) at Wytham, 5 km NW of Oxford.

A. 1951-58 Slope b = -0·735 ; r = -0·89; P<0·01 (replotted from Varley & Gradwell 1962)

B. 1951-70 Slope b = -0·217; r = -0·69; P<0·01

When the actual summer growth was plotted against caterpillar populations for each tree and each year the points were widely scattered; the scatter was reduced by finding for each tree and each year the percentage of the mean growth and the percentage of the mean caterpillar population. When these were averaged for each year a fairly tidy linear regression appeared (Fig. 1A) which could be extrapolated back to zero caterpillar numbers to estimate that the mean summer growth in the absence of damage would be 160%. Including the fairly constant Spring wood, the average increase in timber production in the absence of damage would be 40%.

Holmsgaard (1955) had found that thc growth of oaks in Denmark could be correlated with variations in temperature and rainfall. Using meteorological data from the Oxford Observatory we investiaged our figures using Holmsgaard's methods. The correlations between growth and temperature, and growth and rainfall were neither significant and both correlations were in the opposite direction to those found by Holmsgaard. However, when the caterpillar figures were put into the multiple regression most of the variance was explained by this figure alone.

This illustrates the danger of using multiple regression analysis on incomplete data. The method was devised by R.A. Fisher to test the relative effects of the addition of nitrates, potassium salts and phosphates on plant yield. Fisher already knew that the effects were linear and additive, so his equation for yield Y was

$$Y = b_0 + b_1 x_1 + b_2 x_2 + b_3 x_3 \ldots$$

where the x terms represented added fertiliser and the b terms constants. The computation found the values of b_0, b_1 ... - to give the best fit. Nowadays multiple regression analysis is often wrongly used in an exploratory role without testing that the formula is valid; but if any of the major factors affecting tree rings are excluded from the regression, the method will inevitably give a misleading result.

The original regression method also assumes that the conditions of the year affect only the growth of that year but lag terms could be introduced if required. We have evidence that heavy defoliation has effects which endure for more than two years.

We set up an experiment on defoliation in 1960, choosing two very similar young oaks under 1m high. One we defoliated by hand in May and allowed to refoliate. The tree was defoliated again in May 1961,. and then left to grow undisturbed until both trees were dug up. late in 1964. The defoliated tree weighed only one seventh as much as the control. The trunks were sectioned and the annual rings of the control showed regular increments. The experimental tree, defoliated in 1960 and 1961 had put on virtually no wood in the five years, although it had had three years without defoliation in which recovery might have been expected. Cambini (1975) reports similar results from single and double defoliations of the cork oak *Quercus suber*, and Mikhailov (1972) has found that oaks in the central Volga region which were attacked by *Tortrix viridana* and other defoliators from 1965 onwards had put on less than half of the increment they achieved in the years before the outbreak. Blais (1965) illustrates similar changes in growth of Spruce in Canada in the budworm outbreaks of 1704 and 1748.

Only if the insect damage to trees is synchronous over wide areas will there be perfect matching of tree rings from different places. Sufficiently detailed information about the patchiness of insect attack is difficult to obtain. Some data for the Forest Tent Caterpillar which feeds on Poplar in North America have been summarised in Varley et al. (1973) which show that the outbreaks are not synchronous within the state of Minnesota, and differ markedly on the opposite sides of the continent.

Even if correlation between tree growth and weather factors can be firmly established, the growth of many kinds of tree, but especially oaks, is also greatly affected by insect damage. Until we can find a way to distinguish the effects of weather and of insect damage, the size of the rings cannot safely be used to infer past climatic changes.

References

Blais, J.R., 1965: Spruce budworm outbreaks in the past three centuries in the Laurentide Park, Quebec. For. Sci. 11:130-138.

Cambini, A., 1975: Effetti delle defogliazioni singole e duplici sull'accrescimento della quercia da sughero (Quercus suber L.) Mem. Staz. Sper. Sughero, Tempio Pausania 42:1-31.

Holmsgaard, E., 1955: Arringsanalyser af danske skovtraeer. Forstl. Forsoksv. Dann. 22:1-246.

Jüttner, O., 1959: Ertragskundliche Untersuchungen in Wicklergeschadigten Eichenbeständen. Forstarchiv. 30; 78-83.

Mikhailov, M.M., 1972: Effect of insect defoliators on ring width in oak. Proc. All Union Conf. on Dendrochronology & Dendroclimatology, Kaunas (Lithuania) 132-4. Edited English Translation of Russian Papers on Dendrochronology and Dendroclimatology, J. M. Fletcher and W. Linnard, Research Laboratory for Archaeology, Oxford University, 1977, 91-2.

Southwood, T.R.E., 1961: The number of species of insect associated with various trees. J. Anim. Ecol. 30:1-8.

Varley, G.C., and Gradwell, G.R., 1962: The effect of partial defoliation by caterpillars on the timber production of oak trees in England. Proc. XI Int. Kongr. Ent. Wien 1960, 2:211-214.

Varley, G.C., Gradwell, G.R. and Hassell, M.P., 1973: Insect population ecology. Blackwell Scientific Publications, Oxford.

Note Added by G. C. Varley and N. B. M. Calton, 1978

The study by Varley and Gradwell (1962) on insect defoliation of five pedunculate oaks in Wytham wood near Oxford (Fig. 1A) has been extended to cover the period 1951-1970. They form part of about 20 trees growing in the north-west corner of the wood. They are about 100 years old and have a breast height circumference of between 1.5 and 2 metres. The total ring widths and the latewood growth were measured on four cores per tree.

The ring width patterns of the five trees were found to match well (82% agreement) although the rings of Tree 5 were consistently wider than for the other four trees. The average ring width for Tree 5 was 2.14 mm (1.53 mm latewood only) compared to an average of 1.55 mm (0.85 latewood only) for the other trees. For each tree estimates of the numbers of all caterpillar species living on the tree were expressed as "winter moth equivalents" after having made allowance for the different size of species.

Linear regression of the mean latewood width and mean caterpillar number for each of the 20 years (Fig. 1B) gave a correlation of r = -0.69, again significant below the 1% level. Multiple regression analysis was carried out between the mean latewood width and (i) caterpillar numbers, (ii) previous year's latewood width, and (iii) the mean monthly temperature and precipitation for Oxford. While the previous year's growth had little influence, caterpillar numbers plus six (monthly) climatic variables explained 88% of the variance.

When the regression analysis was repeated excluding the caterpillar numbers, a totally different result was obtained. Although the total variance explained by the first six climatic factors was almost 80%, the significant months indicated by the analysis differed as follows:

	Temperature	Precipitation
With caterpillar numbers	Oct.,* Feb., Mar., May	Oct.,* June
Without caterpillar numbers:	Nov.,* Apr., July, Aug.	Dec.,* Aug.

* Months of the previous year

This implies that results obtained from multiple regression analysis will be misleading when no account is taken of insect defoliation unless proved to be absent.

Insect attack is indicated in years of heavy defoliation by the occurrence of minima in local ring-width chronologies for oak (see Fig.). The years 1952 and 1954 appear as minima for both sessile and pedunculate oaks in Bagley Wood as does 1948, a year in which heavy defoliation was observed. There is little climatic explanation as to why these should be years of such distinct minima. In other years the effect of insect attack can be interpreted as being combined with climatic factors, e.g. in 1965 when defoliation was combined with several months of high soil moisture deficit to give the narrow ring.

Overall, insect damage is widespread; Kulman (Annual Review of Entomology, 1971, 16) cites 174 studies of growth loss due to this cause.

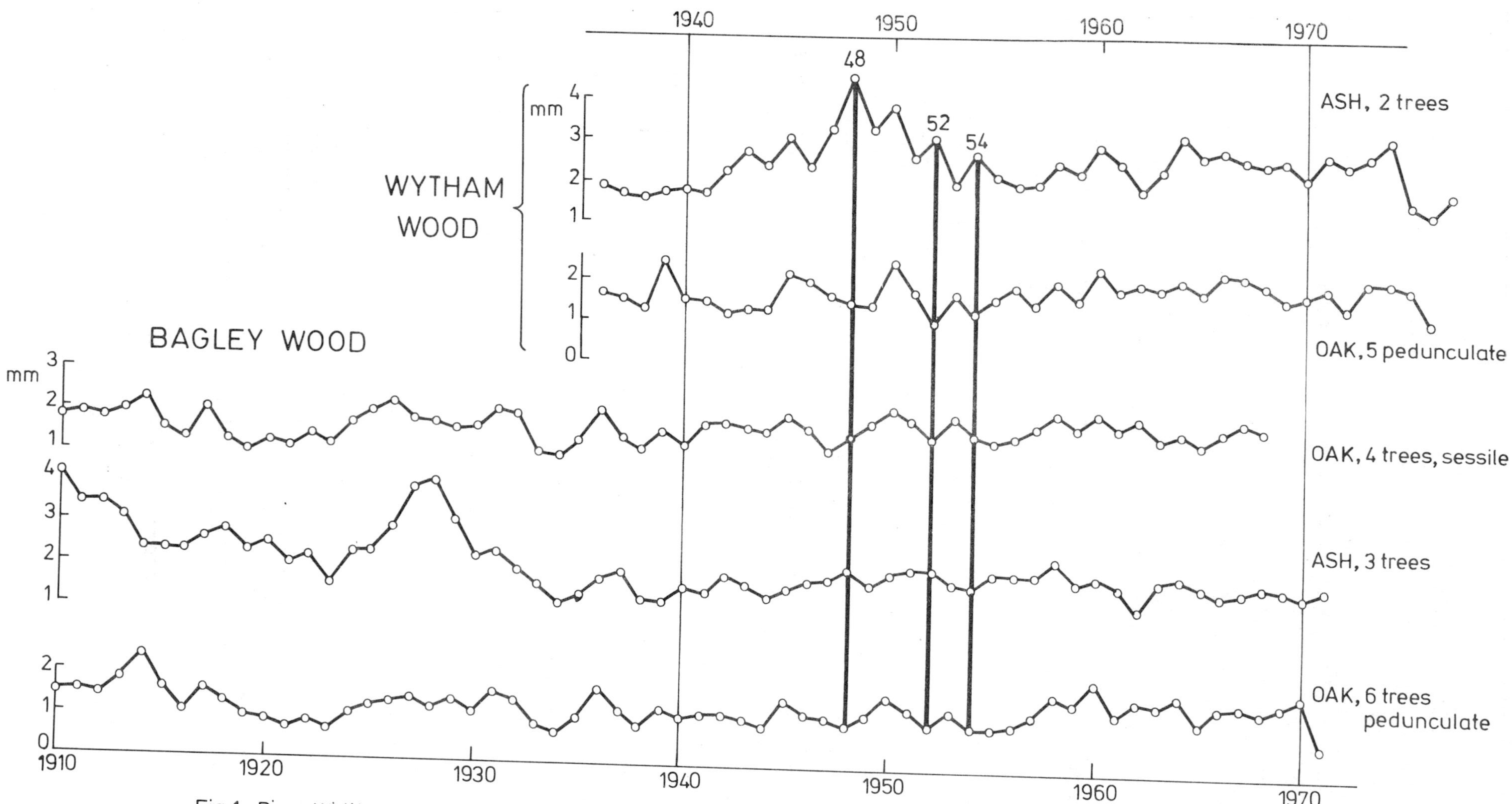

Fig.1. Ring Widths of Mean Curves for Oak and Ash in Two Woodlands near Oxford.

4.3 A COMPARISON OF THE RING WIDTHS OF OAK AND ASH

N. Calton and J. Fletcher,
Oxford University, 1978

Prof. Varley suggested to us that a comparison of the ring width curves of oak and ash growing in the same woodland might separate reduced growth in oak due to insect damage from that due to climatic factors, since ash is much less affected by insects. Results by one of us (J.F.) for discs from three ash, felled with several oaks in 1971/72 in the Underwood compartment of Bagley Wood to make way for the new A34 road, showed little agreement (W, 57% for the period 1837-1971) between the mean curves for the two species (Fletcher, 1974); the trees were of comparable age (about 150 years).

This study has been extended by the other author (N.C.) to cores from two ash in Wytham Wood in association there with the research on caterpillar defoliation of oak (this volume, page 183). Table 1 gives results for the agreement (a) between the single curves for the same species, (b) between each species in the two woodlands (separated by about 8 km), and (c) between the two species in the same woodland. The lowness of the agreements for (c) implies that the growth patterns of oak and ash are different, even for trees growing close to one another.

TABLE 1. Agreement (W) between Annual Increments of Pedunculate Oak and Ash

			W	Years Concerned
(a)	Individual Trees, with one another	5 oaks, Wytham Wood	Mean 82%	1893*-1977
		3 ash, Bagley Wood	Mean 64	1812-1971
(b)	Mean Curves, same species	5 Wytham oaks v 6 Bagley oaks	Mean 73	1893*-1974
		2 Wytham ash v 3 Bagley ash	Mean 69	1936*-1977
(c)	Mean Curves, different species	5 Wytham oak v 2 Wytham ash	Mean 54%	1936*-1977
		6 Bagley oak v 3 Bagley ash	Mean 51%	1882-1971

* The date of the first year is relatively late because cores are involved.

Fig. 1 compares the annual ring widths of the oak and ash in Wytham Wood and in Bagley Wood, the curves for the latter showing sessile and pedunculate oaks separately. The heavy connecting lines apply to 1948, 1952 and 1954, years when caterpillar defoliation of oak was marked (page 183). The curves show that the change in the growth pattern of ash and oak in these years differed considerably.

Regression analysis by Holmsgaard (1955) on Danish ash and oak showed that the growth of ash was correlated with the rainfall in the months of May, June and July to a greater extent than oak; ash was independent of the summer temperature whereas oak was considerably affected. Elling (1966) found that the ring widths of sequences for the two species in the Upper Danube at an altitude of 365 m matched poorly. From regression analyses it appeared that high precipitation in May favoured ash and in June favoured oak; high temperatures in April, May and June were damaging to the growth of ash but tended to favour the growth of oak.

References

Elling, W., 1966. Flora, 153, 459-508.

Fletcher, J. M., 1974. In The British Oak, p. 86.

Holmsgaard, E., 1955. Tree-Ring Analyses of Danish Forest Trees, Copenhagen, Table 22, page 161.

4.4. A MODERN OAK CHRONOLOGY FROM NORTH WALES AND ITS INTERPRETATION

P. Leggett, M. K. Hughes and F. A. Hibbert, Liverpool Polytechnic

Summary

A chronology for slow-grown sessile oak from a typical western oakwood at Maentwrog, North Wales, has been established. Material from living trees and those felled in earlier decades has been cross-dated by both visual and statistical means and summarised as a site chronology (35 trees) for the period 1710 to 1974. The Maentwrog chronology matches both the Belfast regional chronology and the Rostrevor site chronology from Northern Ireland.

INTRODUCTION

The state of a tree ring is determined by many interacting factors, both intrinsic and extrinsic to the individual tree and to the stand. Nevertheless, that there is synchroneity of change in tree ring measurements over wide geographical areas is witnessed by the existence of the discipline of dendrochronology. Whilst the first major thrusts in the development of the field took place in continental Europe and North America, there has been a growth of interest in tree rings in the British Isles in recent years.

Berger, Giertz and Horn (1971) considered that as a climatically complex island Britain would be interesting to investigate. They concluded that 'England, ..., will require a massive onslaught from many different regional centres of study....'. It was in this spirit that the present study was undertaken.

By choosing a site on the west coast of North Wales from a typical western oakwood (Steele 1974) which is an area of equable, moist climate, the most stringent conditions have been imposed for this test of dendrochronology. Oak has been selected as the material for study because of its relative longevity and ubiquity, its extensive use in ancient artefacts, the rareness of double or missing rings and the existence of a considerable body of literature on its dendrochronology (Eckstein 1972).

MATERIALS

The site at Maentwrog, North Wales (Map reference SH661412) is a sessile oak (Quercus petraea (Mattuschka) Leibl.) woodland, with the steep slope and acidic soil characteristic of such situations (Figure 1). The slope faces the south and the trees sampled were at altitudes of 50 to 100 m. Material for tree ring measurement was obtained from three types of source. Firstly, two increment cores were taken at breast height from each of eleven living trees. Cores were, wherever possible, taken in such a manner that the sampled radii were opposite to each other and parallel to the contours of the hillside. Secondly, discs were cut with a chain saw from the stumps of ten trees felled a few days before sampling. Thirdly, discs were similarly cut from nineteen stumps of trees felled at earlier, initially unknown dates. It was believed, at the time of sampling, that these latter had been felled in the last part of the nineteenth century or the first part of the twentieth. However,

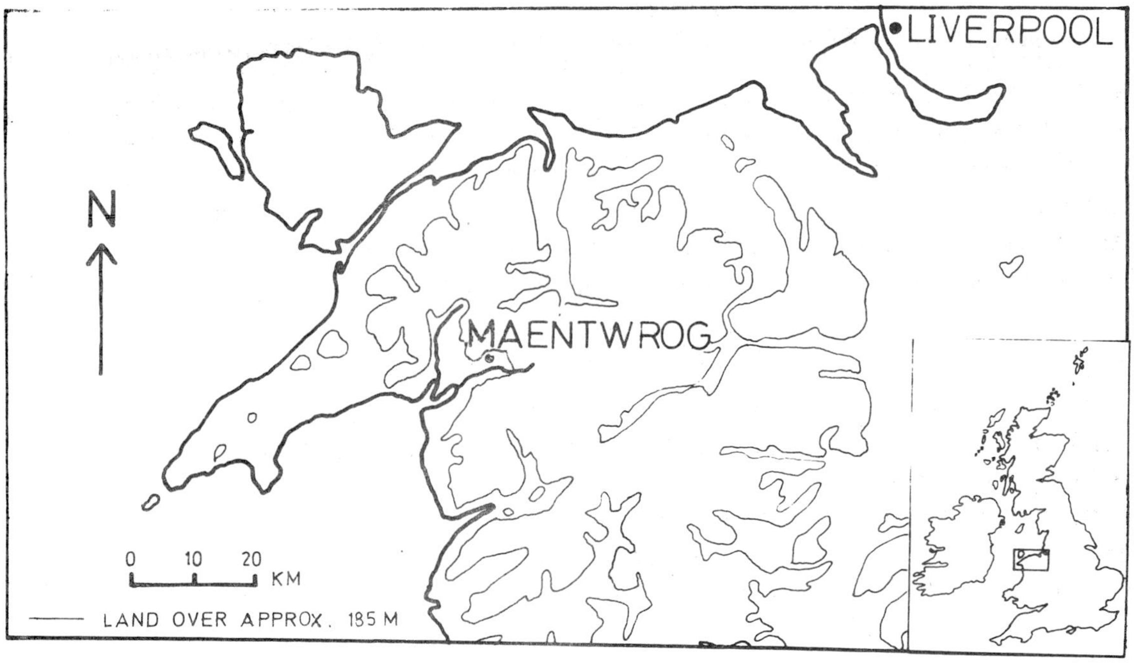

Fig. 1. Site Location

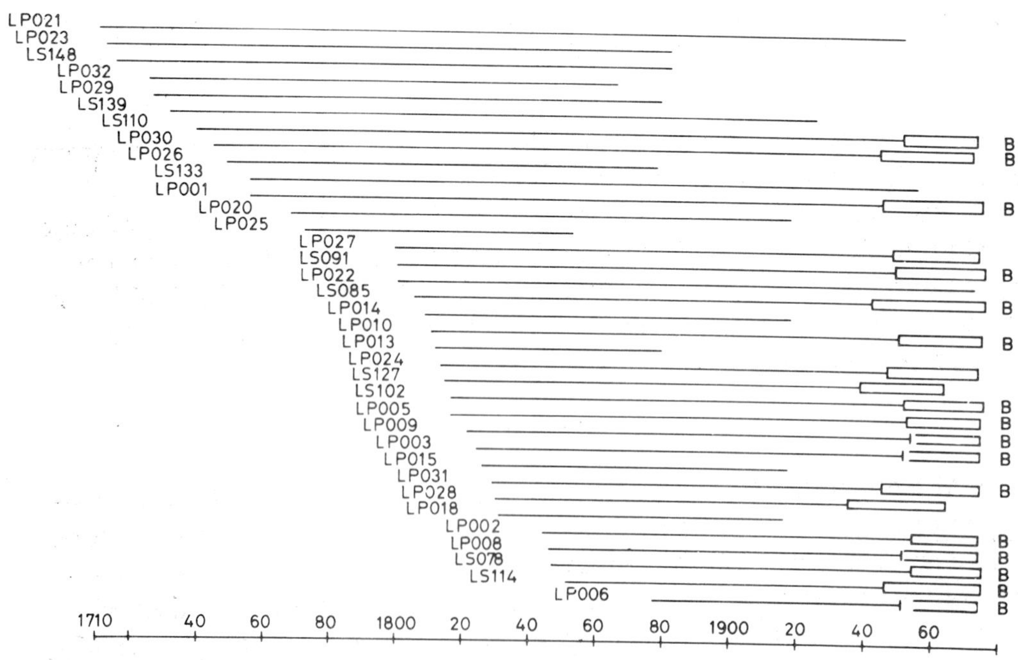

Fig 2. Components of Maentwrog Chronology

fire damage had resulted in distortions in some of these older stumps that prevented measurement of outer sections of the tree ring record. Figure 2 shows the lengths of the records obtained from the samples used in the final chronology. Where bark was present at time of sampling the notation 'B' is used. The number of sapwood rings is shown by a block at the distal end of the record. Where the two radii measured for an individual tree had differing numbers of sapwood rings, the earlier sapwood boundary is shown by a vertical bar. It can be seen that as most of the trees sampled by increment corer or recently felled have a complete record up to and including the bark, the materials present an opportunity both to study the variation in number of sapwood rings at this site and to produce a chronology anchored at a known felling date.

METHODS

Preparation: The preparation of cores for tree ring measurement followed the procedure described by Stokes and Smiley (1968). Slices were stored for at least 10 weeks in an unheated, very well ventilated room, stacked so that air could move freely around them. Sections were cut from the slices so that two opposite radii, analogous to those sampled as cores, could be prepared and measured. All samples were prepared for measurement by sanding to a fine polish. Extra care was required when sanding sapwood so that a good surface could be achieved on this softer material.

Measurement: Ring widths were then measured to 0.01 mm using a Bannister Incremental Measuring Machine (supplied by F. C. Henson Co., Laguna Miguel, Calif.). The various methods by which these primary data may be presented have been outlined by Baillie (1974). In the present study, ring widths were plotted on transparent semi-logarithmic graph paper. The horizontal scale of years in linear, whereas the vertical scale of ring width is logarithmic. As a result the maxima of the curve are smoothed and the minima accentuated whilst the length of the curve is independent of ring width.

Within-tree cross-dating: Cross dating between the two radii measured from each tree was tested using the computer programme CROS (Baillie & Pilcher, 1973). The curves were also compared visually by superimposition of the semi-logarithmic traces described above. The curves were moved relative to each other until a good visual fit on the basis of the general form of the curve and of characteristic features or signatures was achieved. For cross dating to be established between the two radii from a tree both the statistical and the visual criteria had to be satisfied.

In the event of this not occurring it was often possible, using the visual criteria, to identify an area of doubt, for example a group of very narrow rings consisting largely of early wood. In such cases a careful re-examination and re-measurement of the two radii was made in the hope of detecting errors. This procedure was found to be a valuable one in establishing the reliability of ring identification and width measurement. It demonstrates the usefulness of within-tree replication in the production of reliable tree ring series.

Between-tree cross-dating: In those trees where cross dating was established between the two radii measured, mean curves were prepared for each tree. The same procedure of cross dating, based on both statistical and visual criteria, described above for the within-tree case, was then applied to pairs of tree mean curves. For this purpose any measurements which came from only one of the two radii were omitted from the mean curve for that tree. A group of thirteen trees all of whose mean curves cross-dated with one another on both criteria were combined into a working master against which the other materials were tested. The production of this working master was done using the computer programme INDXA (Fritts *et al.*, 1969) modified for use on Liverpool Polytechnic's ICL 1903A computer. The ring-widths for each radius were converted to indices using the polynomial curve fit option also used by Pilcher (1976) for a similar oak site in Ireland. The index values for all radii to be included in the working master were then combined to produce that master.

The materials not already included in the working master were tested for cross dating against it using the procedures described above. Where both within-tree and tree-working master cross dating was shown to exist, the two measured radii for that tree were included in the final Maentwrog chronology, which was obtained in the same manner as the working master.

Sap-wood measurements: In order to estimate a mean and variance for the number of sapwood rings in the lower part of the boles of the sampled trees it was necessary to use samples where bark was present or where absolute date of felling and of sapwood boundary are known. The sapwood boundary was identified using the criteria of colour change and early wood cell characteristics.

RESULTS

Within-tree cross dating was established in 38 of the 40 trees sampled. Thirteen of these were combined into a working master chronology. Figure 3 shows the thirteen-tree working master and the mean curves for the trees from which it was constructed. A number of characteristic years namely, 1839, 1882, 1906, 1966, are marked by vertical lines. The output of Program CROS is shown in Table 1 as probabilities for each pair of tree mean curves (Baillie & Pilcher, 1973). These trees represent a strongly cross-dated group on both statistical and visual grounds.

A further 24 mean curves cross dated with the working master. 35 of this total of 37 were then combined to produce a standard Maentwrog chronology for the period 1710-1975 (Table 2). The temporal distribution of the components of this chronology is shown in Figure 2. (Two trees which did cross-date with the working master were not included in the standard Maentwrog chronology because of a limit on the number of series the Liverpool version of INDXA can summarise.)

The standard Maentwrog chronology, obtained by summarising indices arrived at using a polynomial curve-fit option of INDXA, has a serial correlation of 0.579, a standard deviation of 0.235, a mean sensitivity of 0.175, an average ring width of 1.59 mm and a span of 266 years.

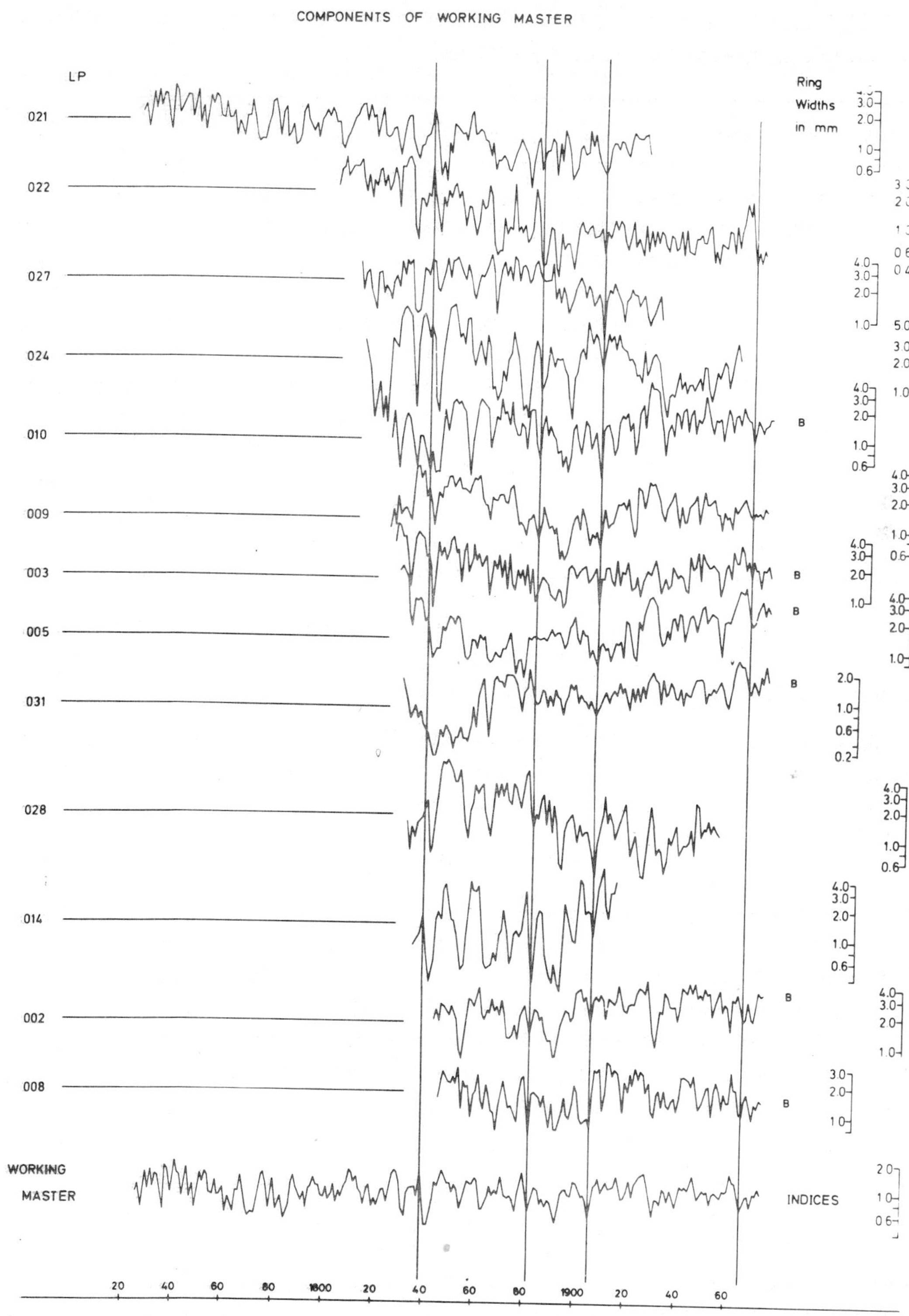

Fig. 3. The Ring Widths, 13 trees, plotted on a semi-logarithmic scale. The MASTER curve is produced from their indices, see text. B, Bark present. Vertical lines mark 1839, 1882, 1906 and 1966.

TABLE 1 Probabilities of observed values of r between components of the working master arising by chance

LP No.	003	005	008	009	010	014	021	022	024	027	028	031
002	xxx	xxx	xxx	xxx	xxx	xxx	x	xx	xx	xx	xx	xx
003		xx	xxx	xxx	xxx	xx	xxx	xxx	xx	xxx	xxx	xx
005			xxx	xxx	xxx	xxx	x	xxx	xxx	xxx	xx	xx
008				xxx	xxx	xxx	xxx	xx	xx	xx	xx	xxx
009					xxx	xx	xx	xxx	xxx	x	xx	x
010						xxx	xx	xx	xx	xx	xx	xxx
014							xx	xxx	xxx	xx	xxx	xxx
021								xxx	xxx	xxx	xxx	xx
022									xxx	xxx	xx	xxx
024										xxx	xxx	xxx
027											xxx	xxx
028												xxx

Key

x ≤ 0.02

xx ≤ 0.001

xxx ≤ 0.0001

The mean of 38 measurements of the number of sapwood rings was 27.21, the standard error 0.8. There was no evidence of either missing rings or apparent double rings in any of the radii measured.

The ring-widths for the 70 radii used in the Maentwrog master chronology are to be deposited in the International Tree-ring Data Bank. They are also available directly from the authors on request.

DISCUSSION

The results presented demonstrate the practicability of establishing adequately replicated, absolutely dated ring-width chronologies in conditions totally different to those of the American south-west. It should be emphasised that this site is very close to the western coast of Great Britain, and at low altitude. Total annual precipitation at Aberystwyth has been within the broad range 800 to 1300 mm in the last six decades. Rainfall at the sample site will, almost certainly, have been significantly greater.

The general conclusion that dendrochronology is feasible even under those conditions that might appear to be unfavourable (Godwin 1940) reinforces the findings of the Belfast group (Baillie 1973a, Pilcher 1976). The validity of this conclusion is further underlined by the existence of cross dating between the Maentwrog chronology and the Belfast master chronology, established for an area some 200 km to the north-west across the Irish Sea. A value of Student's 't' of 3.35 with 261 years overlap ($p < 0.00023$) was obtained. An even stronger crossdating with 't' 5.09, 225 years of overlap ($p < 0.00001$) was found with the Rostrevor chronology (Pilcher 1976).

A comparison of the chronology statistics of the Maentwrog chronology with two other oak chronologies (Table 3) shows its broad similarity with them. In particular it shares a large first-order serial correlation with the Rostrevor chronology (Pilcher, 1976). The mean sensitivity is similar in all three cases, the values being comparable to those given by Fürst (1963).

TABLE 3 Comparison of chronology statistics for oak between Maentwrog and two other sites

	Chronology		
	Maentwrog	Rostrevor[1]	Schleswig[2]
Mean ring width (mm)	1.59	1.00	1.94
Serial correlation with previous years	0.58	0.56	0.36
Mean sensitivity	0.18	0.22	0.19
Standard deviation	0.23	0.30	0.21
Years of analysis	266	226	158

1: Pilcher, 1976 2: Eckstein and Schmidt, 1974

The figure of 27.21 ± 0.8, as mean number of sapwood years (± standard error) compared with 31.8 ± 1.5 for Irish oaks (Baillie 1973).

More extensive statistical analysis of the Maentwrog chronology will be presented elsewhere. The chronology is seen as having three main uses in the work of the Liverpool laboratory. Firstly, as a modern anchor for a longer regional chronology to be used in dating timbers. Secondly, in the examination of the dendrochronological properties of oak on the western coast of Great Britain. Thirdly, as a component of a network of regional chronologies to be used as proxy records of meso-scale climatic change.

Acknowledgements

P. Leggett was supported by a Polytechnic Research Assistantship and part of the research was supported by N.E.R.C. Grant GR3/2244 in which M. K. Hughes was principal investigator. S. J. Milsom provided measured samples from ten trees.

Site facilities were provided by: the Snowdonia National Park, in particular we are indebted to Mr. D. Towler, the Forestry Officer; the North Wales Naturalists' Trust; and the Nature Conservancy.

Dr. H. C. Fritts made programme INDXA available. The Computer Unit of Liverpool Polytechnic gave a great deal of help.

References

Baillie, M. G. L., 1973. A dendrochronological study in Ireland with reference to the dating of medieval and post medieval timbers. M.S. Doctoral dissertation, Queens University, Belfast.

Baillie, M. G. L., 1973a. Tree-ring Bull. 33:15-28.

Baillie, M. G. L., 1974. Ulster Folklife 20:1-23.

Baillie, M. G. L. and Pilcher, J. R., 1973. Tree-ring Bull. 33:1-5.

Berger, R., Giertz, V. and Horn, W., 1971. Vernacular Architecture, 2: 3-6.

Eckstein, D., 1972. Tree-ring Bull. 32:1-18.

TABLE 2 Maentwrog master chronology

	Indices (x 100)										
Date	0	1	2	3	4	5	6	7	8	9	n
1710	90	133	79	72	71	78	66	67	70	69	1
1720	126	113	115	104	152	121	99	113	96	93	4
1730	117	89	92	97	96	134	122	87	137	125	9
1740	94	95	84	119	134	90	110	154	107	104	12
1750	89	90	110	103	95	102	100	102	124	122	14
1760	100	89	85	114	91	74	97	91	114	127	17
1770	94	52	64	48	57	83	121	126	105	80	20
1780	109	139	103	93	97	66	53	43	83	109	23
1790	101	97	85	92	71	87	80	111	104	113	25
1800	101	78	83	82	73	101	101	137	142	149	28
1810	144	171	162	115	120	96	87	73	104	112	36
1820	77	81	78	93	105	111	79	113	140	159	43
1830	159	114	59	50	81	97	103	109	108	126	54
1840	97	58	52	73	76	117	107	142	155	129	59
1850	113	113	89	112	100	73	71	92	110	103	62
1860	101	112	119	62	63	71	85	107	85	93	63
1870	89	139	120	97	94	91	81	70	94	139	62
1880	159	107	69	86	100	115	111	77	92	69	56
1890	82	81	72	57	73	100	111	114	104	102	53
1900	139	128	109	95	96	80	56	91	112	114	54
1910	134	100	108	124	105	108	113	146	109	87	53
1920	88	93	114	82	102	109	128	133	134	107	46
1930	75	58	79	86	80	109	103	83	96	84	43
1940	69	86	109	107	113	106	113	135	112	81	40
1950	90	99	98	98	80	96	104	96	86	101	39
1960	108	123	146	136	143	112	76	73	89	94	36
1970	81	104	91	97	79						29

n is the number of radii for the first year in each line.

Eckstein, D. and Schmidt, B., 1974. Angew. Bot. 48:371-383.

Fritts, H. C., Mosimann, J. E. and Bottorf, C. P., 1969. Tree-ring Bull. 29:15-20.

Fürst, O., 1963. Flora 153:508.

Godwin, H., 1940. Report on timber from the Sutton Hoo Ship-Burial. Antiq. Journ. 200-201.

Pilcher, J. R., 1976. Tree-ring Bull. 36:21-27.

Steele, R. C., 1974. Variation in Oakwoods in Britain. In: The British Oak. ed. M. G. Morris and F. H. Perring. B.S.B.I. London. 130-140.

Stokes, M. A. and Smiley, T. L., 1968. An introduction to tree ring dating. University of Chicago Press, Chicago.

4.5. MEDIEVAL AND RECENT ELMS IN LONDON

D. W. Brett, London

Summary

Results are given of a study of the elm piles which form part of the Thames waterfront at Trig Lane, London and similar piles recovered from the foundations of the Jewel Tower at Westminster. The ring-widths of elm trees grown in recent times are important as a record of summer rainfall.

Elm Piles (Plate 1)

As a traditional source of timber for construction, elms rank second only to oak in England. Their resistance to decay in wet conditions has led to their wide use in shipbuilding, for the construction of drains and conduits, for coffins and for the footings of buildings. Although most of the timbers recovered from the Trig Lane site are oak (our work on them is included in the paper by Morgan and Schofield, this volume, page 234) the main revetments were founded upon elm piles as were many bridges and buildings on low ground in medieval England (Salzman, 1952).

These are uses most likely to lead to the preservation of ancient timber but elm trees had many other uses besides being a source of timber. In particular the foliage was widely used in former times as fodder as it is in other countries where elms grow. The trees used for these several purposes in England were seldom, if ever, forest trees. Only one British elm, Ulmus glabra, appears to be a native woodland species. Although other types of elm are now found in woodlands, English elms were traditionally cultivated in the hedges along the boundaries of fields and roads; there, they were deliberately planted or allowed to grow from suckers. These elms of the southern English landscape are now in danger of virtual extinction as a result of the serious outbreak of Dutch Elm Disease.

Procedure: Slices of the piles from the Trig Lane waterfront were washed and frozen to consolidate the soft sapwood. They were then surfaced with a knife and razor and ring widths measured whilst the wood was still in a moist state. Measurements were taken along two or more radii using a stereo-zoom binocular microscope with an eyepiece graticule; ring width was measured to 0.05 mm. The well established methods of cross-comparison were employed: the ring widths were plotted on transparent graph paper for direct visual comparison, and the computer programs of Eckstein and Bauch, and of Baillie and Pilcher used for cross-matching.

Most of the piles at Trig Lane had been cut from small trees with fewer than 60 growth rings. The bark had not been removed and a partially formed outer ring (present in some cases) suggests a felling in late April to May. Their ring-width sequences can be cross-matched, see Fig. 1. They may not be directly datable by a tree-ring chronology, for it seems unlikely that an elm chronology can be constructed to cover more than limited periods. Conversely, piles recovered from structures firmly dated on documentary evidence can provide a short but accurately dated chronology. It is hoped

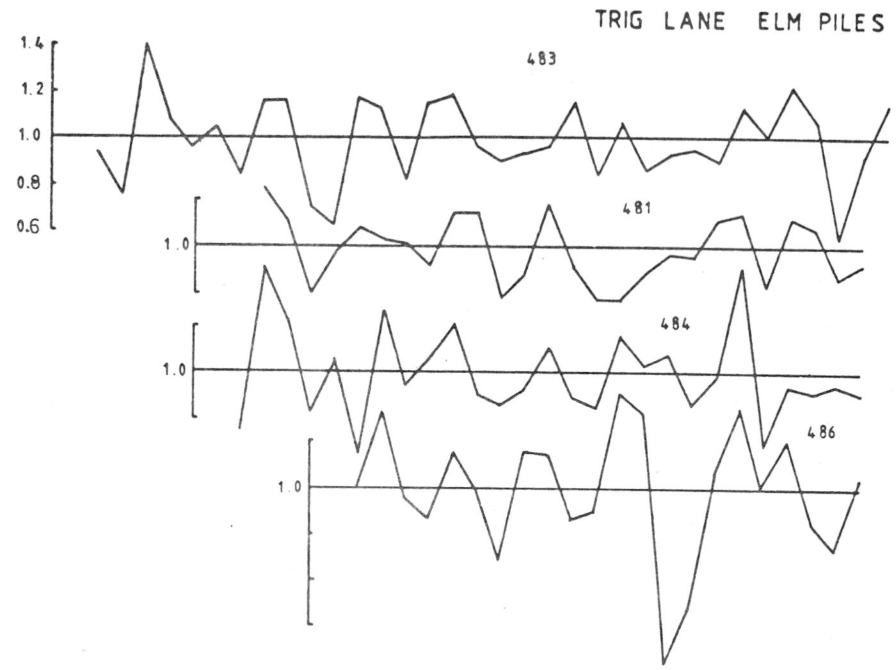

Fig.1. Standardised ring-width indices for the Group 11 piles.

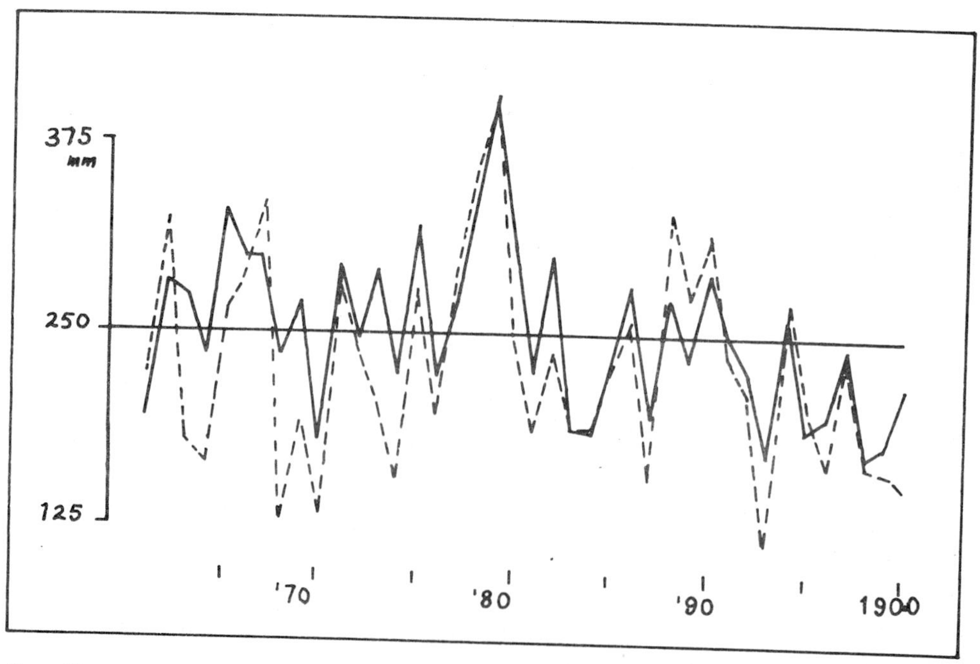

Fig. 2. Summer Rainfall at Kew, near London.
Solid line: Elm ring-width index X average rainfall.
Broken line: Actual rainfall for March - July.

that when more is known of the dendroclimatic relationships of oak and elm in southern England some basis for comparison between the two may be worked out. The presence of bark on elm piles offers the prospect of obtaining the exact year of construction.

Similar small elm piles were recovered from the foundations of the Jewel Tower at Westminster during reconstruction some 20 years ago (Biek, 1963; Wilson and Hurst, 1957, 1965; Green, 1976). These had been allowed to dry out but, despite the considerable tangential shrinkage of the sapwood, ring-width measurements have been possible. Some larger elm piles about 3 m long and fitted with iron shoes were also recovered from excavations close to the Jewel Tower and are thought to be roughly contemporary or a little later in age (Green, 1976). These piles are squared-off trunks with about 90 rings but some inner bark is present in places. The date of construction of the Jewel Tower, 1365 A.D., is firmly fixed on documentary evidence (Taylor, 1965). The ring sequence of the Trig Lane Group 11 piles should therefore overlap the sequence from these Jewel Tower piles. It is hoped that between them they will provide an elm ring record for much of the fourteenth century.

A Modern Elm Chronology

English elms are less long-lived than oaks so are less likely to be useful for extensive chronologies. Some modern trees are about 250 years old but these are rare, most of the largest trees felled at present being about 150-180 years old. Our modern elm chronology has been based on recently felled trees grown in some of the Royal Parks in London. Cross-section discs were taken from the trees within 1-2 m of the base and ring-widths were measured across 2 or 4 opposite radii as described in the previous section.

The agreement for comparison of opposite radii is 71.7% (average for 10 trees) and between-tree comparisons average 64.1%. These figures may be compared with values for oak from sites in southern England. Barefoot et al. (1974) gave examples of agreement from sites in Hampshire: the average of five between-tree comparisons was 64.3%. Fletcher (1974) gives the following average values for between-tree comparisons within three Berkshire woods, 72%, 71%, 71.5%; and 67.5% for a Sussex wood. The trees used by Fletcher were, however, selected for their uniform growth around the central pith. He found the average value for between-radius comparisons within the same trees to be 80%, a value which reflects the circuit uniformity of the well-grown trees that were sampled; while comparison between trees from woods at some distance from one another gave agreement of 65% (at 25 km) and 63% (at 120 km). Thus oak and elm appear to have about the same between-tree agreement and this also holds over a distance, for we find elms grown 18 km apart near London have 67.2% agreement (average of 8 comparisons, Regent's Park with Bushy Park).

A dendroclimatological analysis of this modern elm record has shown that the elm ring width is sensitive to fluctuations in summer rainfall throughout the period studied, 1840-1970. The exact relationship varies according to the serial correlation of the ring record. When that is low, the relationship may be very direct, as illustrated in Fig. 2. The mean ring-width indices based on 8 trees have been multiplied by the average summer rainfall

PLATE I ELM PILES EXPOSED AT THE THAMES WATERFRONT: TRIG LANE SITE, LONDON. Scale = 2 x 100 mm

(250 mm); the mean of the indices is 1.0. These indices provide a reasonable estimate of rainfall during the growing season, as seen from the comparison with the actual March-July total rainfall for Kew plotted in the figure.

One aspect of the importance of medieval elm chronologies therefore lies in their value as a record of summer rainfall. From them a picture of annual climatic fluctuations during the fourteenth century may be inferred (Brett, 1977, 1978).

References

Barefoot, A. C., L. B. Woodhouse, W. L. Hafley and E. H. Wilson, 1974. Developing a dendrochronology for Winchester, England. J. Inst. Wood Sc., 6, 34-40.

Biek, L. 1963. Archaeology and the microscope. Lutterworth, London.

Brett, D. W. 1977. Annual climatic fluctuations in south-eastern England during the fourteenth century inferred from tree-rings. Abstracts, X INQUA Congress, p. 54.

Brett, D. W. 1978. Elm tree rings as a rainfall record. Weather, 87-94.

Fletcher, J. M., 1974. In The British Oak, M. G. Morris and F. H. Perring (eds), Classey, 80-97.

Green, H. J. M. 1976. Excavations of the palace defences and Abbey precinct wall at Abingdon Street, Westminster, 1963. J. Brit. Arch. Ass., 129, 59-76.

Salzman, L. F., 1952. Building in England down to 1540. Clarendon Press, Oxford.

Taylor, A. J. 1965. The Jewel Tower, Westminster. Ministry of Public Buildings and Works, Official Guidebook. H.M.S.O.

Wilson, D. M. and D. G. Hurst, 1957. Med.Arch., 2, 159; and 9, 198.

DISCUSSION—Section 4

Ring Widths of Recent Trees in Northern Europe and Their Environmental Implications

In Eckstein's account (Paper 3.1, page 117) of the relationship between the tree-rings of Schleswig oaks with climate and prior growth, there was no mention, said Dr. M. Hughes, of any other biological influence. Was that intentional or was there at Hamburg no evidence of the influence, for example, of caterpillars? ECKSTEIN replied that to look for other than climatic influences on the width of tree-rings depends from a statistical point of view on having a long series. In the first place, therefore, they had concentrated on those factors for which they had long weather records. Other climatic parameters, e.g. sunshine, have not been recorded as long in their region. Influences like caterpillars are sometimes visible but it is difficult to get them into a model and to handle them statistically. They are only checked visually and by the records or diaries of the foresters that have information about such matters. If they have indications that caterpillars have severely attacked the trees, they exclude the material from their analysis.

Also in answer to Hughes, Echstein agreed that it was his view of climatic reconstruction for oak in lowland Western Europe to look for and then exclude the influence of caterpillars. Asked further, Eckstein saw this influence happening uniformly in the same year over the whole or a large part of the region covered by the six chronologies described by him. Eckstein said that as far as he can recognise it, such a severe attack has not taken place over all regions in the same year. It is possible to get enough material from other sites to overcome this obstacle and to collect good material for close analysis. As to caterpillars, observations had been made in Denmark and analysed by Mr. K. Christensen who might be able to answer in detail, especially as his research had included very good observations on oak trees.

Mr. Christensen said that defoliations by caterpillars and cockchafers are definitely important to the ring pattern. The cockchafer especially has a very great influence upon the tree-ring curves for Schleswig-Holstein: it has a four-year cycle in its population dynamics in Northern Germany and in Southern Denmark. This four-year cycle has been recognised as an inflection in the North German and in the Danish chronology for more than 100 years with only a few interruptions. The factor is very important not only in Northern Germany but also in Southern Germany. There we know that the cockchafer has a three year cycle and for nearly 300 years we find an inflection in the ring of every third year. We think we should warn against the climatic interpretation of such chronologies.

Eckstein concluded by saying that their climatic reconstruction (Fig. 7, page 123) is based on 15 oaks which were checked against these influences. He was aware of Mr. Christensen's findings and had discussed them with him. Attacks were well known in the history of their forests.

- - - - - - - - - - - - - - - -

With regard to Paper 4.2, Dr. Fleming asked whether there was any relationship between the density of caterpillars and altitude. Prof. VARLEY in reply said there is evidence in Switzerland that at higher elevations insect attack on larch increases. The larch bud moth is cyclic and defoliates larches in the Engadine at elevations towards the tree-line every nine years (see also Fig. 4 on page 91). At lower elevations these outbreaks are not at all well marked.

Fritts stated he had been told by entymologists in New England that in their view the biggest attack by the spruce bud-worm occurred in the forest interior and became less the higher the altitude. The use of samples from trees over large regions, and sometimes of many species, helped dendroclimatologists to distinguish the effects of insects from those of climate. He did not think it fair to say that dendroclimatologists had ignored the insect effect completely. He thanked Varley for the suggestion that they had better keep their eyes as open as possible. Varley said although one would like the effect to be simple so that just by looking at one or two trees one could get an answer, the facts were different.

Mr. Haddon-Reece asked to what extent the multiplication of insects depended on climate. Varley replied that he had no evidence, though they have looked very hard for it, that weather directly affects the caterpillar numbers. The major mortality of the caterpillars is at the time of bud-burst when the eggs are hatching. The caterpillars are very small and they haven't had their first meal. They try to get into a nearby bud but it may not be open, so many of them emigrate. They blow away on silken threads and distribute themselves widely so there may very well be a relation to wind speeds or rainfall.

Baillie asked if the results on insect attack obtained before 1958 have now been tested over a longer period of time. Varley replied this was being done (see note on page 183, added 1978 by Varley and Calton).

Prof. Liese asked the reason for the different degrees of resistence between the two trees used for Varley's slides. He also asked whether a reduced amount of latewood in one year influenced the growth of early wood in the next. Varley answered that the trees which flushed first suffered the worst damage. Again and again they have found that the trees which flushed late suffered less damage. He had carried out no analysis on the effect of the latewood of one year on the growth of the earlywood of the next, but intended to look into this point.

Hughes asked whether the amount of the previous year's growth had been included in the regression equations and received the reply from Varley that future regression analysis would include this information (see page 183). Hughes also asked whether other species, likely to be growing side by side with oak, such as ash, had been studied in parallel. Varley replied that he had thought of that way round the problem. The result of a preliminary investigation with ash, undertaken at his suggestion had already been published (Fletcher, The British Oak, 1974, 84). [Note: for later work see paper, added 1978, page 184). Within fairly easy reach in Wytham Wood there are ash, beech and poplar, but the only tree which is periodically defoliated is the oak. Usually, neither ash, beech or other species suffer visible damage, so that it might be possible by comparison to determine the component caused by caterpillar damage to oak and to separate it from the weather components. The snag of course is that though oak flushes late in the season, ash often flushes three weeks after oak.

Hughes said he thought that Varley had perhaps presented his strongest case and acted as Devil's Advocate. There was an artificial, but for the time being necessary, separation between the two kinds of work. One was the examination of the interaction between different organisms and their physical environment in relation to tree-ring growth: what they had heard would make them extremely concerned and would require a great deal of effort to unravel. The other is concerned with establishing empirical relationships. If in the first instance these relationships work empirically, as had been established for 'response functions' for a number of sites in Britain, then for the time being that is quite good enough.

Baillie said that although he 'believed' in caterpillars he thought they did not seriously influence master chronologies for the simple reason that oak chronologies agree well over large areas. So he asked Varley or Fletcher to give a clear and specific answer to the following question: if teleconnections exist as between Scotland and Dublin, or between England and Germany, are those correlations due to caterpillar influence or some other systematic factor such as climate?

Varley replied that he could not answer that question in detail. On page 139 of their book (Varley, Gradwell and Hassell, Insect population ecology, 1973, Blackwell Sci. Publ.) measurements were given for caterpillars by species for three places, two in England, one in Holland. The large distance over which defoliation of oak can occur in the same year was exemplified in 1948 at Oxford, Durham and the Forest of Dean. Defoliation of oak is largely by the winter moth and by tortrix. The distribution of the two in Europe is very widespread— from the British Isles, through Germany and Yugoslavia to the Moscow area— but they don't peak everywhere in the same year. As for the effect in England of altitude, the highest oak wood, Wistman's Wood in Devonshire, is defoliated by winter moth almost every year and the population of the moth is enormous.

Fletcher adds the following written communication in reference to Baillie's question. It does not follow that teleconnections between chronologies rule out caterpillars having an influence. The agreement between two well-matched chronologies was often in the range 65 to 70%. That of 69% (with $\underline{t}$ = 7.9) for the years 1354-1550 between his REF 1 and Bauch's Chronology II for the Netherlands provided an example. A figure of 69% implied that 100-69 = 31% of the directional year-to-year changes in width were in opposite directions in the two chronologies. This provided plenty of scope, in spite of the similarities, for differences in certain years both in the climatic factor and in that for insect damage. His answer therefore was that teleconnections between English and German oaks, as well as agreement between areas in Western Europe less far apart, were likely to be due partly to climate and partly to insects and other biological factors with the former generally predominating. The relative proportions could vary with the species of oak, with the site, the altitude, the decade and the century.

Schove thought that the ring-width chronologies of oak were not significantly affected by caterpillars. The main agreement between the German chronologies of Huber and Holstein on the one hand and English ones on the other hand was based on anti-cyclonic years with dry weather in spring and summer. There are interesting differences, however, between the reaction of German and English trees to low winter temperatures. The German upland series of Giertz and of Hollstein have narrow rings after cold winters, but in England he had found that some trees had wide rings after cold winters, many however, are 'neutral'.

— — — — — — — —

Fleming queried, re Paper 4.4, whether there was a need for visual comparisons. Surely, he said, what the eye can do, the computer can do equally as well. Mrs. LEGGETT replied that to give just one example, adjustment of doubtful areas was more easily done by eye. Other speakers, Becker, Barefoot and Baillie also stressed the importance of visual matching.

— — — — — — — —

Varley said that he was concerned about Mr. BRETT's correlation between growth and climate records being divided into calendar months as presented in the studies of oak and elm in London. For instance, this year (1977) the late spring could possibly make trouble. Brett replied that he had also used dendrometers on trees and these confirmed that last year (1976) was very abnormal. Fritts suggested that dendrometry was not the most useful way to approach the problem since it was less a question of cambial activity than of total tree activity. Brett agreed and pointed out that elms can keep their leaves a long time; in abnormal years, such as last year (1976), well into November.

5.1. APPLICATIONS OF DENDROCHRONOLOGY IN SWITZERLAND

O. U. Bräker, Birmensdorf

(with a contribution by J. Lambert and C. Orcel, Neuchâtel)

Summary

The Swiss plateau is rich in lakeside dwellings dating back to the neolithic and bronze ages, with much well-preserved timber (Plate 1) which is excellent for dendrochronological research. Measurements have been made on large quantities of material from excavation sites on the Lakes of Zurich, Neuchâtel and Bienne. Although a continuous chronology is not yet feasible, the many tree-ring measurements of oak samples will help the construction of a standard chronology for the neolithic and bronze ages.

The various lakes, bogs, and alluvial sites formed in Switzerland in the late-glacial and post-glacial period present today a very complex landscape (Fig. 1). Conditions in the neolithic period favoured intensive settlement along lakes, rivers and bogs, and wood remains are excellently conserved in such conditions. So the Swiss plateau is rich in sites (Fig. 2).

Plate 1. Neolithic Swiss Lake Dwelling, Portalban, Lake Neuchâtel. Oak Beams and Posts, formerly floating, partially dried.

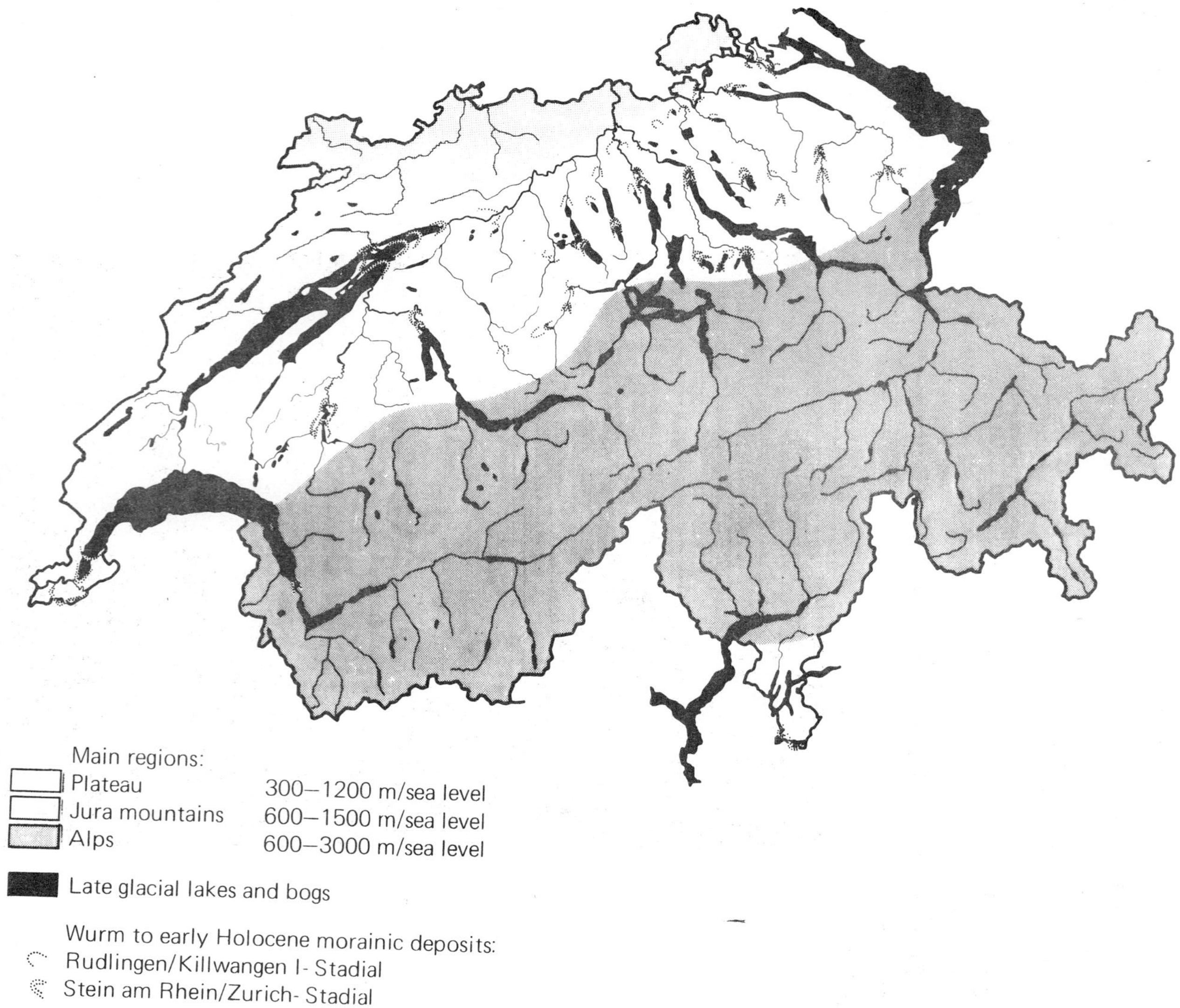

Figure 1 Main topographic features and regions of Switzerland. Source: HANTKE, 1969.

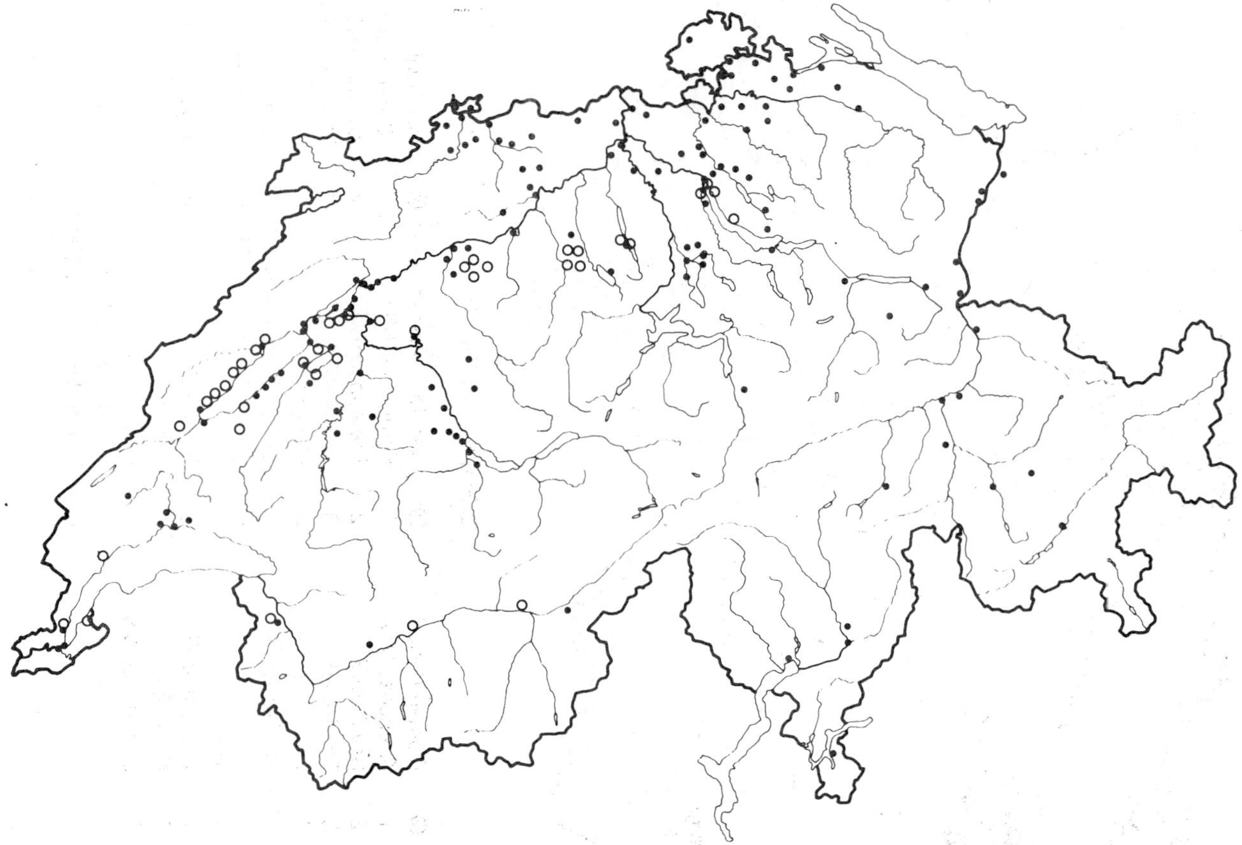

Figure 2 Known locations of prehistoric findings in Switzerland for the Cortaillod techno-complex (circles) and late Bronze Age (points). Source: SAUTER/GALLAY, 1969, and PRIMAS, 1971

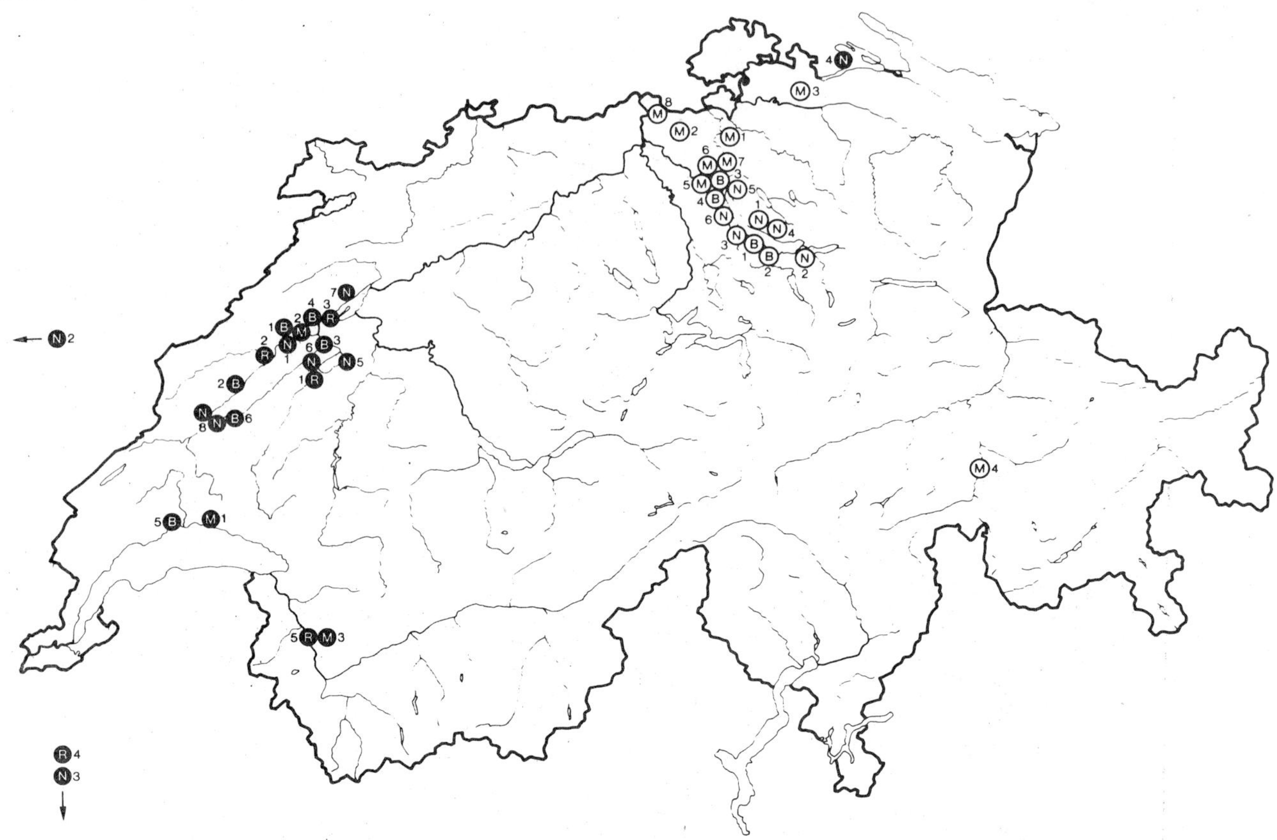

Laboratory of Neuchâtel

● N	Neolithic Period	● R	Roman Period
1	Auvernier/NE	1	Avenches/VD
2	Clairvaux/France	2	Bevaix/NE
3	Charavine/France	3	Thielle/NE
4	Horn/Germany	4	Lyon/France
5	Montilier/FR	5	St Maurice/VS
6	Portalban/FR		
7	Twann/BE	● M	Medieval
8	Yverdon/VD	1	Lausanne/VD
		2	Neuchâtel
● B	Bronze Age	3	St Maurice/VS
1	Auvernier/NE		
2	Concise/VD		
3	Cudrefin/VD		
4	Le Landeron/NE		
5	Morges/VD		

Laboratory of Zurich

○ N	Neolithic Period	○ M	Medieval
1	Feldmeilen/ZH	1	Kloten Rohr/ZH
2	Freienbach/SZ	2	Lengnau/AG
3	Horgen/ZH	3	Stammheim/ZH
4	Meilen/ZH	4	Zillis/GR
5	Zurich Pressehaus/ZH	5	Zurich Altstetten/ZH
6	Zurich Haumesser/ZH	6	Zurich Höngg/ZH
		7	Zurich Zentrum/ZH
○ B	Bronze Age	8	Zurzach/AG
1	Horgen/ZH		
2	Halbinsel Au/ZH		
3	Zurich Hafner/ZH		
4	Zurich Haumesser/ZH		

Figure 3 Sites studied by dendrochronology in Switzerland

The dating of Swiss material started in other countries. Huber and Merz (1963) at Munich synchronized oak, ash and alder posts from the late Bronze Age settlement Zug Sumpf and then oaks from the early Neolithic settlement of Thayngen with oaks from Burgaschisee and Egelsee. Giertz-Siebenlist at Munich examined posts from the Auvernian site of Saunerie and Munaut at Louvain posts from Niederwil. The sites studied since 1970 by the Swiss laboratories are concentrated in a relatively small area of the country (Fig. 3).

The presence and importance of the species of tree varies with the temporal period and the topographic region. Silver fir and oak are the species most frequently examined but in alpine regions pine and larch occur, while for recent times spruce is important as shown by the following chart:

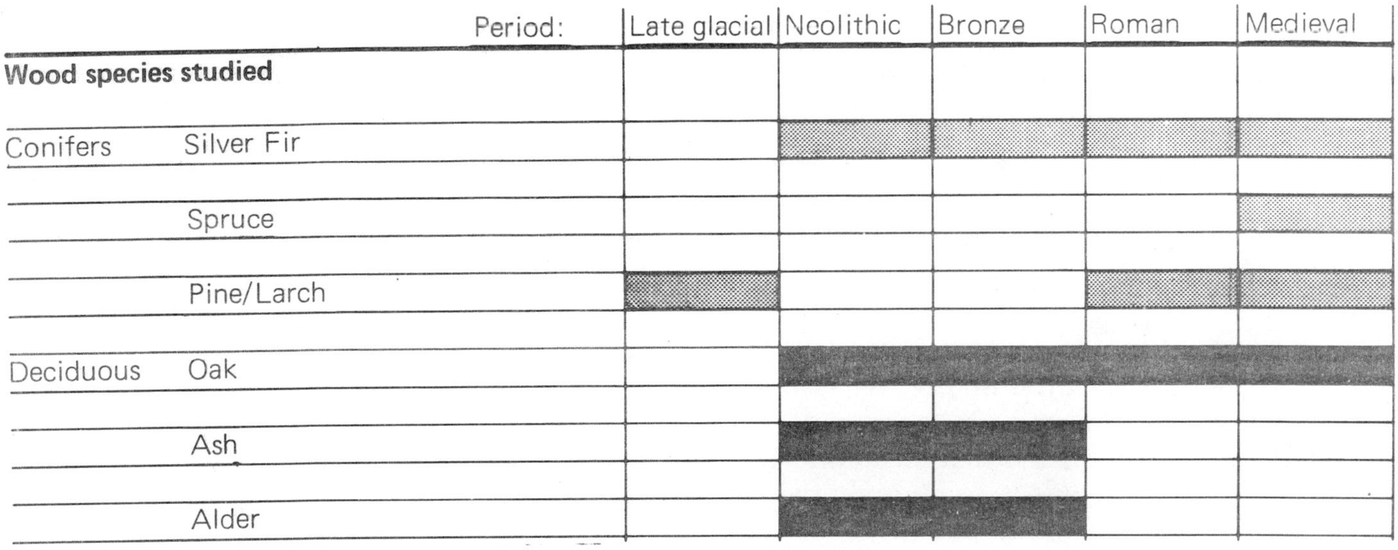

In addition to the relative and absolute dating of wood from cultural layers, archaeologists also expect results which can aid in the interpretation of the structure of houses, of site formation and of the chronological development of a site.

Important problems for the dendrochronologist are therefore:-

a) The separation of contemporary posts where there are dense concentrations of posts;
b) The identification of houses of the same construction period;
c) The identification of replacements and of reused timbers;
d) The establishment of the duration of habitation periods and phases of abandonment; and
e) The establishment of time interval between construction within a site and between different sites.

APPLICATIONS

The various applications that have been carried out in different fields by the Swiss laboratories are summarised below:-

Period	Late Glacial	Neolithic	Bronze	Roman	Medieval
Geomorphology	*				
Archaeology Dating	*	*	*	*	*
Structural Chronology		*	*		
Architectural history					*
Art history				*	*
C14 Calibration/Serial measurements	*	*			

Archaeology: The years spanned by the tree-ring chronologies for oak derived from various sites shown in Fig. 2 and examined in the laboratories at Neuchâtel or Zurich are shown as bars in heavy framing in Fig. 4. This figure also includes other chronologies as a background, indicated by bars with light framing. Those with shading are for Swiss sites investigated prior to 1971 in other countries (see above), and those without shading are for other oaks in southern central Europe as published by Becker (1971).

Prehistoric Settlements in Auvernier (Neuchâtel): by G. Lambert and C. Orcel.

"In connection with the building of a highway skirting the bay of Auvernier, excavations were made south-west of Neuchâtel. These excavations brought to light villages from the Neolithic Age (3,800-3,200 BC), the Chalcolithic Age (3,000-2,300 BC) and the late Bronze Age (1,200-800 BC). The understructure of the houses consisted of oaken posts. The dendrochronological study of such posts has enabled us to establish with great accuracy:-

(a) the period of time during which these villages were occupied (Lambert and Orcel, 1977); and
(b) the relation between these settlements and others in Switzerland (Fig. 5)."

"The tree-ring analysis has also shown

(a) a progressive occupation of the shore, rapidly extending westwardly (2800-2700 BC);
(b) the former existence of a village of the late Bronze Age which had remained undetected up to now (Auvernier-Graviers, late period);
(c) the method of selecting the trees that were suitable for building purposes (1000-900 BC); and
(d) the possibility of synchronizing very short curves with the help of a suitable combinative analysis. However this analysis only applies to certain stratigraphic groups."

Applications to Architectural History and to Geomorphology: The evolution of building types and stylistic details in farmhouses have been dated by dendrochronology. The identification of the wood species alone may indicate the general construction period. Thus during early medieval times in the plateau, oak was commonly used for vertical elements. Later, the oak was replaced by conifers, either fir or spruce, first in the upper parts of the house, then in the ground floor structure. Oak was retained only for the supporting ground-floor beams.

In the higher parts of the lower slopes of the Alps, a similar evolution can be observed: fir was replaced by spruce. Dendrochronology has been able to separate original work from later additions to farmhouses, town houses, churches and city halls.

In the field of geomorphology, Kaiser (1970) created a stratigraphic chronology with 30 pine stems from a clay site near Zurich. Two main horizons from the Bolling and the Allerød interstadials were found. Various C14 measurements indicate the period between 11,200 and 9,260 BP. This wood was also used in 10-year serial radiocarbon measurements for calibration work by Prof. Oeschger at the University of Berne; a similar study was made by Suess and Strahm (1970) with Auvernier material.

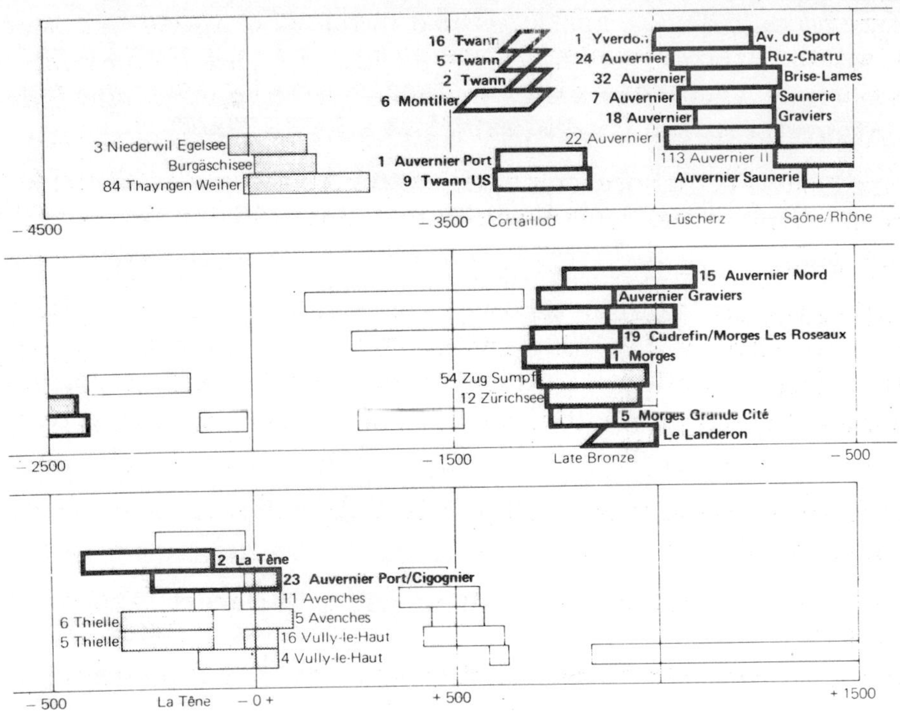

Figure 4 Comparison of the relative oak chronologies of southern central Europe compiled in 1971 by BECKER (ground plan: light print, Swiss chronologies: shaded) with the relative chronologies of Switzerland compiled in 1977 by LAMBERT/ORCEL (superimposed figures: bold print).

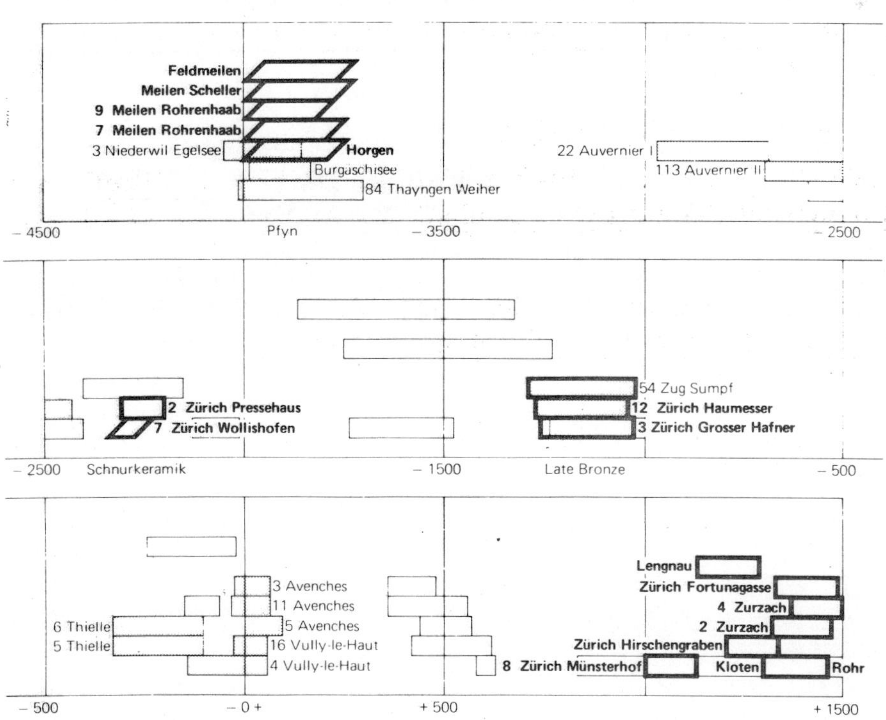

Figure 5 Comparison of ground plan Figure 4 (light print) with the relative oak chronologies of Switzerland compiled in 1977 by RUOFF (bold print).

Future Developments

Correlations within Swiss chronologies are at least one degree of significance higher than their correlations with the standard one for southwestern Germany; this demonstrates the need to establish a Swiss standard.

The Zurich and Neuchâtel laboratories will measure the material found in their own part of the country and construct relative chronologies which will be periodically exchanged.

The most challenging problem for the two laboratories is the construction of a long sequence covering the Neolithic and Bronze Ages. Several relative chronologies are already compiled. Cultural layers of different sites are synchronized, but a complete chronology over the whole period has yet to be achieved.

Subjects that require investigation with the data collected in these dendrochronological studies are:

- biological research; looking for growth trends on different sites and in different periods;
- forestry history; reconstructing forest covers and man's intervention, and
- climatic developments in the past, similar to the densitometric studies of Schweingruber (this volume, page 115).

References

Becker, B., 1971. Zwischenbericht über die dendrochronologische Bearbeitung subfossiler Eichen aus Flussterrassenschottern des südlichen Mitteleuropas. Jahrbuch der Akademie der Wissenschaften und der Literature, 140-145.

Huber, B. and Merz, W., 1963. Germania, 41, 1-9.

Kaiser, K. F., 1970-72. Ein eiszeitlicher Wald im Dättnau. Mitteilungen der Naturwissenschaftlichen Gesellschaft Winterthur, 34, 25-41.

Lambert, J. and Orcel, C., 1977. L'état de la dendrochronologie en Europe occidentale et les rapports entre dendrochronologie et archéologie en Suisse, Archives Suisses d'Anthropologie Générale, Geneva, 41, 2, 73-97.

Suess, H. and Strahm, C., 1970. The Neolithic of Auvernier, Switzerland. Antiquity, 44, 91-99.

5.2. TREE RING STUDIES IN THE SOMERSET LEVELS

R. A. Morgan, J. M. Coles and B. J. Orme

Summary

The well-preserved wood of a variety of species, age-ranges and date found in the Somerset Levels trackways has enabled us to examine unusual aspects of tree-ring analysis for archaeology. In addition to the construction of floating oak chronologies which provide relative dates for various plank-built trackways, younger brushwood material of ash, hazel, alder and birch has been examined.

Young stems used for various functions in the trackways can be related in age by ring-counting, to give indications of woodland management as early as the fourth millenium b.c. Short patterns allow some relative dating within a trackway. Worked wood from different millennia indicate changes in the type and quality of timber available locally.

The peat deposits of the Somerset Levels have yielded many prehistoric finds in the past decade. The removal of peat by commercial interests has created opportunities for archaeological discovery and excavation, and the Somerset Levels Project was formed to undertake this work. Most of the prehistoric finds are of wood, and consist of roads and tracks built out into and across the marshlands (Fig. 1) which have existed since c. 4000 b.c.

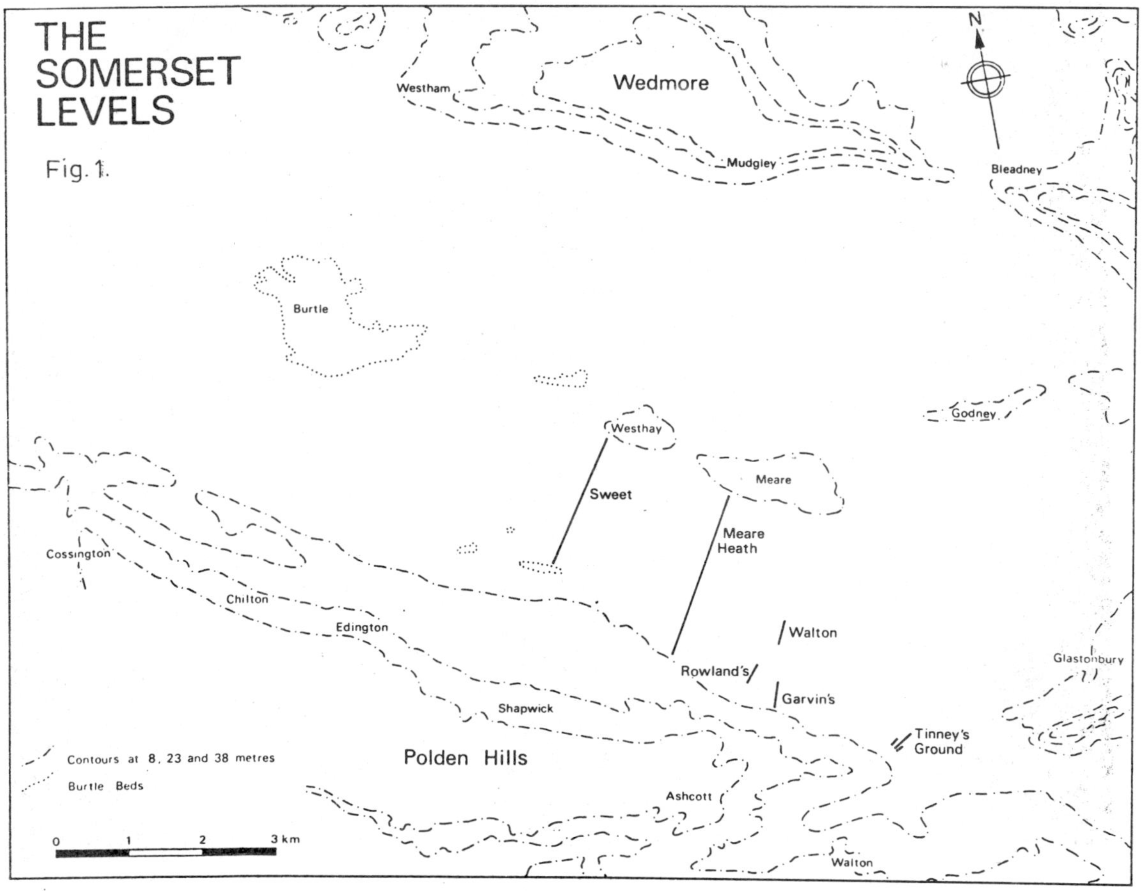

Fig.1. Map showing the physical features and some of the trackways.

Plate I. The Sweet trackway (Factory site)

Other prehistoric structures, such as terminal platforms, and stray finds of equipment, are also discovered, but the roads and tracks are particularly suited to tree ring studies as they provide ample quantities of well-preserved wood, firmly stratified at different levels in the peat and dated by multiple radiocarbon analyses.

In construction, the trackways fall basically into three groups, which have no chronological significance. Firstly, there are those built of substantial timbers, usually oak planks, held in place by pegs, and looking not unlike a catwalk or a railway line, Pl. 1. The second group, only recently discovered, consists of prefabricated woven hurdles laid down in a line and held in place by pegs. Finally, there is a large group of tracks made by throwing down rough bundles of brushwood, again held in place by pegs. The trackways may be continuous from high ground to high ground over a long distance, or they may consist of a short stretch consolidating a particularly wet and boggy area.

Vast quantities of wood were required in the trackway construction, and with such abundant material for analysis a number of new approaches to our understanding of prehistoric man's activities in the Levels have been developed over the last few years. The first tree-ring samples to be examined were some from the neolithic Sweet trackway in 1973. The encouraging results from these led to further thoughts on the application of tree-ring analysis to the interpretation of trackway construction, and studies of wood of varying ages and species have now been undertaken. This work goes hand in hand with the archaeological work and with specialist studies such as the identification of wood species and the examination of growth characteristics and other signs of woodland management.

Analysis is facilitated not only by the very large quantities of wood available, but also by the opportunity to take a second look at individual trackways. The cooperation between the peat-cutting firms and the Somerset Levels Project enables trackways to be examined as each area is cut, possibly over a period of several years, and the questions raised by the initial excavations can be answered perhaps in subsequent ones. The approach to sampling too can be extended and revised according to previous results. The great advantage of such a long-term and intensive project is well illustrated by the examination of the Sweet trackway which has now been exposed at four main sites spread over several hundred metres, and sampling on each occasion has been revised in the light of work done on and problems raised by the previous excavation.

Dendrochronology finds its conventional application in the establishment of floating tree-ring chronologies from the oak timbers used in various trackways. The periods and approximate dates in radiocarbon years of those from which samples of all species have been examined are:

Sweet	Early Neolithic	3200 b.c
Rowland's	Late Neolithic	2300 b.c
Walton Heath	Late Neolithic	2300 b.c
Garvin's	Late Neolithic	2300 b.c
Meare Heath	Bronze Age	1100 ? b.c
Tinney's Ground	Bronze Age	1100 b.c

Three floating chronologies exist so far, two roughly dated by radiocarbon analysis of wood of known relative age. The best known and most informative trackway, still under examination, is the Sweet track, dated by radiocarbon to about 3200 b.c. and consisting of a walkway (Fig. 2) one plank wide supported above the contemporary marsh surface on a sub-structure of rails and crossed pegs (Coles, Hibbert and Orme, 1973; Coles and Orme, 1976a). About 170 oak planks and boards have now been examined from the Sweet track and proved to be extremely suitable for tree-ring analysis, having very sensitive ring-width sequences averaging one to two millimetres growth per year (Fig. 3). Despite age, degree of waterlogging and penetration by roots, the wood structure is still perfectly preserved, and easily exposed by deep-freezing and surfacing with a knife or plane.

As one might expect from a homogeneous structure, many of the oak timbers came from the same very large and aged trees, estimated to be over three hundred years in age and about one metre in diameter; ringwidth sequences are often identical (with correlation values in excess of W = 90 per cent, or Student's t = 13). While the final mean curve of 314 years represents

Fig. 2 Schematic Reconstruction of the Sweet Track (Drawn by S. Goddard) A, underlying peat; B, rail laid on contemporary peat surface; C, rail pegs driven into peat; D, peat-packing around and on top of rail; E, supporting plank peg; F, plank laid on peat packing (D) and on plank peg (E) and on upper part of rail pegs (C); G, plank peg through hole or notch in plank.

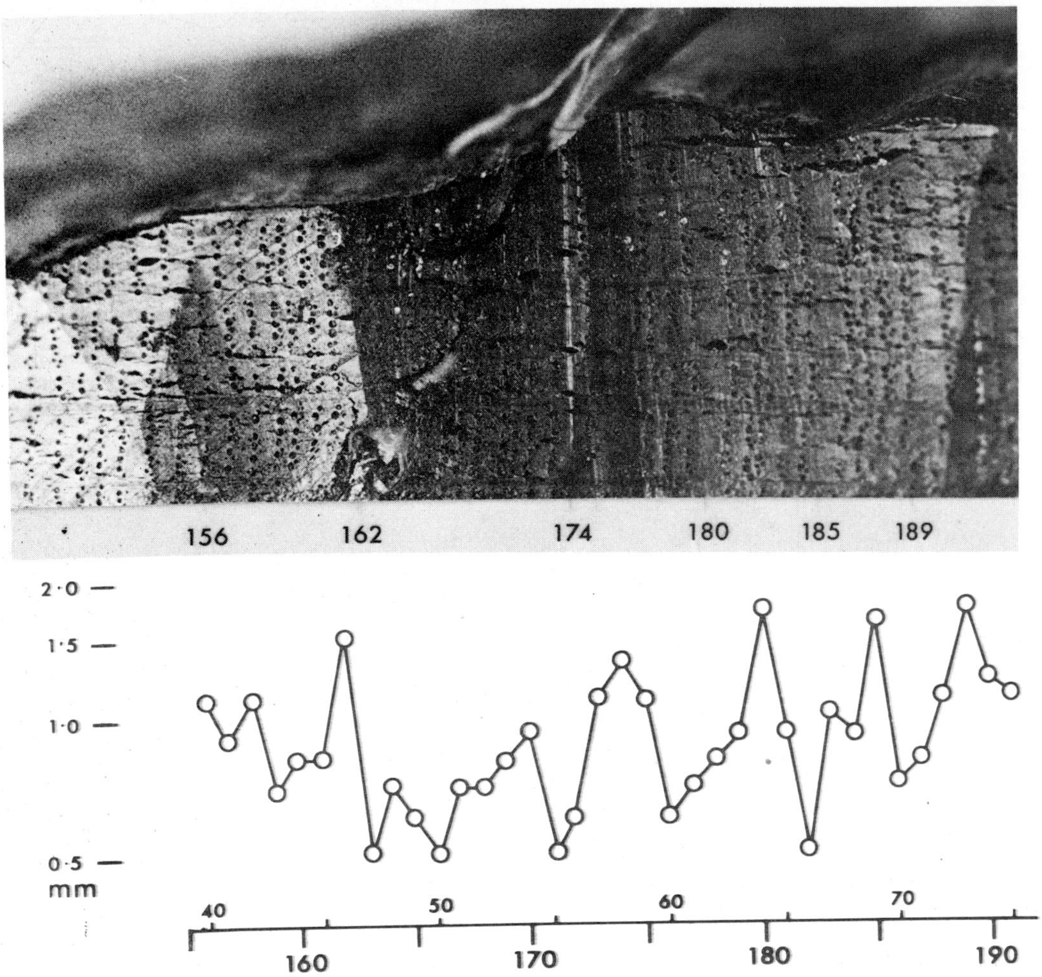

Fig. 3 Photo of the annual rings of an oak board from the Sweet track with, below, the plot of the widths for years 156 to 191 of the 314 year mean curve.

129 oak pieces (Fig. 4) it probably represents only a few trees, and this is confirmed by the distribution of each individual curve within the mean curve, the position of the finishing years being completely random (Morgan, 1976).

Sapwood has never been located on any of the oak timbers until this year, and it was assumed to have decayed; however, excavations in April 77 produced three chips of oak consisting mostly of sapwood, and it is now believed that the qualities of sapwood must have been appreciated and that it was trimmed off the boards deliberately.

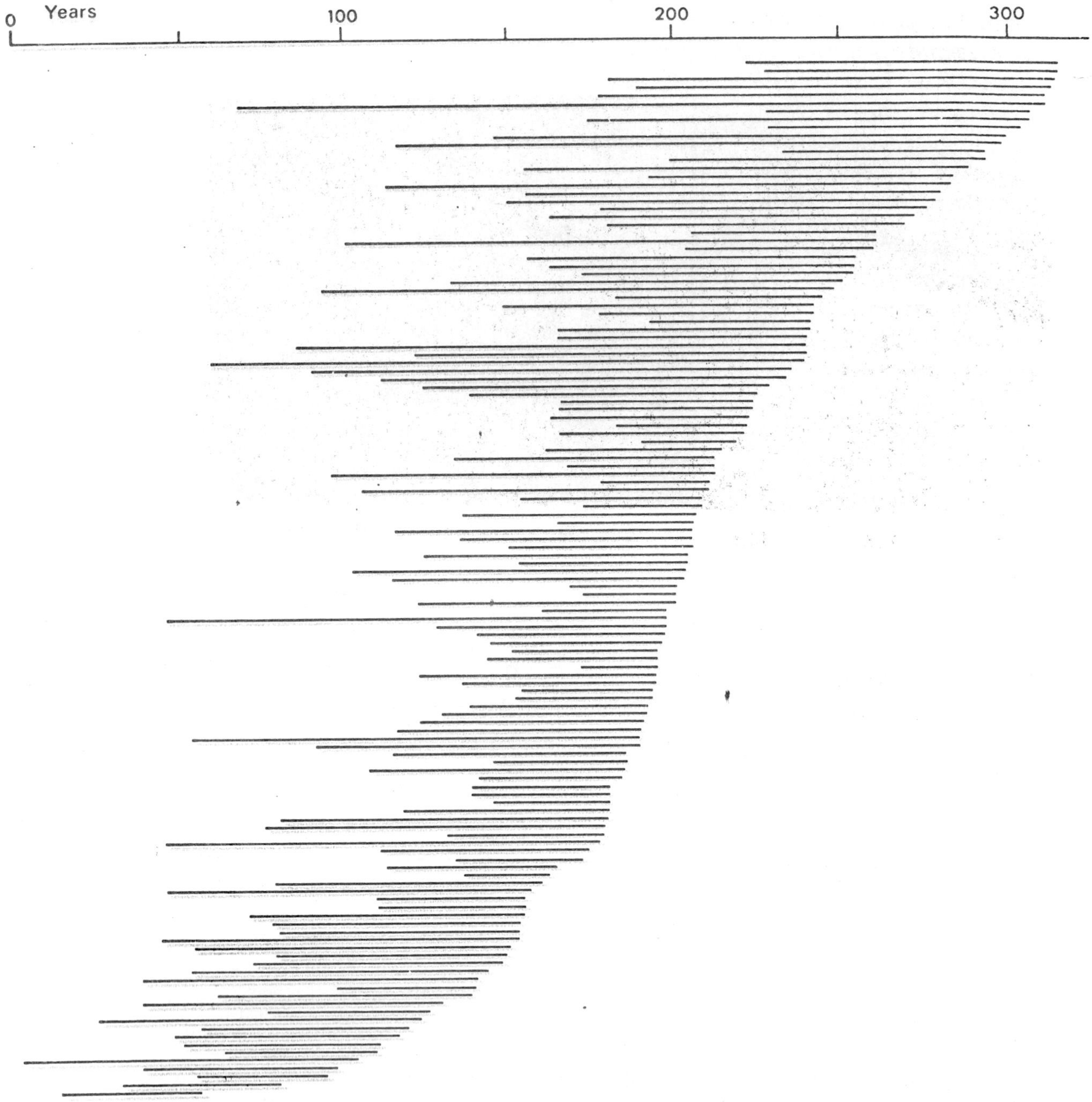

Fig. 4 Years (arbitrary scale) spanned by each of the 129 ring-width sequences included in the 314 year mean curve.

The 314 year floating tree-ring chronology has provided a series of radiocarbon samples of known relative age, which should both closely date the construction of the trackway and give some indication of radiocarbon fluctuations over this period; the dates are currently being examined statistically. They show that the curve is roughly contemporary with the Swiss sites established by Huber for Thayngen and Burgäschisee-Süd (Huber, 1967), and Bräker, this volume, p. 207), and by Munaut for Niederwil.

As important as the oak timbers which contribute to the tree-ring chronology, are those which do not crossdate with others, in this case about thirty of the 170 examined. It is possible that some of these were reused, or that they were not an integral part of the trackway, and here the archaeological context is obviously important; some of the timbers from the Sweet site were probably flood debris. Details like this are important for interpreting the construction methods of the trackway, provided a sufficient length can be excavated.

The Meare Heath track, known since the 1930s (Godwin, 1960; Coles and Orme, 1976b), dates to the late second millennium b.c., and consists of longitudinal planks held in place by pegs. Here, in contrast to the Sweet track, the oak planks are narrow and tangentially split from much younger trees so that the final mean curve extends over only 120 years. Again a number of timbers had been cut from the same tree, and planks from a single tree have been found at sites far apart along the track, suggesting that the treetrunks had been split up on the dry ground and the planks randomly mixed before being carried down into the marsh to build the trackway (Morgan in Coles and Orme, 1976b).

The trackway in Tinney's Ground consists of alder and birch brushwood bundles laid on oak planks which may have had a prior function as they have a number of holes and notches which are not constructional (Coles et. al, 1975). The structure is dated by radiocarbon to about 1100 b.c. As with the previous sites, many of the planks come from a single tree, and if reused must have come from the same dismantled building or the like. Ringwidth curves exceed 250 years but the growth pattern is erratic and disturbed. These timbers are still being examined, and whilst no correspondence can be found with the possibly contemporary Meare Heath tree-ring sequence, it is hoped eventually to link the Bronze Age structures and to establish a framework for relative dating. Such a project may be confused by the absence of sapwood on the timbers and by the apparent reuse of planks, but it may be possible as more evidence accumulates.

One useful aspect of tree-ring studies is the opportunity to compare woodworking techniques from the different prehistoric periods, and to glimpse the effect man was having on the local woodland through extensive felling. The contrast is particularly strong between the Sweet and Meare Heath trackway, the former having wide thin planks of fine quality slow grown oak, radially split using only stone tools and wedges, and the latter, some 2000 years after, having mostly thick and narrow tangentially cut timbers liable to split along their length (i.e., along the rays), and only cut with difficulty from a smaller trunk than those of the Neolithic. The Bronze Age trees in fact show much younger and wider-ringed trunks being used, suggesting that there had been a history of human exploitation of the local woodland, and that forest growth as well as composition had altered since the first Neolithic felling of primary Atlantic forest, which yielded the slow-grown and sensitive trees used in the Sweet track.

Apart from the oak timbers, there is a vast quantity of complete trunks and stems of various species of tree which were used as pegs, posts and brushwood, which could hardly fail to provide some useful information from tree-ring analysis. Many species are represented, mainly ash, alder, hazel and birch; elm, holly, lime, willow and apple also occur in small quantities, but they have not been examined to any extent. These pieces are almost invariably complete stems, with the bark still preserved, and they are usually only worked at one end.

150 thick stems of ash, alder and hazel used as posts, pegs and rails in the Sweet track were examined first, to see how informative such wood might be (Morgan, 1976). The posts were very finely facetted at one end for driving into the marsh, and they were usually about 10 cm in diameter and some 25 years old. This uniformity of age suggests that an even-aged stand of growing timber had been cut, to give large quantities of poles of similar size, which might have been deliberately grown by means of coppicing. There is other evidence from the Sweet track to support this suggestion from the tree-ring studies, and we can thus claim with some confidence to have demonstrated the practice of forest management in the fourth millennium b.c.

Stem-ageing is useful to determine the degree of selection of particular timbers to serve certain functions in a track. A typical ash peg from the Sweet track produced a sensitive ringwidth pattern of about twenty five years, which proved to be extremely similar to 80 others out of almost 100 examined; most of these were felled in the same year, whilst about one third were cut 11 years previously and were evidently reused.

A recently excavated hurdle trackway of late Neolithic date, Rowland's track (Coles and Orme, 1977a; Morgan, 1977), has been carefully sampled for stem-ageing, Pl. II. The central poles and the pegs holding the hurdles in place were made from 14-15 year old hazel, whilst the sails of the hurdles were about 9 years old, and the rods only about 5; the histogram shows the relationship between stem age and function, the young rods being on the left, Fig. 5. A conspicuous peak occurs at the age of 15 years among the hazel pegs, which strongly suggests the selection of suitable stems from a coppiced woodland. The uniformity of stem age is only one feature diagnostic of coppicing; the long straight hazel rods with curved heels could only be produced by this method (Rackham, 1977). There is likely to be information on coppice cycles and growth rates in the annual rings, but the effect of coppicing on growth is not yet clearly understood, and awaits tree-ring examination of modern hazel coppice of similar age.

Apart from simple ageing, the ringwidth patterns of older stems are in many cases sensitive enough to merit comparison within individual trackways; it is reasonable to assume that most stems in an apparently one-period track are contemporary, and this can be checked in the growth pattern. Often it is evident that many stems came from the same tree or stool.

Plate II **Part of a hurdle trackway showing the principal components of Walton Heath and Rowland's tracks; the size and age of stems used for each are given in the table.**

		Rowland's track			Walton Heath track		
number	function	species	age	diam mm	species	age	diam mm
1	central pole or rib	hazel	10-20	30-50	hazel	av. 15	40
2	rod	hazel	2-6	10-20	hazel	3-5	8-20
3	sail	hazel	5-15	10-30	hazel	10	12-55
4	peg	hazel	10-20	20-40	hazel	av. 13	34
		birch	20-25	30-50			
5	stray	birch	20	35-55	birch	22-40 +	55-90

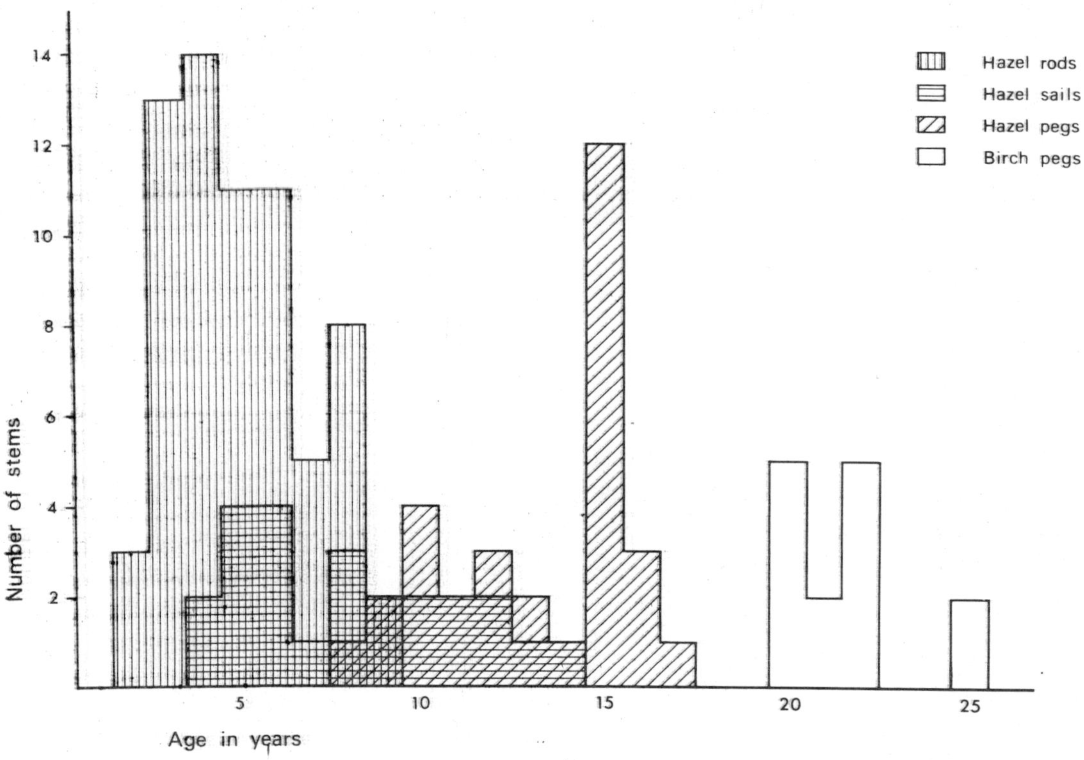

Fig. 5. The age distribution of hazel and birch stems used in the Rowland's track.

Many of the older hazel stems from Rowland's trackway also showed an identical pattern of growth with each being cut in the same year; all 37 stems examined had characteristic narrow rings in arbitrary years 18 and 26. Probably these stems came from the same coppice stools; it is interesting to know that stems used in weaving the hurdles are contemporary to those used for pegs, suggesting that the hurdles had not served a prior purpose but were intended from the start for use in the trackway.

Not far away on Walton Heath another hurdle trackway was thought to be broadly contemporary with Rowland's on the basis of structure, alignment, peat stratigraphy, level and radiocarbon dates (Coles and Orme, 1977b). Among a great variety of growth types in the stems, confirming some repairs to the Walton track, were a small group of hazel stems with identical growth patterns to those in the Rowland's track (Fig. 6), which strongly suggests the contemporaneity of the two structures, given the supporting evidence from other fields of investigation

While we are dealing here with very short ringwidth series, it can be demonstrated particularly on ash and hazel that the same growth pattern is reproduced in most of the stems from a trackway, though of course there is no question of using this material by itself for cross dating between tracks. It is only when track construction, wood selections, alignments, peat stratigraphies and radiocarbon dates, all suggest a close relationship that tree-ring counts can help to support archaeological reasoning of exact contemporaneity.

The felling season of track timbers is often determinable from the stage of growth of the outermost ring of complete stems, particularly of the ring-porous woods such as oak, ash and elm. Stems cut in winter have the final ring completely formed, whilst those cut in summer terminate with the fast-grown and large-celled wood of the major growth period in the late spring. Most of the Sweet ash pegs can be shown to have been felled in winter.

The felling season is less easily determined in the diffuse-porous woods like hazel, due to the homogeneous nature of the annual ring, but the outer ring of the Rowland's hazel stems was invariably much narrower than the inner ones, and suggested the possibility of summer cutting.

Finally, the brushwood bundle trackways normally contain very young stems, but their identification and ageing gives us a useful guide to the selection of material practised by prehistoric man, whether the wood was being coppiced or merely randomly collected from a natural woodland. Usually the brushwood consists of one or two species, birch, hazel or alder, of consistent size. The Tinney's track above the oak planks was made entirely of alder stems about 7 or 8 years old with very regular growth rings. The Garvin's track (Coles and Orme, 1977c), of Neolithic date, contained some coppiced hazel rods and many rough birch stems and branches of varied age which were probably randomly gathered.

Continuing studies of tree-rings from the many wooden structures in the Somerset Levels should allow more precise correlations to be made, within and between the sites, and more information about woodland growth to be accumulated. These studies, allied to the examination of pollen, peat and

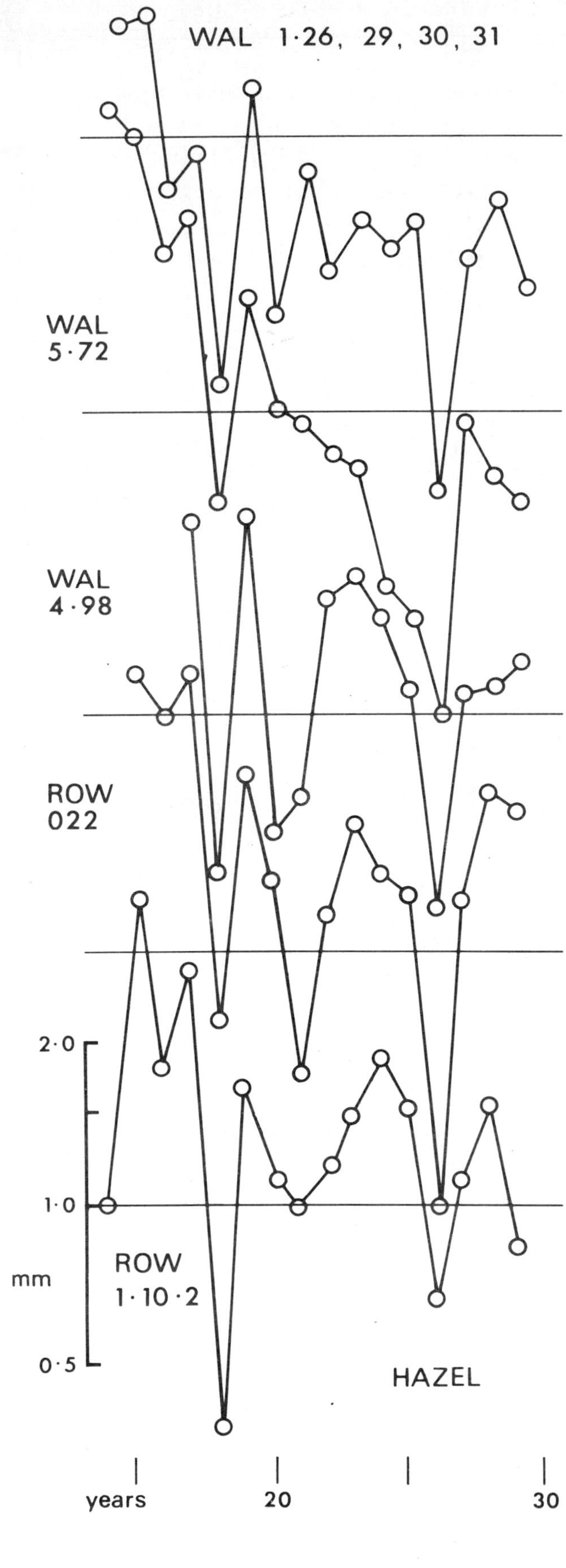

Fig. 6 Ring-width curves of six hazel stems from Walton Heath (the first is an average of four) and two hazel stems from Rowland's; the ring in arbitrary year 29 represents the year of cutting.

beetles, will permit archaeological work to move in new directions in assessing the impact of prehistoric man upon the Levels.

References:

Coles, J. M., Hibbert, F. A. & Orme, B. J. 1973, Prehistoric roads and tracks in Somerset: 3. The Sweet track. Proc. Prehistoric Soc. 39, 256-293.

Coles, J. M., Orme, B. J., Hibbert, F. A. & Jones, R. A. 1975, Tinney's Ground, 1974. Somerset Levels Papers 1, 43-53.

Coles, J. M. & Orme, B. J. 1976a, The Sweet track, Railway site. Somerset Levels Papers 2, 34-65.

Coles, J. M. & Orme, B. J. 1976b, The Meare Heath track: excavation of a Bronze Age structure in the Somerset Levels. Proc. Prehistoric Soc. 42, 293-318.

Coles, J. M. & Orme, B. J. 1977a, Rowland's hurdle trackway. Somerset Levels Papers 3, 39-51.

Coles, J. M. & Orme, B. J. 1977b, Neolithic hurdles from Walton Heath, Somerset. Somerset Levels Papers 3, 6-29.

Coles, J. M. & Orme, B. J. 1977c, Garvin's Tracks. Somerset Levels Papers 3, 73-81.

Godwin, H. 1960, Prehistoric Wooden trackways of the Somerset Levels. Proc. Prehistoric Soc. 26, 1-36.

Huber, B. 1967, Seeburg, Burgäschisee-Süd — Dendrochronologie. Acta Bernensia 2, 145-56.

Morgan, R. A. 1976, Tree-ring studies in the Somerset Levels: the Sweet track. Somerset Levels Papers 2, 66-77.

Morgan, R. A. 1977, Tree-ring studies in the Somerset Levels: the hurdle tracks on Ashcott Heath (Rowland's) and Walton Heath. Somerset Levels Papers 3, 61-65.

Rackham, O. 1977, Neolithic woodland management in the Somerset Levels: Garvin's, Walton Heath and Rowland's tracks. Somerset Levels Papers 3, 65-71.

5.3. TREE-RINGS AND THE ARCHAEOLOGY OF THE THAMES WATERFRONT IN THE CITY OF LONDON

Ruth Morgan and John Schofield,
(with a contribution by D. Brett and G. Milne)

Summary

Timber quays that formed part of the Roman and Medieval waterfronts to the River Thames at London have been excavated at four sites. They extend either side of London Bridge over a total distance of about 1.2 km.

For the Roman timbers from three sites, a mean curve, as yet floating, of 282 rings has been formed. 20 medieval timbers from the Seal House site have been matched and dated by the reference curve from southern Germany and have helped to form a reference chronology for southern England. They also enable three phases of waterfront development around 1200 AD to be identified. At the Trig Lane site a large number of 14th century timbers from fast-grown trees have also been recovered and have been examined.

Introduction

Since 1973 the Guildhall Museum and its successor, the Museum of London, has excavated four sites along the north bank of the River Thames in the City of London. The relative position of the sites are shown by numbers in Fig. 1.

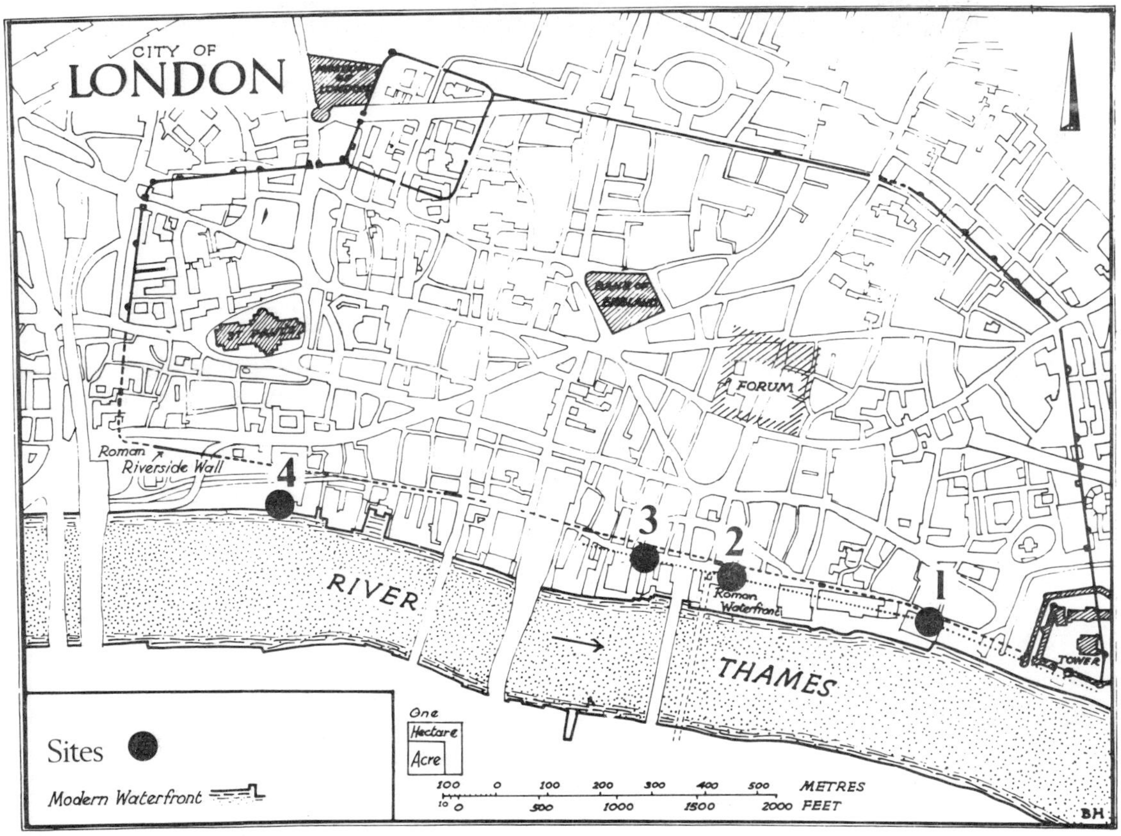

Fig. 1. City of London with position of sites excavated. 1, Custom House; 2, New Forest Wharf; 3, Seal House; 4, Trig Lane.

Plate I New Fresh Wharf: the Roman waterfront, Area III, looking E.

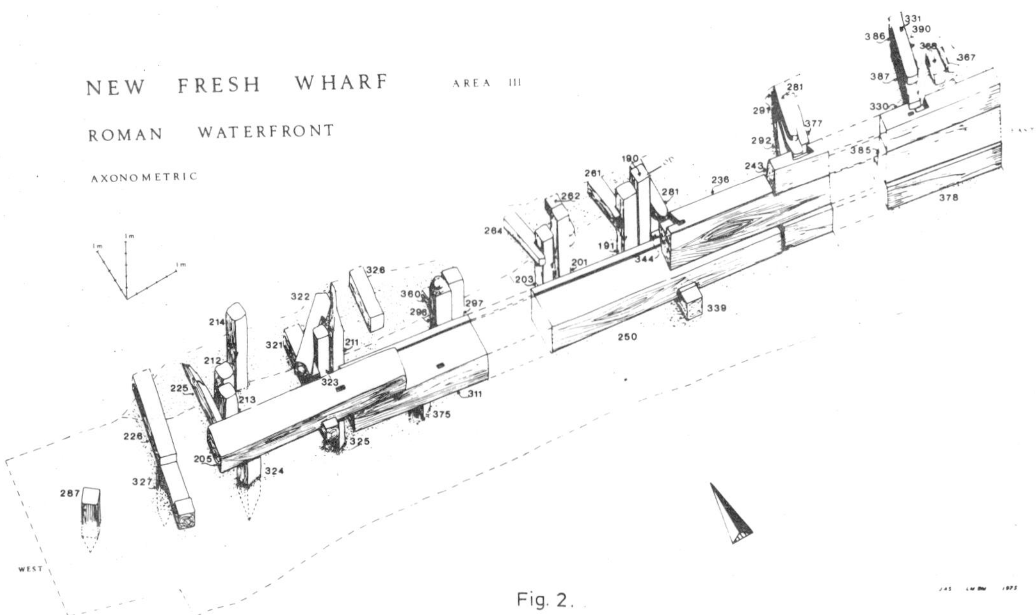

Fig. 2.

The results of dendrochronological studies of the many oak timbers from the waterlogged and well preserved wooden structures which formed the Roman quay are given in Section I, and the corresponding results for the Medieval quays are given in Section II.

I. ROMAN WATERFRONT

Custom House Site (Fig. 1, site 1)

In 1973 a Roman waterfront was located on this site, which lies just upstream of the Tower of London. It was traced for over 26 m (Tatton-Brown, 1974) and consisted of a series of timber boxes jointed at the front to a quay wall of four or five tiers of horizontal oak beams. These became smaller in section as they rose and were jointed vertically with false tenons. Inside the boxes, but not connected directly to them, were scattered vertical piles perhaps for buildings. Pottery in the layers beneath the structure indicated a late 2nd century date.

Eight samples of the oak timber from the waterfront formed the basis of dendrochronological analysis by Fletcher (1974). A curve for 218 years was built up for the five largest beams. Tentative cross-dating with the curve for oak timbers from a well at Wederath (Hollstein, 1972), and including 20 ± 6 years for sapwood (all absent), gave the last quarter of the 2nd century as the likely period of erection of the waterfront. The feasibility of a match with Hollstein's curve was based on the trees used for the four largest beams having maintained an average to slow growth (ring widths 1.2 to 2 mm for more than 100 years). Pottery evidence indicated the second half of the second century A.D.

New Fresh Wharf and Seal House (Fig. 1, sites 2 and 3)

New Fresh Wharf lies west of Billingsgate Market and about 300 m upstream of the Custom House. In 1974 and 1975 two trenches (Areas II and III) located 23 m of the Roman waterfront roughly on the same alignment, but of significantly different construction (Schofield, Miller *et al.*, forthcoming; interim, Schofield and Miller, 1976, and Hobley and Schofield, 1977). A quaywall of three (originally four or five) beams was found resting on timbers at right angles which gripped the sillbeams via trenches cut in their upper surfaces (Fig. 2 and plate I). These cradling timbers were in turn supported on piles. The sillbeams, which were up to 7.95 m long, butted at their ends, allowing for slight deviations in alignment. At the west end of Area III, where extensive robbing had taken place in the Saxon period, a beam with rounded upper edges overlapped the normal sillbeam and carried the waterfront to the west at a higher level. The quaywall was braced internally with horizontal timbers jointed to all levels of the wall beams and running back *c.* 3 m to large squared piles.

Five pairs of piles had been driven immediately behind the sillbeam in Area III, forming an east-west structure of four bays of *c.* 2 m each, either a building or a crane on the edge of the quay. As at Custom House, no trace of flooring was found, so the level of the top of the quay could not be suggested. Pottery under the structure indicated a date of *c.* 150-180 A.D.

At Seal House about 200 m upstream from New Fresh Wharf and on the upstream side of the line of the Roman bridge across the Thames, the waterfront was located in 1974 and 1975 (Schofield 1975; Hobley and Schofield, 1977). 21 m of the quaywall, of which three beams survived, were excavated during building development. The structure seemed to be a lighter version of the New Fresh Wharf structure, without groups of piles for buildings. A fragmentary section of the top was recorded, and its level at +0.4 m O.D. agrees with inferences drawn from the structures at New Fresh Wharf and Custom House for a general level of the quayside. No accurate archaeological date was obtained due to the small amount of structure uncovered and the problems of working in salvage conditions.

Tree-Ring Analysis

This aimed to date the timber structures of the New Fresh Wharf and Seal House sites by the matching of their tree-ring sequences, and so to evaluate the time relationship between the two stretches of waterfront. It was hoped to produce a reference curve for the area.

From the New Fresh Wharf site 13 major timbers of the Roman waterfront were sampled by power saw, and grouped on the basis of average- ring-width and number of growth rings. The first group (timbers 321, 378, 236, 311, 205 and 322, see Fig. 2) included sillbeams and first-row beams above, most

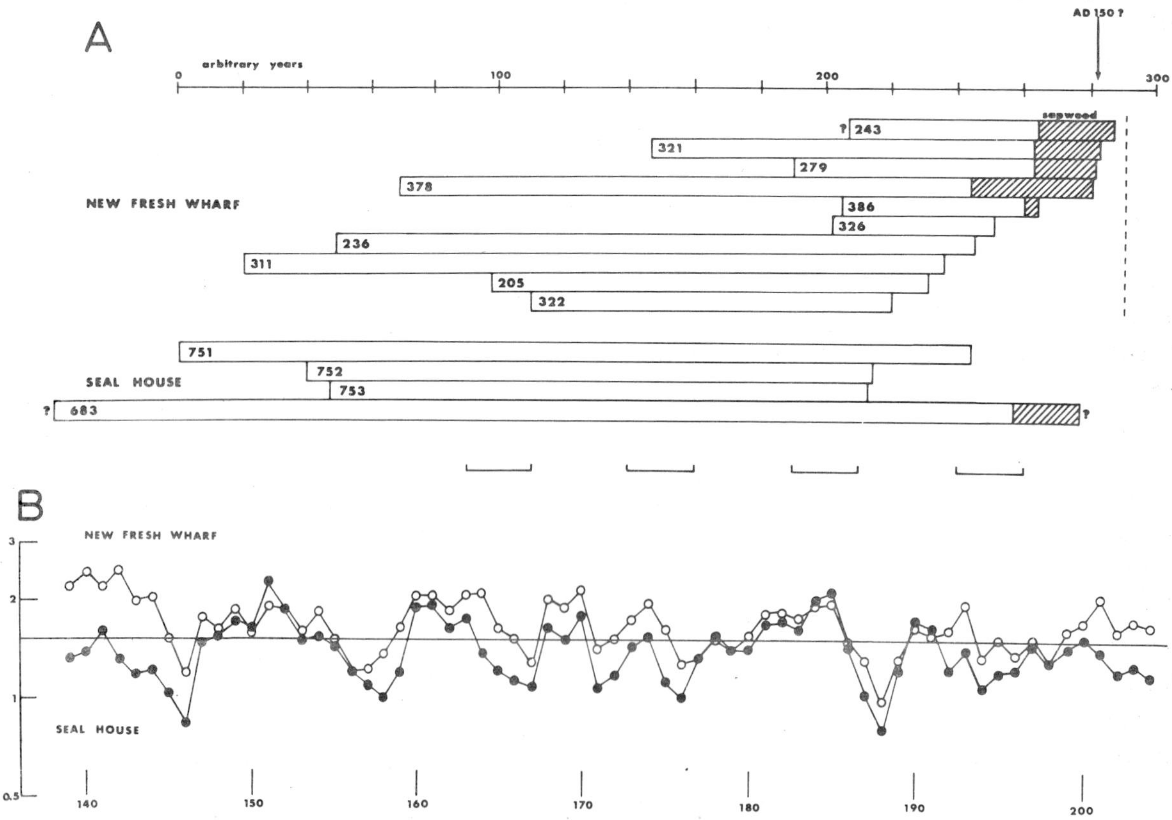

Fig. 3. A. Arbitrary years spanned by samples numbered as in Fig. 2.
B. Matching of the Roman mean curves for the two sites.

of which were oak trunks split in half and trimmed to a rectangular cross-section. They came from mature trees with 100 to 200 growth rings of 1-2 mm average width. The ring-width patterns were synchronised in the relative positions shown in Fig. 3A. The sillbeams 311 and 378 and the first-row beam 236 may have come from the same tree; if so, the tree must have reached a diameter of over 800 mm and an age of more than 260 years with a clear bole some 13 m long. The ring-widths and sensitivity suggest that it must have grown under forest conditions on well-drained land. Timbers 378 and 321 have retained most of their sapwood, so that the year of their respective final rings should fall very close to the year of felling. The corresponding ring-widths for each year in the six timbers were averaged to give a mean curve of 262 years, part of which is shown in Fig. 3B.

The second group (seven timbers) consisted of complete trunks hewn to square cross-section from younger and faster-grown oaks. These acted mainly as piles and braces, for which strength rather than size was required. The timber came from trees up to 100 years in age and about 400 mm in diameter. The ring-widths were in the range 2-5 mm. Timbers 279, 326 and 386 of the seven sequences were matched to provide a mean curve of 92 years. This matches the 262 year mean curve between arbitrary years 190 and 281. A fourth timber (243) was tentatively positioned in relation to the same mean curve. Three of these timbers had sapwood which indicated felling in about the same year (the vertical dotted line in Fig. 3A).

The remaining three tree-ring sequences matched neither with these two groups nor with each other; Fig. 2 shows that all three (timbers 190, 212 and 213) are vertical piles immediately behind the waterfront, perhaps supporting a building or crane. The tree-ring evidence suggests that they are not contemporary with the main waterfront timbers, and are therefore not integral to the primary structure. A radiocarbon age determined on timber 213 gave A.D. 320 ± 70 (HAR 1421; uncalibrated). Nothing during excavation indicated secondary insertions.

Seal House: Five Roman timbers were analysed. Four provided sequences of up to 294 years, the oldest having some sapwood remaining. The three without sapwood (751 and 752, sillbeams; 753, a second row beam) matched well, giving an average curve of 244 years; subsequent comparison with the 262 year New Fresh Wharf curve showed a very good match (t=9.96) and indicated that both groups had come from the same forest (Fig. 3B). The nine sequences were averaged to give a 282 year mean curve.

The key as to whether the two sections of the waterfront were built in the same year lies with Seal House timber 683, which has sapwood, is much slower grown but has a growth pattern which does not match the others well. When placed (Fig. 3) in the position relative to the others which gave the best visual match, the two groups of timber would have been felled at the same time.

A comparison of the Seal House/New Fresh Wharf mean curve with that for the Custom House timbers established by Fletcher (1974) showed a reasonable match (t=4.12) with Custom House year 218 corresponding to New Fresh Wharf/Seal House year 185. This suggests that the former (in the absence of sapwood) was built slightly earlier or at the same time as the latter. The ring-widths for the 282 year mean curve (oak) for these Roman timbers from the three sites are given in Table 1.

Dating of these curves is not yet possible. No English reference curve extends as far back as the Roman period, but there is a continental one for oak timbers used at Wederath in Belgium (Hollstein, 1972) with which a very tentative match ($\underline{t}$=4.08) suggested a date of A.D. 151 for the final ring of the New Fresh Wharf/Seal House curve, with felling and use about A.D. 155, in general agreement with archaeological evidence.

However, the results in Table 2 from radiocarbon analyses of samples from New Fresh Wharf beams 311 and 378 and the cradling timber 321 conflict with the other evidence. The samples were cut in positions to lie at 50 year intervals on the tree-ring curve (the positions are marked by brackets at the base of Fig. 3A). The growth allowances, assuming felling in arbitrary year 290, are included in Table 2; the felling dates are uncalibrated. The mean of the four results is 295 ± 35 a.d.

Table 2. Radiocarbon dates derived from samples from the Roman Waterfront at New Fresh Wharf (Site 2)

Analysis Number HAR	Rings of 290 mean curve	Growth allowance, years	Radiocarbon Age+	Radiocarbon Date+	Felling Date (with growth allowance)
1867	90-110	190	1840 bp	110 ad	300 ad
1865	140-160	140	1800	150	290
1864	190-210	90	1660	290	380
1868	240-260 (partially sapwood)	40	1760	190	230

\+ Each ± 60 years

Discussion

The dating of the Roman waterfront is of paramount importance for the archaeology of the Roman City of London. In various constructional details the structures compare closely with the late 1st/early 2nd century wharf at Xanten on the Rhine (von Petrikovitz, 1952), and now cover a length of 500 m. Their building must have been an operation of immense size and of considerable civic, if not military, organization.

The various parts of the wharves may have been zoned according to merchandise. Samian and imported wares were found in great abundance around the New Fresh Wharf quay, with very little pottery at Seal House, indicating that perhaps unbreakable commodities such as timber or animals were being unloaded at the latter site. Being upstream of the bridge, it may also have handled domestic traffic coming down the river. This zoning may be reflected in the significantly different construction of the wharves at the three sites. A date for the quay would also suggest the date for a large-scale development of waterfront facilities, warehousing and presumably the development of the quayside area south of the Forum, most of which is now inaccessible to modern archaeological investigation. Further, the quayside area may be seen as an index of the magnificence of the city it served. These are factors which affect the history of the whole Roman city, and the discovery and excavation of the waterfront since 1973 may be seen as the greatest contribution to the study of Roman London since the discovery of the Cripplegate fort in the immediate post-War period.

II. MEDIEVAL STRUCTURES

Sites and Land Reclamation

Medieval developments in the waterfront area have been excavated on all four of the sites shown in Fig. 1.

At the Custom House (Site 1) two medieval waterfronts were found in 1973, one directly above and one about 7 m south of the Roman waterfront and up to 2 m higher. Both structures were of Seal House III type with diagonal braces; the first incorporated re-used planks from the hull of a clinker-built boat, up to 4.5 m long and with up to five strakes articulated (Tatton-Brown, 1974, 128). The second, and presumably later structure, was better built with new materials and good carpentry.

The waterfronts were dated by a large amount of associated debris in their structure (mostly pottery) to the 14th century, and on documentary grounds quite possibly to the second third of that century (Tatton-Brown, 1975, 105). Samples were taken for dendrochronological analysis by Fletcher (1974), but were found to be nearly all fast-grown and containing less than 75 rings, making dating at that time impossible. However the 158 rings on the one slow-grown sample from a post (all but the first thirty were less than 1 mm wide) have been dated by one of us (R.M.) to the years 977 to 1135; it had no sapwood.

New Fresh Wharf (Site 2) shows most clearly the first stage of land reclamation in the medieval period. A marine transgression silted up the Roman structure in the late Roman and early Saxon periods. The quay was probably standing, in a decayed condition, until sometime between the 7th and 9th centuries when extensive robbing took place. Rows of straight oak branches, averaging 4 m long and 0.16 m in diameter, were inserted behind and in front of the submerged Roman quayfront to form a rough grid of upright pointed stakes, projecting 2 m out of the silt. This seems to be part of a defensive work to protect the vulnerable shore downstream of the bridge.

The stakes were unsuitable for tree-ring dating because they were too immature. A C14 date from one stake of a.d. 870 ± 60 (HAR 1422; uncalibrated) was obtained, and the refurbishing of the city by King Alfred in 886 is a possibility. It is unfortunate that dendrochronology could not place the date more precisely. A bank of clay and rubble on a base of cut branches and worked timbers was formed around the stakes at a later period, in stages

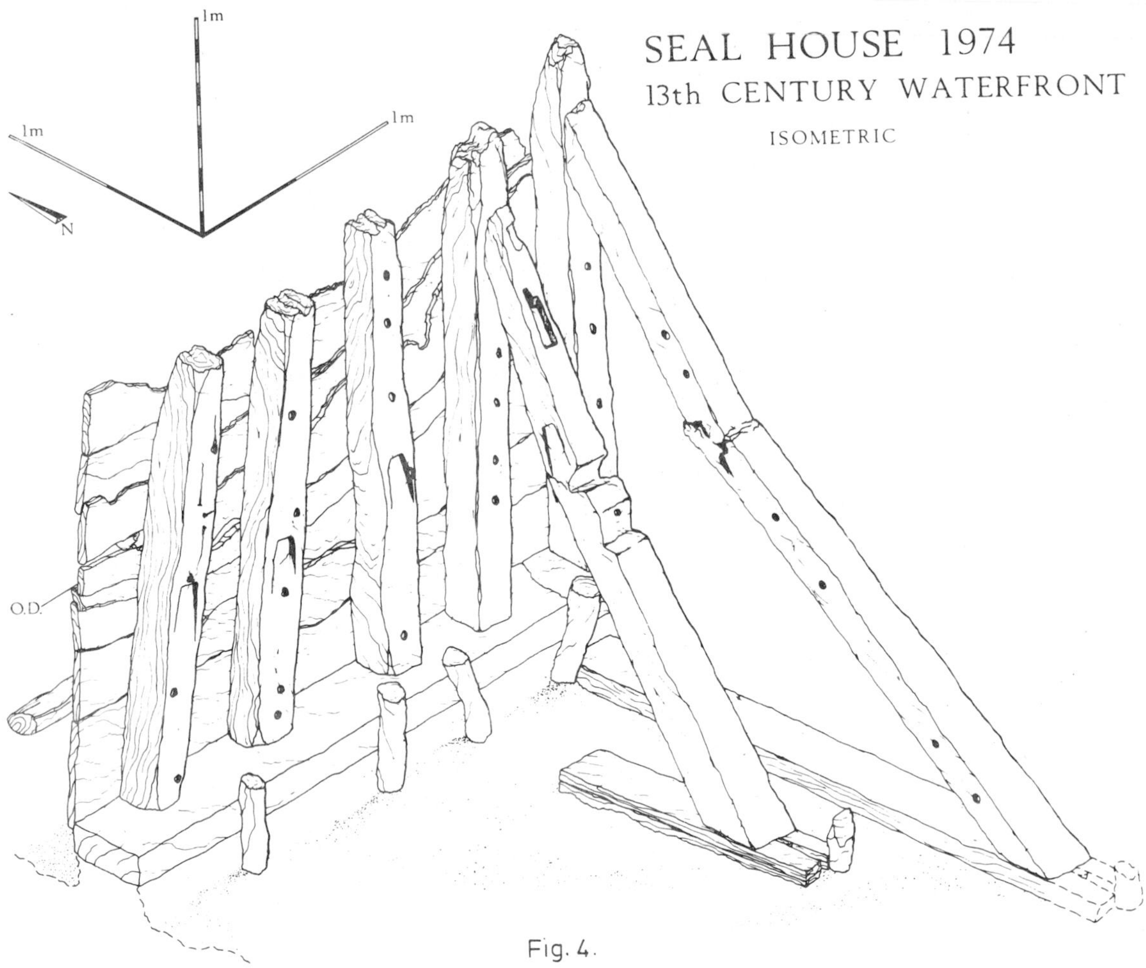

Fig. 4.

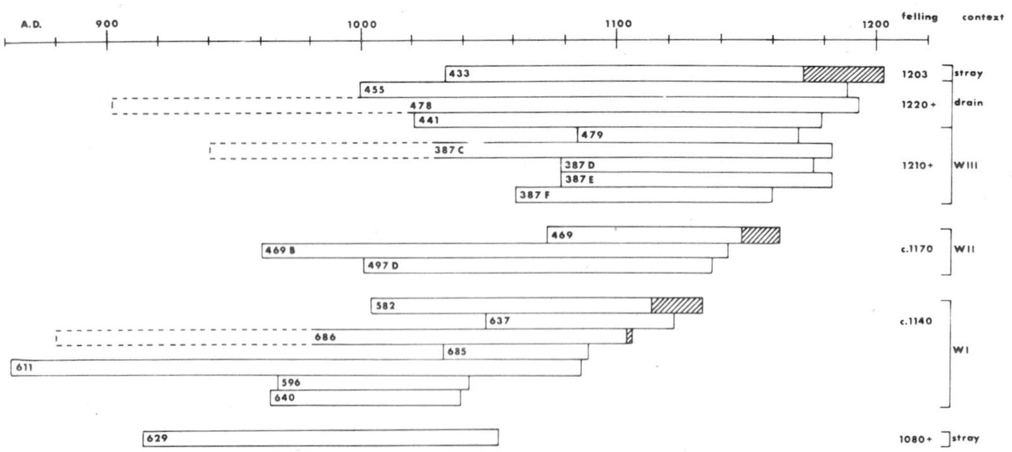

Fig. 5 Calendar years spanned by timbers from medieval Waterfronts I, II and III at Seal House.

which culminated in the 12th century in a surface, about 1 m above the Roman waterfront, probably behind a wharf to the south of the Area III trench. This was the first step in a process of land reclamation which is continuing in the 20th century quaywall, some 75 m to the south of the Roman line.

Seal House (Site 3): Waterfronts or revetments marking early stages in land reclamation were excavated here in 1974, in a trench only 3 m wide. Three waterfronts extended the quayside into the river from the mid-12th century onward. These structures (Waterfronts I-III) were all robbed to varying degrees as each was superseded by the next stage, but survived to show three different designs. Waterfront I comprised a sill-beam which probably held vertical boarding, braced back with cruciform braces and diagonal riverwards supports down to wedges in the silt. Its successor Waterfront II comprised horizontal weatherboarding slotted into a large vertical post, with a compartment behind and rough logs in front. Waterfront III, the best preserved, was of horizontal planks pegged onto verticals which jointed at regular intervals of c. 0.45 m with a sillbeam held by piles, and were supported by diagonal braces going down to piled soleplates (Fig. 4).

Provisional dating of the associated pottery indicated dates for the three waterfronts at around 1200, around 1225 and sometime in the 3rd quarter of of the 13th century respectively.

Tree-Ring Dating (Seal House)

Since the publication of the tree-ring dating of oak timbers from this site in *London Archaeologist* (Morgan, 1977), a number of additional samples have been examined by Jennifer Hillam. Further comparisons have also enabled more original timbers to be dated, while tentative matches have been withdrawn. The block diagram (Fig. 5) has been modified, although the dating of each waterfront is not significantly altered.

Out of 43 timbers of 12th to 14th century date, 20 have been accurately dated by comparison with Hollstein's (1965) German reference curve and Fletcher's (1977) curves for southern England. Interpretation of the felling dates of timbers used in each waterfront is based on Fig. 5. Four timbers retained some or all of their sapwood, so felling dates can only be estimated with due sapwood allowance.

The seven dated timbers of Waterfront I (W.I in Fig. 5) are scattered in time. Timbers 582 and 686 have some sapwood remaining, indicating felling in about 1140 (686 is an old and narrow-ringed tree with probably 30+ sapwood rings). Timber 637 is young and wide-ringed and could have been felled around the same time, or a little later as the archaeological evidence may suggest: it was from the support to a quaywall brace, possibly added to the structure as a repair.

Three timbers, integral members of Waterfront II (W.II) were dated. 469, the main timber of the excavated section was, from sapwood evidence, cut about 1170. This date is also likely for timbers 497D and 469B which end in 1137 and 1143 respectively (Fig. 5). Plate II shows the rings, after preparation, of one of the timbers, undated, from this waterfront.

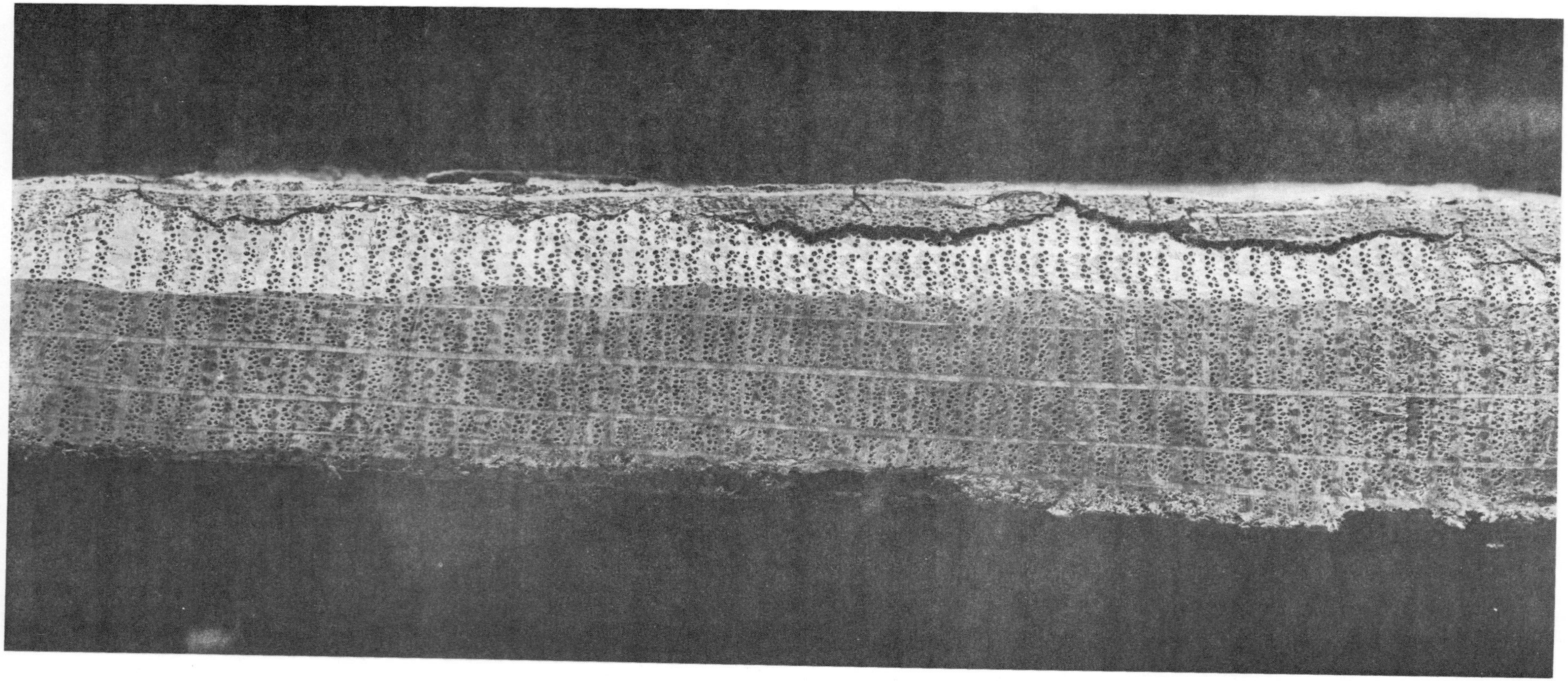

Plate II Seal House: oak growth rings of board 503 from Waterfront II. The transverse section was surfaced, while frozen, with a surform plane. Shows large early wood vessels and dense latewood. Direction of growth left to right. Average width, 1-2 mm. Rings lack sensitivity and the 160 year curve was not dated with certainty.

Waterfront III (W.III) has been dated only by five timbers (four trees) without sapwood. Timber 479 a brace, and boards 387 C, D, E (D and E from the same tree) and F are the only integral members, and with sapwood allowance must have been cut after about 1210.

Three other dated timbers, 455, 441 (Fig. 6) and 478, come from a drain which can be associated with Waterfront III archaeologically. The age and slow growth of the latest, 478, suggests a felling date after about 1220. Finally, timber 433, cut in the summer of 1203, was a stray find on the surface of the waterfront. So evidence from the drain and Waterfront III itself suggest a felling date of, or later than, 1210-1220.

It is difficult to accept construction dates as late as those, 50 to 70 years later, suggested provisionally by the pottery, associated and foreign. Although some of the gap may be explained by the medieval builders using seasoned oak it seems more likely that the provisional pottery dates will eventually be revised to reduce or eradicate the gap.

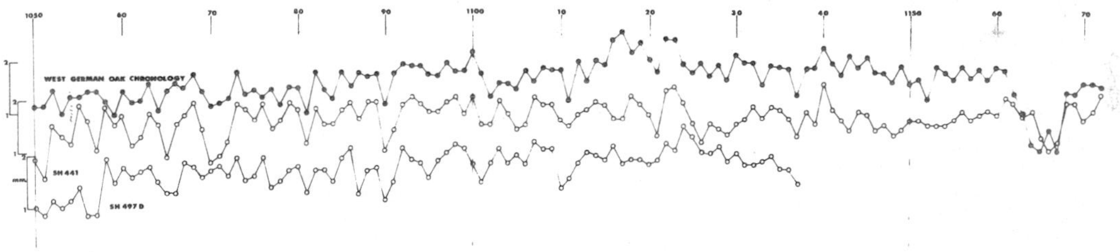

Fig. 6 West German oak chronology compared to those for two Seal House timbers. 1050 to 1172 A.D.

Trig Lane (Plate III), by D. W. Brett and G. Milne

At this site (Fig. 1, site 4), 800 m upstream of the medieval London Bridge, excavation of the medieval waterfronts began in 1974 and has continued until 1977. An area of 500 m² has revealed a series of nine timber revetments and two stone riverwalls based on timber piles; several of the timber structures were 2 m high and up to 16 m long.

The revetments varied in their construction. The earlier type (C in Plate III) is front-braced with horizontal planking laid edge to edge against regularly spaced vertical posts set in a base plate founded on piles. This type is very similar to the early 13th century waterfront (W III) at Seal House (Fig. 4). A later type (B in Plate III) comprised a vertically set stave wall, similarly housed, supported by back braces. With the exception of the elm piles the constructional timber was oak.

There were only a few oak timbers of a size suitable to use for dating. Most have fewer than 100 growth rings. 94 timbers have been sampled, 76 analysed, and some 40 timbers placed on a relative time scale which spans 200 years.

A prominent phase (Group 11) consists of a box-like projection (Plate III) with vertical staves and large back braces. Some of the best samples were provided by these braces since they were halved tree trunks with centre and sapwood present. Samples 420 and 425 show a high level of agreement. W = 73% and t = 11.2. They are clearly the two halves of a single trunk about 350 mm in diameter. Sample 424 (Fig. 7) matched these two at 69% and 74% respectively.

The large corner post (sample 408, Fig. 7), a squared trunk about 2 m in length and 400-500 mm across, matched one brace, sample 424, at 71% and another, sample 425, at 81%. Its ring widths (Fig. 7) are larger (mean width, 3.86 mm) than those of the brace timbers (mean widths from 1.7 to 2.6 mm); this suggests that the post was cut from nearer the base of the tree. It is of interest that the rings of this post show a higher sensitivity (0.335) than the narrower rings of the other samples (0.183-0.221).

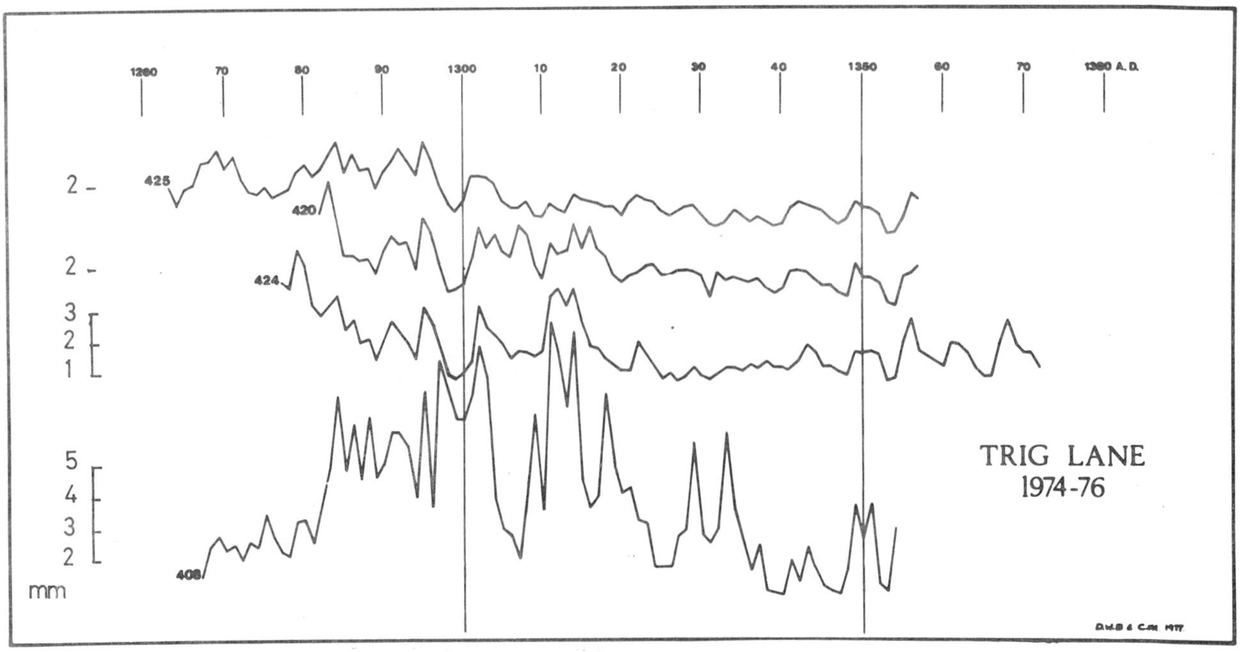

Fig. 7 Ring width sequences of four oak timbers from Group 11—the projection shown at A in Plate III. The dating is that suggested by agreement with the Southern German oak chronology.

Plate III Trig Lane: View across the site, looking N.W.

A. Box-like projection (G 11) with back braces and vertical stave wall.
B. Staves capped by top-plates of a later revetment (G 12).
C. Earlier, front-braced revetment (G 3) on same alignment as B.
The stone wall is the foundation of a later building and encases horizontal planking of the revetment.

These four Group 11 timbers form the basis of a putative single-tree mean chronology which spans 110 years. Comparison with the oak chronology for southern Germany (Huber and Giertz-Siebenlist, this volume,) indicates a matching position at 68.8% agreement. This would place the latest ring of this mean curve at 1372 A.D. and indicate a building phase with timbers cut between 1375 and 1385 A.D., a period in broad agreement with ceramic and coin evidence.

The analysis of timbers from the main construction phases has resolved some queries. For example, the top plate and back braces of the Group 12 structure are contemporary (the top plate of G 12 is visible on top of the vertical staves labelled B in Plate III), and were added to the earlier Group 10 revetment. A detailed recording of the marked structural diversity (Milne and Milne, forthcoming) will provide for the first time a tentative typology for medieval revetments.

The elm piles (Plate III) upon which the main revetments rest are described elsewhere in this volume (Brett, page 195).

DISCUSSION AND CONCLUSIONS

Re-Use of Timbers

This presents an unknown factor except when, as at the Custom House and Trig Lane sites, a re-used timber can be identified by peg holes or joint matrices which do not fit in with its context in the waterfront. Thus there is a problem for Waterfront III of Seal House (Fig. 4). The vertical posts bear tapered mortices or scotches, which may well be not for rearing but for diagonal braces which went down to an excavated baseplate in the silt just to the south of the main structure. A similar scotch, a different mortice and superfluous peg holes may be seen on the adjacent diagonal braces, where they were clearly not used in the waterfront; the farther brace (479) appears from Fig. 5 to be contemporary with the other unmarked timbers of Waterfront III, and may be a vertical post turned round. Analysis of the nearer brace was not possible. There are almost no signs of re-use on the carefully prepared main timbers of all four waterfronts, though ancillary members, such as stakes and piles, are more likely to be re-used.

Experience from the Trig Lane site, where there were many different structures, indicates that primary timbers comprised the sillbeam, the planks, and probably the verticals, and that these tend not to have sapwood; the supports and piles would be more likely to be re-used timbers.

Conclusions

When dates are obtained for the series of waterfront structures, the consequences will be dramatic and far-reaching. The dated extensions to the quays and reclamation zones, each associated with a different stage of the developing house and property plan, will be of immense importance. The various buildings and quays already tie together closely, especially when, as at Seal House and Trig Lane, a building with a subsequent waterfront disturbs or intentionally re-uses the immediately preceding structure.

For both the Roman and Medieval waterfronts, objects associated with the dated structures will become dated objects. The medieval wooden quaywalls each revetted a separate land extension comprised of hard-packed domestic rubbish, including large groups not only of pottery but of organic objects well preserved in anaerobic conditions: shoes, sheaths, leather clothes, bone objects, as well as cloth, hair and seeds. Thus a typological series of each object may well be formed from each series of waterfronts. There are also important examples of dated imports, attesting to the city as an acknowledged centre of international trade and commerce.

The preservation of both Roman and medieval timbers is a result of water-logged conditions which were brought on by rising river levels; these in turn necessitated rebuildings and repositioning the waterfront structures. The existing top of each succeeding structure is therefore a record (though imperfect) of the level to which the Thames normally rose at high tide (Willcox, 1975) and these levels can be tied by dendrochronology into a temporal sequence. In terms of ecology the timbers may be able to interpret and verify palaeo-climatic studies, while on a more local level, individual timbers provide information on conditions of growth, such as whether or not the timbers were from managed woodlands.

Acknowledgements: The authors would like to thank Louise Miller, Charlotte Harding and George Willcox for their criticism of the paper. They and the authors of the paper on the Somerset Levels (page 211) are grateful to Mr. J. W. G. Musty and the Ancient Monuments Laboratory of the Department of the Environment for financing the research.

References

Fletcher, J. 1974: The Dendrochronology, in Tatton-Brown (see below), 211-5.

Fletcher, J. M. 1977: Tree-ring chronologies for the 6th to 16th centuries for oak of southern and eastern England. Journal of Archaeological Science 4, 335-352.

Hobley, B. and Schofield, J. 1977: Excavations in the City of London, 1974-5: First Interim Report, Ant. J. 57, 31-66.

Hollstein, E. 1972: Dendrochronologische Datierung von Hölzern aus Wederath (Belgium), Trierer Zeitschrift 35, 123-125.

Milne, G., Milne, C. et al. (forthcoming): The Advancing Waterfront: Excavations at Trig Lane, City of London, 1974-6, Med. Arch.

Morgan, R. A. 1977: Tree-ring dating of the London waterfronts, London Archaeologist 3, 40-45.

von Petrikovitz, H. 1952: Die Ausgrabungen in der Colonia Traiana bei Xanten, Bonner Jahrbuch 152, 145-57.

Schofield, J. 1975: Seal House, Current Archaeol. 5, no. 2, 54-7.

Schofield, J. and Miller, L. 1976: New Fresh Wharf: 1, The Roman Waterfront, London Archaeologist 2, no. 15, 390-5.

Schofield, J., Miller, L. *et al.* (forthcoming): Excavation at New Fresh Wharf, City of London, 1974-6, *Special Paper Series, Lon. Middx. Arch. Soc.*

Tatton-Brown, T. 1974 and 1975: Excavations at the Custom House site, City of London, *Trans. Lon. Middx. Arch. Soc.* 1974, 25, 211-5: and 1975, 26, 103-170.

Willcox, G. 1975: Problems and possible conclusions related to the history and archaeology of the Thames in the London region, *Trans. Lon. Middx. Arch. Soc.* 26, 285-292.

	0	1	2	3	4	5	6	7	8	9	n
0		15.0	14.0	12.0	15.0	15.0	17.0	16.0	22.0	22.0	1
10	19.0	18.0	15.0	21.0	26.0	20.0	24.0	24.0	30.0	32.0	1
20	22.0	21.5	24.5	27.2	32.2	31.0	30.2	30.5	27.5	25.6	4
30	25.8	22.9	27.1	21.5	22.7	26.6	25.0	18.7	26.8	20.5	5
40	25.4	20.7	22.5	22.9	24.6	22.2	21.4	27.3	32.9	28.2	6
50	21.9	19.6	17.5	12.4	11.7	15.7	20.7	22.9	21.2	21.8	8
60	23.0	23.3	21.2	17.6	21.2	24.4	20.9	23.6	26.0	21.1	9
70	25.9	21.5	18.5	14.8	19.1	23.6	22.1	21.2	21.9	27.8	9
80	25.9	21.7	26.8	20.7	16.3	14.6	16.7	18.2	21.7	21.1	9
90	23.1	22.5	21.5	20.7	14.7	12.3	10.3	17.7	17.6	20.4	10
100	21.8	20.7	24.0	15.9	16.8	14.7	14.9	14.4	17.0	15.7	10
110	18.1	23.5	20.5	17.8	19.8	18.2	16.3	18.4	19.7	18.1	11
120	19.2	19.7	20.3	16.6	18.7	22.2	18.8	20.2	19.6	17.6	11
130	17.4	16.8	17.3	17.2	21.0	17.2	18.4	18.0	15.5	18.4	11
140	19.4	20.3	19.4	16.5	16.3	13.0	11.1	16.2	15.4	17.3	11
150	15.1	19.6	18.4	15.9	17.3	14.7	12.1	12.0	11.8	14.5	11
160	19.3	19.7	16.8	18.5	17.1	14.7	13.6	11.4	17.4	17.0	11
170	20.0	13.5	14.1	16.8	18.2	15.4	11.8	12.8	16.2	14.5	11
180	15.6	16.7	13.1	16.4	18.3	19.4	14.7	12.2	9.1	13.0	9
190	16.5	15.6	14.7	17.3	12.4	13.9	13.1	14.7	13.0	15.5	9
200	16.4	18.2	14.5	15.5	14.9	15.5	15.1	16.8	17.9	18.3	9
210	15.1	12.2	13.7	12.0	12.7	15.4	16.2	15.6	13.5	10.9	7
220	15.3	14.3	12.5	16.8	17.2	15.5	14.0	10.3	13.7	12.3	6
230	12.7	12.0	12.6	12.0	12.4	12.8	12.2	10.8	11.6	10.0	4
240	13.7	14.5	14.2	12.2	11.2	13.3	14.5	17.5	12.5	17.0	2
250	15.0	14.0	12.0	14.5	12.5	14.0	14.5	14.0	13.0	12.0	2
260	12.0	14.5	13.0	13.0	11.5	9.0	11.0	12.0	12.0	17.0	2
270	15.5	13.0	12.0	11.5	12.0	12.5	11.0	9.0	9.5	14.0	2
280	18.5	14.0	13.0								1

Table 1. Ring-width values for a floating mean curve of 282 years based on 12 Roman timbers from the waterfronts at New Fresh Wharf (6 timbers), Seal House (3 timbers) and Custom House (3 timbers, measured by Fletcher). It is impossible to estimate the number of trees involved. n is the number of timbers.

5.4. DATING ANCIENT WOODEN BOATS

Sean McGrail, Greenwich

Summary

The application of dating techniques to boat finds is considered, and it is shown that radiocarbon assay and dendrochronology are often the only feasible methods. It is suggested that dendrochronologists could use logboats to establish regional chronologies, for these are essentially local boats and they are known from recent C.14 dates to have been used in the medieval as well as the prehistoric period.

INTRODUCTION

There are three principal ways of attempting to date boat finds:

(a) by consideration of the boat's form and structure,
(b) by dating the context of the find, and
(c) by dating the materials used in the construction of the boat.

The special nature of boats means that only rarely can all these potential methods be used. Wooden boat remains survive only in special circumstances: wood, skin, fibre, and most other materials used in ancient boatbuilding, are readily destroyed by bacteria and fungi unless they are in anaerobic conditions. In addition, the natural forces of wind and tide and the depredations of man, may well remove a boat's contents and destroy or remove much of the boat's structure, before the site is covered by a protective layer of silt.

Those boat structures which do survive relatively intact have usually been deposited in isolation, on sites such as inshore waters, estuaries, river beds, lake bottoms, and former water courses which, until recently, have seldom been investigated by archaeologists. Thus, boats are likely to be chance finds, revealed by mechanical means to the drainage contractor, the bridge builder, and the gravel extractor; or they may be found by the diver. In these circumstances, evidence which might be used for dating has, in the past, almost invariably been disturbed or ignored during this initial, non-archaeological investigation, because its relevance was not appreciated. Even when material from a boat find does get into archaeological hands there are problems, for much of it is ephemeral unless conserved, and thus dating may become impossible because of inadequate or inappropriate treatment. With this array of difficulties in mind, we may consider each dating method in turn.

DATING METHODS

A. Typology

This method aims to produce a seriation based on some criteria such as functional improvement, change in fashion, or increasing complexity, which is then equated with a development through time, producing a tentative, relative chronology. This may be turned into an absolute chronology by dating representative members of the series by an independent method, for example, by association, or by radiocarbon assay. Fox (1926:130-1; 132, fig. 3) published a typological classification of the logboats of England and Wales, using mainly morphological characteristics, but his subsequent evaluation of the dating evidence (1926:127/8) is unsatisfactory. In addition, it is arguable that many of the characteristic hull forms associated with logboats (and indeed, plank boats) may have had very long phases of use, and thus a sufficiently narrow date range cannot be obtained from them to make their use as bench marks practicable. Fashion appears to have played only a small part in boat design, and it may be that form is more closely correlated with function and with environment, than with age. For the present, this method of dating boats is unsatisfactory and, in any event, it is doubtful whether attempts to base chronological series principally on hull form will be successful.

A priori, a more satisfactory way of dating boats would seem to be by technological changes rather than merely by morphological ones. We have evidence to show in broad terms that boat building techniques and the available tool kits did change through the centuries: for example, split planks were replaced by sawn planks. Thus, whilst form may have remained relatively unchanged for a considerable time, satisfying the required function and matching the operating environment, the methods of fastening fittings to boats, or of repairing them, or the types of joints used, may well have altered through time, as boat builders sought to enhance the strength, the watertight integrity, and the capacity of their boats. Olsen and Crumlin-Pedersen (1958, 1967) have attempted to use such technological changes as a basis for dating Viking Age plank boats of the western Baltic and adjacent regions. But the well argued and plausible development sequence they proposed some 20 years ago (1958: 171-3), cannot yet be claimed as a chronological series, as none of their key finds has been precisely dated (McGrail, 1977B:250-1). However, there is now a reasonable number of well documented boat finds in this research field and, if precise dating can be achieved, Crumlin-Pedersen's current research could result in a soundly based chronological series.

Elsewhere, and in other periods, there are insufficient well-recorded boat finds to support even provisonal typological dating, except in very broad bands indeed. Listed in Appendix 1 and 2 are the few published radiocarbon dates for European log and plank boats. It should be noted that these lists probably include boats built for different functions and these may not be readily comparable with one another in any theory of development. In addition, and regrettably, several dated boats do not have a traceable archaeological publication, and thus these dates lose much of their value. The difficulties of dating boats in Britain by typology were well high-lighted by Valerie Fenwick (1972:9) when she recorded how the Graveney boat, when first found, could only be dated to the broad range 4th to 13th centuries A.D.

As additional boat finds are recorded and precisely dated, it may prove possible to progress from dating by broad stages of technological advance, to more precise dating based on the recognition of the techniques used in one 'boatyard'—what Crumlin-Pedersen (1977:164) has called, 'a certain local practice at a given time and place'. A comprehensive, standardised recording system, possibly involving the use of electronic data storage and retrieval, will be essential before this precision becomes more than a pious hope (McGrail, 1977A:131/2). Individual constructional features may have had long and widespread usage. For example: flat, double-dovetail clamps were used to repair splits in the Bronze Age logboat from Appleby, Lincolnshire

(Dudley, 1943); in the undated logboat from Vaaler Moor, North Germany (Paret, 1930:106/7); and also to fasten the planking of the 12th dynasty Egyptian Dahshur boats (Hornell, 1970:193), and the 19th-century dragneurs of the R. Adour, France (Beaudouin, 1970:92 fig. 13). Thus, only if many features are recorded and dated, can there be any possibility of obtaining satisfactory combinations of diagnostic features which may subsequently enable other boat finds to be more precisely dated and their 'boatyard' identified.

B. Dating by Context

Consideration of an artifact's spatial relationship to other dateable artifacts or to dateable horizons can provide a range of dates within which the artifact must have been deposited. There are intrinsic problems, however, in the application of these methods to boats. Firstly, artifacts associated with the boat at deposition may subsequently be removed by man or by natural forces; conversely, but nevertheless confusingly, artifacts not associated with the boat during its period of use may be deposited in apparent close association, in or under the boat by tide and stream before the boat find is truly sealed. Secondly, in tidal, especially estuarine, conditions, scouring out and redeposition of deposits can greatly change the stratigraphy, and may result in a very complex or an indeterminate stratigraphy, both of which may be very difficult to interpret. It is not inconceivable that a boat, deposited and sealed in one era, may be shifted by the scouring action of a river in full flood in a subsequent era, to be redeposited and subsequently resealed, but now out of primary context. Thirdly, a boat may settle and sink through an unconsolidated deposit, coming to rest in a stratigraphically earlier deposit than appropriate to its deposition. In certain circumstances, meticulous excavation techniques, such as used by van der Heide (1970:24/5) in his excavations from the reclaimed lands of the Zuider Zee, (fig. 1) can detect and quantify these effects, but on other sites these techniques may be impracticable.

(i) Association with other artifacts: Before association can be used to date a boat find it is essential to demonstrate that the artifact was unambiguously associated with the boat at the time of its initial deposition. A coin trapped in the mast step of the Blackfriars 1 plank boat (Marsden, undated), or the logboat found in Jutland with inscribed early medieval runes (Wilson, 1966:85), are examples of the valid use of this method. On the other hand, there are many examples in the literature of wrongly deduced associations: one will suffice. Fox claimed that logboat Glastonbury 1 was "undoubtedly associated with the lake dwellings, therefore c 100 BC" (1926:150), but this is not substantiated by the earliest reports (Bulleid, 1893:121; Bulleid, 1906:52) which refer to the boat being found in a ditch "a few fields distant" from Glastonbury Lake Village. Conversely, logboat Glastonbury 2 was excavated from peat in the sub-structure of this village (Bulleid, 1906:52; Bulleid and Gray, 1911: 333), and thus its date may be related to that of the village.

Ironically, it is often less difficult to date boat timbers which have had a secondary non-nautical use, as at Wood Quay, Dublin (fig. 2), where they are associated with dateable artifacts, and are excavated under controlled conditions. But here, because of the dispersed nature of the fragments, the evidence for boatbuilding techniques is not so straightforwardly interpreted, and the date obtained only gives a terminus ante quem for the boat's primary use.

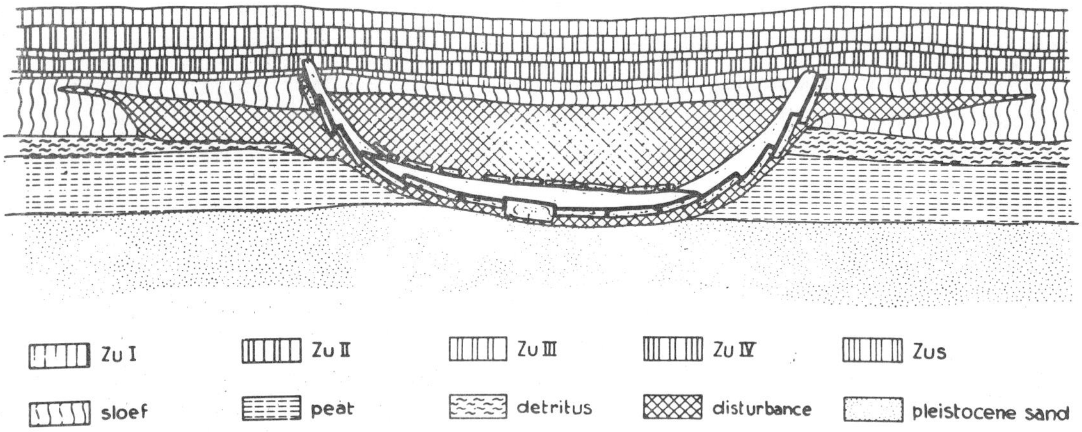

Fig. 1 Diagram of the stratigraphy in and around a Zuider Zee wreck. G. van der Heide

Fig. 2 The National Museum of Ireland's medieval site at Wood Quay, Dublin, with re-used ship's timbers in the foreground.

(ii) Dating by stratigraphy: Another indirect method of dating boats is by dating the layer within which the boat is stratified. This date may be established by several means: contained dated artifacts; a peat layer dated by radiocarbon assay; or from a dating attributed to that horizon in the local sequence. The problems caused by the natural disturbance of the site discussed above, apply with great force to this method of dating. Depth below the present surface *per se*, means little, although this statistic has been quoted, with the implication that the deeper the boat the older it was. Rates of deposition vary tremendously (see Holmes, 1902:11), and only where boats are found unambiguously vertically above one another, can anything be inferred about their relative age.

The stratigraphy of boat finds has only infrequently been recorded in the past. All too often the remains were extracted before either geologist or archaeologist was on site, and thus only hearsay or other indirect evidence is now available. For example, the oft-quoted mesolithic date for the logboat found at Perth, Scotland (Geikie, 1879; Clark, 1936) may not be as reliable as has been assumed. Geikie did not see the boat until 'a number of years' after it had been found, and the identification of the find spot depends on the memory of the manager of the brickworks at Perth, and not on an *in situ* identification by Geikie (McGrail, forthcoming).

The hazards in dating a boat find solely by its apparent relationship to a waterborne deposit are well illustrated by the boat remains found partly buried in a clay sediment, *c* 4 m deep in Lago Maggiore, Italy in 1955. By reference to these deposits the boat was believed to be *c*. 1st century A.D.; subsequent radio-carbon assay (R-847 α) gave a date of *post* A.D. 1800 (*Radiocarbon*, 18, 1976, 323).

(iii) Dating by pollen analysis: During the 1920s Paret (1930) dated several logboats from Lake Federsee,, and since then other analyses have been published, in Scandinavia, the Netherlands, and Britain, in which boats were assigned to one of the broad climatic phases, or to an interface between such phases. In at least two cases, this broad date band has subsequently been confirmed and refined by radiocarbon assay of wood from the boat; the Shapwick, Somerset, logboat (Godwin, 1967); and the Pesse, Netherlands logboat (van Zeist, 1957). It is now realised that there is no simple relationship between climatic phase and date, and that pollen dating cannot be sufficiently precise.

C. Dating the Raw Material of the Boat

For the reasons given above, dating boats by typology or by context is, for the present at least, often impracticable. Thus we are left with dating the raw material of the boat; and the two methods most evidently applicable to wooden boats are radiocarbon assay, and dendrochronology.

(i) Radiocarbon assay: The principles for the radiocarbon dating of logboats discussed by McGrail and Switsur (1975), generally apply to all boat finds. The sample (preferably several) should be taken from the outermost growth rings of the parent log so that a close approximation to the felling date may be obtained. However, where there is surface contamination by a wood preservative such as Stockholm tar, this may not be possible. A sample from a fitting fashioned from a shortlived timber such as a branch, or a sample of moss or hide which we may assume to have had a short useful life, should give a precise date, providing these are truly associated with the boat. Dating a repair or a secondary fitting should help to define the period within which the boat was used. With present techniques, and even if we discount the difference between radiocarbon years and calendar years, the smallest date bracket to be expected is c. 200 years, using the range of two standard deviations. An increasing number of boats, both log and plank, has been dated in recent years, and these dates are tabulated in Appendices 1 and 2.

(ii) Dendrochronology: With the right sort of material, potentially the most reliable and precise method of dating boats is by dendrochronology. I know of only three plank boats that have been dated by this method: the Bremen Cog (Bauch, 1969); the Blackwater boat from the north of Ireland (Farrell and Baillie, 1976:53); and the Graveney boat (Fletcher, forthcoming) (Fig. 3). However, we have in the National Maritime Museum, Greenwich, a letter dated 12th February 1948 from one of the pioneers of dendrochronology, A. W. G. Lowther, offering to date one of the Ferriby boats (Fig. 4). Regrettably, I can trace no further action, but we now know from radiocarbon assay that these sewn plank boats can be dated to the Bronze Age (McGrail and Switsur, 1975:192/4). We have recently initiated a dendrochronological investigation of our other prehistoric boat, the Brigg 'raft' (Fig. 5), which has been dated to c. 600 bc by radiocarbon; dendrological investigation of Ferriby boats 1 and 2 may prove impossible however, because of the early methods of conservation.

Logboats have not yet attracted much dendrochronological interest but I believe that they are well worth consideration, particularly in view of the value already derived from work on log coffins (Becker, this vol. page 62 and Bartholin, this vol. page 275). Several logboats have more than the minimum number (c. 100) of growth rings required to be dateable, either at the ends (Fig. 6, 7) or across an integral bulkhead (Fig. 8). Although it is customary in the 20th century to remove all sapwood from timber destined for boats, this was not always the case in antiquity, and small patches have been found—see, for example, the Short Ferry, Lincolnshire, logboat (Baker, 1952:18). Where the station with the optimum number of growth rings is not at the broadest part of the logboat, it should be possible to increase the measurable growth rings to a maximum by relating the optimum growth ring pattern to the pattern at other stations where later growth rings survive. Where no sapwood is present it is reasonable to assume that the boatbuilder would have made the maximum possible use of his log and, therefore, that the latest measureable heartwood growth ring in a sound logboat is close to the heartwood/sapwood transition. Thus, with the addition of one of the standard allowances for missing sapwood (Baillie, 1973:10, and Fletcher, 1978), a close approximation to the felling date (with an appropriate confidence limit) can be obtained. Similar considerations apply, when dating the radially split, or diametrically hewn, elements of plank boats.

The archaeologist must then consider what relationship this felling date bears to the dates he is most interested in: the building date, and the in-use

Fig. 3 The Graveney boat in situ, 1970.

Fig. 4 Ferriby boat 1 on the foreshore of the R. Humber, 1946. E. V. Wright.

Fig. 5 The Brigg 'raft' first exposed in 1888 and re-excavated by the National Maritime Museum in 1974.

Fig. 6 The logboat from North Stoke (now in the British Museum) made from half a log split longitudinally.

Fig. 7 Logboat Hardham 2 (now in Worthing Museum) made from a whole log.

Fig. 8 The logboat from the R. Thames at Kew (now in the Museum of London) with an integral midships bulkhead.

dates of the boat; and the date of the earliest and latest use of specific woodworking and boatbuilding techniques. He must consider whether there should be an allowance for the time taken to haul the timber from forest to 'boatyard'. Was there a lengthy seasoning time, or was the timber used 'green'? (McGrail, 1974:40-42). Were partly worked logs and crooks stored underwater for an appreciable time before use? There is evidence for this in north-west Europe during the medieval period (McGrail, 1974:39/40), and the practice may have been widespread. The possibility of repairs and reused timbers must also be considered. The archaeologist must quantify the answers to these questions, and use any other dating evidence available, so that he can arrive at the closest approximation to the dates of most importance to the history of technology (Fig. 9).

As Farrell and Baillie (1976:45/8) have pointed out, only rarely can the country of origin of an ancient boat be unambiguously identified. Dendrochronologists have thus been wise to assume that timber from a plank boat may be foreign to the region of deposition and excavation, and have therefore avoided using boats' timbers when establishing their master chronologies (Baillie, 1973:13). On the other hand, European logboats are intrinsically local boats, most probably built from local logs preponderantly of oak (Quercus Sp.), and used and deposited within a particular river catchment area. Thus they could profitably be used to assist in establishing a regional oak chronology. (In passing, we may note that this also applies to log coffins and troughs.) Obtaining suitable samples from a recently excavated oak logboat should be relatively simple, providing sufficient remains have survived. Logboats which have been excavated for some years may present more of a problem, for they may have split, fragmented or distorted during drying out or during conservation treatment, making it difficult to find a suitable sample. Any conservation treatment must be noted so that its possible contaminating effects may be evaluated before ancillary radiocarbon dating.

The chronologies for Britain are as yet far from finalised, and no logboat, apart from a curious reference to the dating of the Scotswood boat in Dodds (1964:287), has been dated by dendrochronology; nor has one been used to establish a chronology. The few radiocarbon dates that are available for logboats show that their use spans the medieval as well as the prehistoric period in Britain, and dates from other European countries (see Appendix 1) confirm this. Some 40 or so additional logboat dates for England and Wales are shortly to be published by Dr. Roy Switsur of the Radiocarbon Dating Research Laboratory at Cambridge. When these are available it would seem eminently worthwhile to examine these boats dendrologically to see whether reliable samples for dendrochronology can indeed be taken.

Notes

1. In this paper no distinction is made between boat and ship.

2. In Appendices 1 and 2 lower case 'bc', 'ad' are used to denote radiocarbon years: 'bp' means radiocarbon years before AD 1950.

Dating by Dendrochronology

The Parent Log

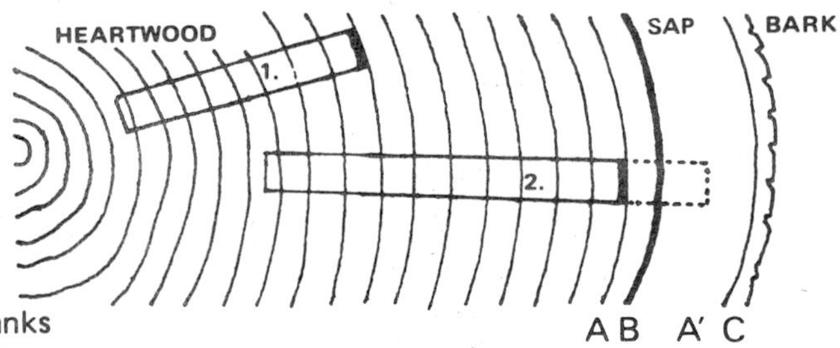

1 & 2	radially cleft planks
A. A'	latest surviving growth ring.
B.	heartwood/sapwood transition.
C.	outermost growth ring = felling year.

After Felling

C.	Felling Date	Intermediate Period (C.–D.)	Haulage time? Underwater storage? Recently felled? or Seasoned?
D.	Boat Building Date	(D.–E.) or In-Use Period (D.–F.)	
(E.	Secondary Use)		This may include a dateable repair
F.	Deposition Date		

Fig. 9 Dates of importance to the dendrochronologist and the archaeologist.

References

Baillie, M., 1973, Dendrochronology: an exercise in archaeological dating, in Scott, B. G., (ed) Perspectives in Irish Archaeology: 6-16.

Baker, F. T., 1952, A prehistoric 'dugout' boat from Short Ferry near Fiskerton, Lincolnshire, Lincolnshire Poacher, 1:17-19.

Bauch, J., 1969, Die Bauzeit der Bremer Kogge, Wittheit zu Bremen, 8:123-6.

Beaudouin, F., 1970, Les Bateaux de L'Adour Musee Basque Bayonne.

Bulleid, A., 1893, Ancient canoe found near Glastonbury, Somerset and Dorset Notes and Queries, 3:121.

Bulleid, A., 1906, Prehistoric boat found at Shapwick, 1906, Proceedings of the Somersetshire Archaeological and Natural History Society, 52:51-54.

Bulleid, A. and Gray, H. St. G., 1911, Glastonbury Lake Village, 1, Glastonbury.

Burleigh, R., forthcoming, Radiocarbon dating, in Fenwick, V., (ed), The Graveney Boat.

Clark, G., 1936, Early navigation in N.W. Europe, Proceedings of the Prehistoric Society, 2:146.

Crumlin-Pedersen, O., 1965, Cog-Kogge-Kaage, Handels og Sjøfartsmuseet pa Kronberg Arbog: 81-144.

Crumlin-Pedersen, O., 1977, Some principles for the recording and presentation of ancient boat structures, in McGrail, S., (ed) Sources and Techniques in Boat Archaeology, 163-177, British Archaeological Reports S29, Oxford.

Dodds, W., 1964, Ryton dugout canoe, Archaeologia Aeliana, 4th Series 42:285-8.

Dudley, H. E., 1943, One tree boat at Appleby, Lincolnshire, Antiquity, 17: 156-61.

Farrell, A. W., and Baillie, M. G. L., 1976, Use of dendrochronology in Nautical Archaeology, Irish Archaeological Research Forum, 3(2):45-55.

Fenwick, V., 1972, Recent work on the Graveney boat, in Three Major Ancient Boat Finds in Britain: 9-17, National Maritime Museum Monograph No. 6.

Fletcher, J., 1978, Tree-ring studies, in Fenwick, V., (ed) Graveney Boat, 115, British Archaeological Report 53, Oxford.

Fox, C., 1926, Dugout canoe from South Wales, Antiquaries Journal, 6:121-151.

Geikie, J., 1879/80, Discovery of an ancient canoe in the old alluvium of the Tay at Perth, Scottish Naturalist, 5:1-7.

Godwin, H., 1967, Discoveries in the peat near Shapwick Station, Somerset, Proceedings of the Somerset Archaeological and Natural History Society, 111:20-3.

Holmes, T. V., 1902, Geological notes on the new reservoirs in the Valley of the Lea near Walthamstow, Essex Naturalist, 12:1-16.

Hornell, J., 1970, Water Transport, Newton Abbot.

Lieger, A. and Marguet, R., 1974, Découvertes récentes dans les Dragages, Revue Archéologique de l'Est et du Centre Est (Dijon) 25:215-234.

McGrail, S., 1974, Building and trials of the replica of an ancient boat: the Gokstad faering, National Maritime Museum Monograph No. 11 (part 1), Greenwich.

McGrail, S., 1977A, Searching for pattern among the logboats of England and Wales, in McGrail, S., (ed) Sources and Techniques in Boat Archaeology: 115-135, British Archaeological Reports S29, Oxford.

McGrail, S., 1977B, McKee's Timbers from Old Ships—A comment, International Journal of Nautical Archaeology, 6:248-252.

McGrail, S., forthcoming, The Logboats of England and Wales.

McGrail, S. and Switsur, R., 1975, Early British boats and their chronology, International Journal of Nautical Archaeology, 4:191-200.

Marsden, P. V., undated, Ship of the Roman period from Blackfriars in the City of London, Guildhall Museum, London.

Nielsen, H., 1973, Baden ved Vejene, Skalk, 6:12-15, Højberg.

Olsen, O. and Crumlin-Pedersen, O. 1958, The Skuldelev Ships, Acta Archaeologica, 29:161-175.

Olsen, O. and Crumlin-Pedersen, O., 1967, The Skuldelev Ships II, Acta Archaeologica, 38:73-174.

Olsson, H. and Sjöberg, J. E., 1971, Låssby Skäggered Göteborg Stockbåtar Bronsålder/Järnålder, Göteborg Museum.

Paret, O., 1930, Die Einbäume im Federseeried und im übrigen Europa, Prahistorische Zeitschrift, 21:76-116.

Philipsen, J. P. W., 1965, The Utrecht Ship, Mariners' Mirror 51:35-46.

Sjöberg, J. E., 1970, Hypotetiska båtar—och verkliga, Göteborgs Arkeologiska Museum Årstryck: 41-6, Göteborg.

van der Heide, G., 1970, Ship-archaeological investigations in the Netherlands, in Aspects of the History of Wooden Shipbuilding, National Maritime Museum Monograph No. 1, Greenwich.

van Zeist, W., 1957, De Mesolithische boot van Pesse, Nieuwe Drentse Volksalmanak (1957):4-11, Assen.

Wilson, D. M., 1966, Medieval boat from Kentmere, Westmorland, Medieval Archaeology, 10:81-8.

APPENDIX 1

PUBLISHED RADIOCARBON DATES FOR EUROPEAN LOGBOATS

Laboratory Ref. No.	Reference	Logboat, and material used for dating	bp	bc/ad	±
(A) Britain					
Q-288	McGrail & Switsur, 1975:196	Branthwaite, oak (*Quercus* Sp) from "backboard"	3520	1570 bc	100
BM-213	*Radiocarbon*, 11 (1969):285	Chapel Flat Dyke, wood	3450	1500 bc	150
Q-80	McGrail & Switsur, 1975:196	Appleby	3050	1100 bc	80
Q-79	*Radiocarbon*, 3 (1961):72	Short Ferry, oak (*Quercus* Sp)	2796	846 bc	100
Q-78	*Radiocarbon*, 3 (1961):73	Brigg, oak (*Quercus* Sp)	2784	834 bc	100
Q-357	*Radiocarbon*, 2 (1960):62-3	Shapwick, oak (*Quercus* Sp)	2305	355 bc	120
Q-821	McGrail & Switsur, 1975:196	Poole	2245	295 bc	50
Birm-132	*Radiocarbon*, 12 (1970):395	Holme Pierrepont 1, wood from top edge of boat	2180	230 bc	110
Q-827	McGrail & Switsur, 1975:196	Hardham 2	1655	ad 295	50
Q-828	McGrail & Switsur, 1975:196	Amberley 3	1310	ad 640	70
Q-857	McGrail & Switsur, 1975:196	Llyn Llangorse	1136	ad 814	60
Birm-269	*Radiocarbon*, 15 (1973):9	Gate Warth (Warrington 11)	950	ad 1000	90
	NOTE: The validity of some of these dates is discussed by McGrail & Switsur (1975:195-6).				
(B) Denmark					
K-1473	*Radiocarbon* 15 (1973):96	Praestelyngen 1, lime (*Tilia*) from bow of boat	5010	3060 bc	100

Laboratory Ref. No.	Reference	Logboat, and material used for dating	bp	bc/ad	±
K-1650	Radiocarbon, 15 (1973):96	Praestelyngen 1, willow (Salix) stake	4960	3010 bc	110
K-1651	Radiocarbon, 15 (1973):96	Praestelyngen 1, moss from stern of boat	4890	2940 bc	110
K-1649	Radiocarbon, 15 (1973):99	Praestelyngen 2, wood	4420	2470 bc	110
?	Nielsen, 1973:15	Varpelev	2980	1030 bc	100
(C) France					
Ly-792	Radiocarbon, 18 (1976):73	Les Baigneurs à Charavines A, wood	4190	2240 bc	150
Ny-313	Radiocarbon, 16 (1974):122	Chaudeney sur Moselle 1, oak (Quercus Sp)	1850	ad 100	60
?	Lieger & Marguet, 1974:217	Ancenis	1820	ad 130	150
Ny-314	Radiocarbon, 16 (1974):122	Chaudeney sur Moselle 2, wood	1750	ad 200	70
Ly-68	Radiocarbon, 15 (1973):140	Parc de la Tête d'Or, wood	1500	ad 450	110
(D) Italy					
R-359α	Radiocarbon, 15 (1973):166	Lago di Fimon, wood	4580	2630 bc	50
R-786α	Radiocarbon, 15 (1973):166	Bande di Cavriana, wood	3520	1570 bc	50
R-375	Radiocarbon, 15 (1973):165-6	Lago Lucone, oak (Quercus Sp)	3360	1410 bc	50
R-375α	Radiocarbon, 15 (1973):165-6	Lago Lucone, oak (Quercus Sp)	3160	1210 bc	50
R-855α	Radiocarbon, 18 (1976):322	Lago Monate, 2a, wood fragments	1820	ad 130	50
R-2	Radiocarbon, 15 (1973):166	Valle Isola	1810	ad 140	140
F-63	Radiocarbon, 16 (1974):13	Lago Monate, 2, wood	1580	ad 370	105

Laboratory Ref. No.	Reference	Logboat, and material used for dating	bp	bc/ad	±
R-854α	Radiocarbon, 18 (1976):322	Lago Monate, 2, wood	1430	ad 520	50
R-917α	Radiocarbon, 18 (1976):323	Selvazzano 1, oak (*Quercus* Sp)	1210	ad 740	50
R-918α	Radiocarbon, 18 (1976):323	Selvazzano 2, oak (*Quercus* Sp)	1200	ad 750	50
R-920α	Radiocarbon, 18 (1976): 324	Selvazzano D, wood	1130	ad 820	50
R-919α	Radiocarbon, 18 (1976):323	Selvazzano C, wood	1020	ad 930	50
F-62	Radiocarbon, 16 (1974):13	Lago Monate 1, wood	940	ad 1010	75
Pi-84	Radiocarbon, 3 (1961):103	Lago Trasimeno, wood	744	ad 1206	100
R-894α	Radiocarbon, 18 (1976):324	Laghi di Monticolo, wood	710	ad 1240	50
(E) Netherlands					
Gro-486	van Zeist, 1957:11	Pesse	8265	6315 bc	275
(F) Norway					
T-1429	Radiocarbon, 17 (1975):394	Moen, pine (*Pinus* Sp)	740	ad 1210	110
(G) Russia					
Le-654	Radiocarbon, 12 (1970):131-2	Peschanoye	1120	ad 830	100
(H) Sweden					
?	Sjöberg, 1970, 43	Kvillehed Bergsgård	2735	785 bc	?
St-3550	Olsson & Sjöberg, 1971:41	Låssby	2215	265 bc	100
St-3551	Olsson & Sjöberg, 1971:41	Skäggered	2485	535 bc	100
?	Sjöberg, 1970:44	Vo Frölunda	2025	65 bc	?
St-27	Science, 126 (1957):496	Kattegat, 40 m NE Christiansö, wood	650	ad 1300	150

APPENDIX 2

PUBLISHED RADIOCARBON DATES FOR EUROPEAN PLANK BOATS

Laboratory Ref. No.	Reference	Boat, and material used for dating	bp	bc/ad	±
(A) Britain					
Q-837a	Radiocarbon, 8 (1966):396	Ferriby 2, yew (Taxus) stitches, and a slat from underneath boat	3393	1443 bc	210
Q-837b	Radiocarbon, 8 (1966):396	Duplicate oxidation sample Q-837a	3506	1556 bc	110
Q-1197	McGrail & Switsur, 1975:193	Ferriby 1, oak (Quercus) from main planking	3380	1430 bc	100
Q-1217	McGrail & Switsur, 1975:193	Ferriby 1, yew (Taxus) stitches	3312	1362 bc	100
Q-715	Radiocarbon, 7 (1965):211	Ferriby 3, alder (Alnus) branch directly beneath boat	3120	1170 bc	105
Q-1199	McGrail & Switsur, 1975:194	Brigg 'raft', oak (Quercus Sp) from planking	2630	680 bc	100
Q-1200	McGrail & Switsur, 1975:194	Brigg 'raft', oak (Quercus Sp) from cleat	2543	593 bc	100
BM-688	Radiocarbon, 18 (1976):22	Sutton Hoo, wood, possibly from burial chamber	1256	ad 694	45
BM-957	Radiocarbon, 18 (1976):25	Graveney, oak (Quercus Sp) from strake P4D	1126	ad 824	76
BM-956	Radiocarbon, 18 (1976):25	Graveney, oak (Quercus Sp) from strake S6C	1283	ad 667	51
BM-715	Radiocarbon, 18 (1976):25	Graveney, brush-wood from underneath boat	1003	ad 947	40
BM-661	Radiocarbon, 18 (1976):25	Graveney, loose fragments of oak (Quercus Sp) planking	1064	ad 886	54

Laboratory Ref. No.	Reference	Boat, and material used for dating	bp	bc/ad	±
BM-660	Radiocarbon, 18 (1976):25	Graveney, oak (Quercus Sp) from strake P6 or P7	1080	ad 870	40
BM-1137	Burleigh, (forthcoming)	Graveney, oak (Quercus Sp) from strake S3D	1178	ad 772	38
BM-1138	Burleigh, (forthcoming)	Graveney, oak (Quercus Sp) from strake S3D	1095	ad 855	37
D-71	Radiocarbon, 3 (1961):37	Kentmere, oak (Quercus Sp)	650	ad 1300	120
BM-961	Radiocarbon, 18 (1976):39–40	Walthamstow, wood	346	ad 1604	54
(B) Cyprus					
P-1622	Radiocarbon, 13 (1971):363-4	Kyrenia, wood from hull	2222	272 bc	43
(C) Denmark					
K-876	Radiocarbon, 8 (1966):229	Skuldelev 1, hair and wood-tar caulking	940	ad 1010	100
K-583	Olsen and Crumlin-Pedersen, 1967:163	Skuldelev 2, oak (Quercus), twigs inside ship	-	ad 940	100
K-584	Radiocarbon, 2 (1960):22	Skuldelev 2, wool caulking	1040	ad 910	100
K-908	Radiocarbon, 8 (1966):229	Skuldelev 2, willow (Salix) treenails	960	ad 990	100
K-877	Radiocarbon, 8 (1966):229	Skuldelev 3, willow (Salix) treenail	920	ad 1030	100
K-878	Olsen and Crumlin-Pedersen, 1967:163	Skuldelev 3, hazel (Corylus Sp) fascines by wreck	-	ad 980	100
K-875	Radiocarbon, 8 (1966):229	Skuldelev 5, hair and wood-tar caulking	990	ad 960	100
K-872	Radiocarbon, 8 (1966):230	Skuldelev ?, worked wood, perhaps from ship	870	ad 1080	100

Laboratory Ref. No.	Reference	Boat, and material used for dating	bp	bc/ad	±
?	Crumlin-Pedersen, 1965:130	Wreck from Sound near Amager, juniper (Juniperus) treenail	-	ad 1250	75
(D) Egypt					
BM-332	Radiocarbon, 13 (1971):161	Cheops boat, grass rope	3990	2040 bc	105
(E) France					
Gif-1862	Radiocarbon, 16 (1974):32	Bon-Porte Bay, wood	2520	570 bc	100
Ny 315	Radiocarbon, 16 (1974):122	Chaudeney sur Moselle 3	840	ad 1110	50
(F) Italy					
P-1436	Radiocarbon, 12 (1970):585	Torre Sgarrata, wood	2027	77 bc	43
St-103A	Science, 126 (1957): 494	Lake Nemi, wood from innermost treerings	1940	ad 10	70
St-103B	Science, 126 (1957): 494	Lake Nemi, wood from innermost treerings	2090	140 bc	75
P-1435	Radiocarbon, 12 (1970):585	Pantano Longarini, wood	1328	ad 622	48
R-1080α	Radiocarbon, 18 (1976):326	Argentario 1, wood	< 150	ad 1800	-
R-1081α	Radiocarbon, 18 (1976):326	Argentario 2, wood	< 150	ad 1800	-
R-847α	Radiocarbon, 18 (1976):323	Gravellona Toce, wood	< 150	ad 1800	-
(G) Netherlands					
GrN-5646	Radiocarbon, 14 (1972):100	Kapel Avezaath, wood	1820	ad 130	30
?	Philipsen, 1965:37	Utrecht	1160 or 1030	ad 790 or ad 920	45 45

Note: the data given by Philipsen is ambiguous.

Laboratory Ref. No.	Reference	Boat, and material used for dating	bp	bc/ad	±
(H) Norway					
T-944	Radiocarbon, 14 (1972):442	Klåstad, hazel (Corylus Sp) twig associated	1150	ad 800	80
T-1349	Radiocarbon, 17 (1975):394	Verne Kloster, hair caulking	730	ad 1220	100
T-1148	Radiocarbon, 17 (1975):394	Isegran 1, treenails	320	ad 1630	100
T-1348	Radiocarbon, 17 (1975):394	Hellesö, treenail	290	ad 1660	100
T-1149	Radiocarbon, 17 (1975):394	Isegran 2, treenails	210	ad 1740	100
(I) Sweden					
St-150	Science, 126 (1957): 496	Badelunda, wood from ship grave No. 75	1225	ad 725	65
St-661	Radiocarbon, 4 (1962):129	Äskekärr, cattle hair	1120	ad 830	75
St-204	Science, 126 (1957): 496	Galtabäck, cattle hair caulking	880	ad 1125	65
St-137	Science, 126 (1957): 496	Falsterbo, cattle hair caulking	850	ad 1155	70
Lu-745	Radiocarbon, 16 (1974):327	Slädö 2, wood	590	ad 1360	50
Lu-744	Radiocarbon, 16 (1974):327	Slädö, oak (Quercus Sp) probably from rib of wreck 1	290	ad 1660	50
(J) Yugoslavia					
Z-256	Radiocarbon, 17 (1975):150	Bosanka Gradiska, wood	1759	ad 191	55
Z-298	Radiocarbon, 17 (1975):152	Nin, rib fragments	1177	ad 773	80
Z-296	Radiocarbon, 17 (1975):152	Nin, hull fragments	961	ad 989	64
Z-297	Radiocarbon, 17 (1975):152	Nin, hull fragments	937	ad 1013	71

Laboratory Ref. No.	Reference	Boat, and material used for dating	bp	bc/ad	±
Z-251	Radiocarbon, 17 (1975):150	Hrvatska Dubica, wood fragments	541	ad 1409	60
Z-255	Radiocarbon, 17 (1975):150	Hrvatska Dubica, wood fragments	417	ad 1533	60
Z-236	Radiocarbon, 17 (1975):150	Hutovo Blata, oak fragment	430	ad 1520	60

Sean McGrail examining re-used boat timbers on the Custom House site in London.

5.5. DATING OF SOME SHIPS' TIMBERS FROM WOODQUAY, DUBLIN

M. G. L. Baillie, Belfast

Between 1969 and 1973 the opportunity existed to examine large numbers of oak timbers from the excavations in Dublin conducted by Mr. B. Ó Ríordáin on behalf of the National Museum of Ireland. By 1973 a floating chronology of 452 years had been constructed covering the approximate period A.D. 850 to 1300 (Baillie, 1973). Despite considerable efforts to complete a continuous and independent Dublin area chronology it proved impossible to extend this floating chronology across the 14th century due to a serious depletion in local material for this period. More recently it has been possible to define the dates of the Dublin chronology as A.D. 855 to 1306 by comparison with 1000 year Scottish and North Irish chronologies.

The timbers used in the original construction of the Dublin site master were selected on the basis of local origin. Any timbers which showed signs of having served in the construction of a boat or ship were ignored for the purposes of chronology building. In this way it was hoped that the possibility of including exotic material could be avoided. At that time it seemed reasonable to assume that ships or boats broken up or repaired at Dublin might well have distant origins, since trade was an important facet of the life of this port. In addition, in 1973 it was not considered worth looking for cross-agreements between stray ships' timbers and non-Irish master chronologies for the 12th and 13th centuries, since only the basic German chronologies of Huber and Hollstein were available.

The situation has changed considerably since 1973 and recently it has been possible to compare ring patterns with master chronologies (Fig. 1) from Dublin, S.W. Scotland, Germany and S.E. England, the latter chronology kindly supplied by J. Fletcher.

Since 1973 further excavations have been conducted at Wood Quay, Dublin by Mr. P. Wallace also for the National Museum of Ireland. The site, on the 12th-13th century waterfront, proved to be rich in ship derived timbers. This pilot study was conducted on stray ships' timbers built into a series of early 13th century revetments and mid-13th century drains. Apart from dating the timbers, a doubtful exercise in itself, since few had sapwood and most would have derived from vessels 20 to 50 years old at the time of breaking, the hope was that some indication of area of origin could be deduced (Farrell and Baillie, 1976).

In the first instance 9 timbers, S1-9, were investigated. These were unquestionably planks derived from clinker built vessels, still exhibiting clench nails and hair caulking. Of these, 6 cross matched with the existing Dublin gies. Of the 6 dated timbers the highest correlation value in each case occurred against the Dublin chronology (see Table 1A). In order to boost the number of timbers tested, a further batch of 6 ships planks, Z 1-6 were measured. The results from these are shown in Table 1B. In each case the individuals matched extremely well with Dublin and much less well with the other chronologies, with the exception of that for S.W. Scotland. The quality of the cross-correlations can be gauged from Fig. 2, which shows a portion of the ring pattern of timber Z 1 against the Dublin Index chronology, and from the 't' value, 10.2, using the Belfast Program CROS (Baillie and Pilcher, 1973).

What can we make of these results? Obviously any conclusions must be tentative as the sample studied was so small. However, we can infer that the timbers are likely to be of Dublin origin and can hardly have come from an area outside the Irish sea coast. So it appears that we are not dealing with exotic material in the sense of coastal European or Mediterranean but with vessels which may have been built or extensively repaired in Dublin. If these were merely fishing boats or lighters rather than trading vessels then the local origin is highly reasonable. Unfortunately, most of the ship remains at Dublin are disarticulated planks and only in a few cases will it be possible to arrive at an estimate of original size. That is the work of the nautical archaeologists. It needs to be stressed that there may well have been a considerable fishing industry out of Dublin, if the numbers of fish remains throughout the archaeological deposits are any measure of the importance of that commodity. Thus perhaps we have to expect boat remains from such sites to be predominantly local in origin.

While the number of samples studied was rather small to support generalisations, it is of interest that 12 out of 15 timbers appear to be of local origin while the remaining three do not give indications of agreeing with any other area. Since these timbers were selected randomly, on the sole basis of nautical origin, it is suggested that the majority of ships' timbers from the Dublin sites will ultimately prove to be of Dublin origin. It now seems worth repeating the experiment on some of the larger articulated samples from the same sites to test this theory.

On the present rather slight evidence it looks as though ships or boats tend to be broken up or repaired in their home port. This supports the generalisation that any vessel, not lost at sea, will ultimately return to its point of origin or ownership.

References

Baillie, M. G. L. A Dendrochronological Study in Ireland, PhD Thesis, Queen's University, Belfast, 1973.

Baillie, M. G. L. and Pilcher, J. R. A Simple Cross-Dating Program for Tree-Ring Research, Tree-Ring Bulletin, 33, 1973, 7-14.

Farrell, A. W. and Baillie, M. G. L. The Use of Dendrochronology in Nautical Archaeology, Irish Archaeological Research Forum, 3, 1976, 45-55.

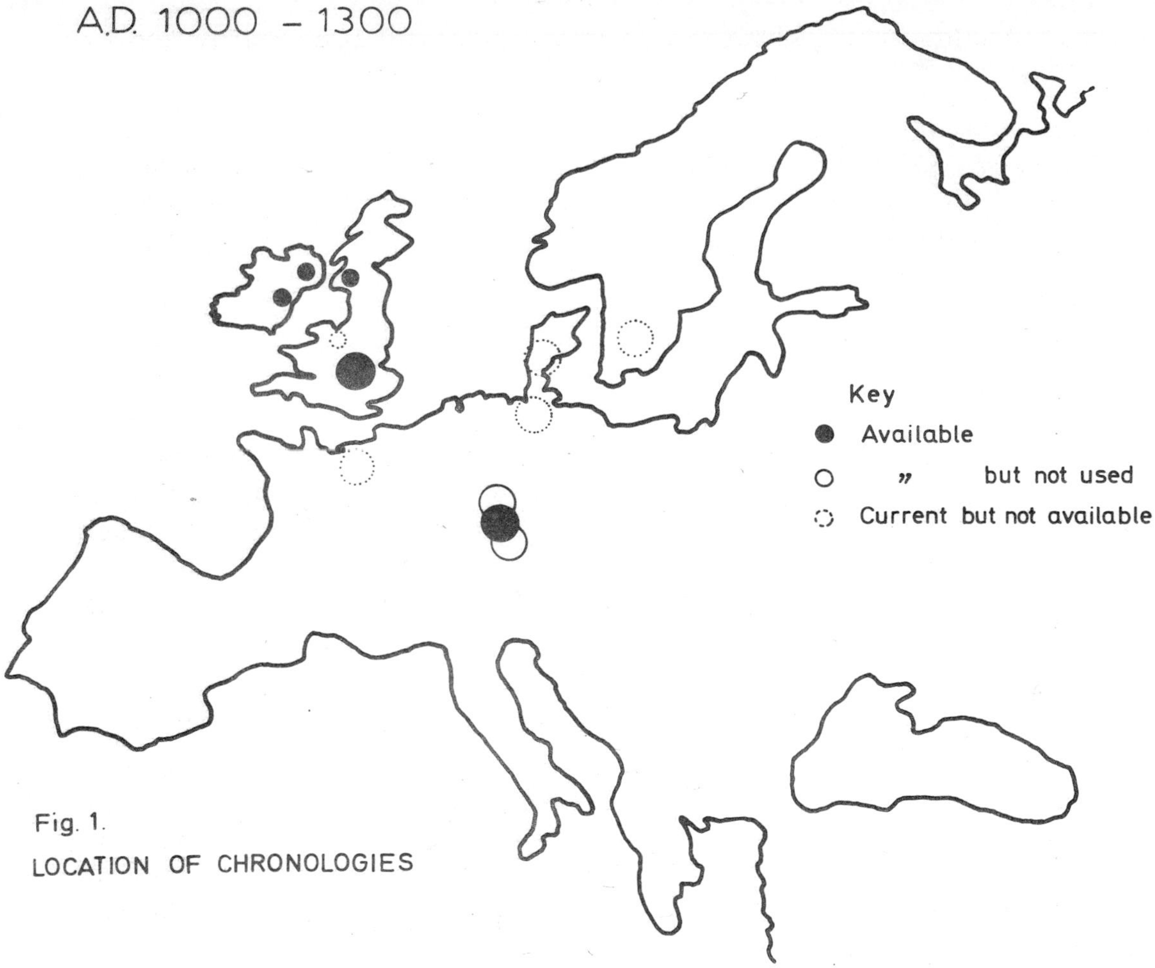

Fig. 1.
LOCATION OF CHRONOLOGIES

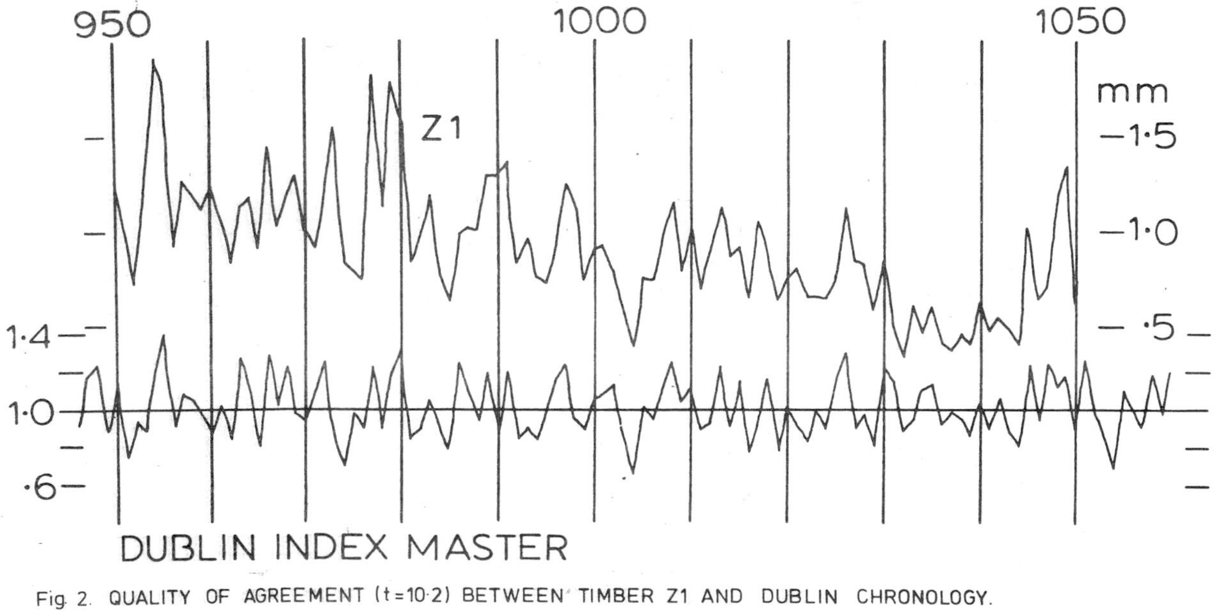

Fig. 2. QUALITY OF AGREEMENT (t=10·2) BETWEEN TIMBER Z1 AND DUBLIN CHRONOLOGY.

	DATE	DUB.	BEL.	SCO.	ENG.	GER.
S1	1096	4·6	1·5	3·5	2·4	1·8
S2	1156	5·8	2·2	4·2	3·3	0·2
S3	1180	3·6	1·2	2·9	3·1	1·2
S4	1121	5·4	0·5	·9	0·5	—
S5	1157	4·8	0·4	3·8	—	0·7
S6	1143	7·0	1·1	4·2	1·3	0·7

A

	DATE	DUB.	BEL.	SCO.	ENG.	GER.
Z1	1089	10·2	1·4	3·8	3·3	1·5
Z2	1162	8·6	4·1	5·4	2·7	0·5
Z3	1155	7·3	2·3	2·8	2·2	0·8
Z4	1115	9·7	—	4·5	2·8	—
Z5	1123	8·6	2·1	3·5	4·0	3·0
Z6	1209	10·0	1·6	3·2	3·8	2·4

B

't' correlations

Tables 1A & B

Dates of Outer Rings of Ships Timbers. Agreements of Ring-Width Pattern with Master Chronologies for Dublin, Belfast, Scotland, England, and Germany. A. First Batch B. Second Batch

5.6. DENDROCHRONOLOGICAL DATING FOR ARCHAEOLOGICAL RESEARCH IN THE GDR

Manfred Jährig, Berlin (GDR)

The main task of dendrochronology at the Central Institute of Ancient History and Archaeology is to investigate timber finds dating back to the Slavonic settlement (Ed. Early Medieval) period. As is generally known, Slavonic archaeology was strongly neglected for decades. For this reason major excavations started in the GDR in the 1950s to complement the history of the Slavonic tribes west of the rivers Oder and Neisse. This research raised fundamental dating problems which made it seem expedient to apply dendrochronological methods for more exact chronologization.

Thanks to large-scale excavations and due to the fact that the settlements and strongholds are frequently located at rivers or lakes and in inaccessible lowlands, much wooden material suitable for dendrochronological examinations has survived. This will make it possible to elaborate a standard curve for the GDR as far back as the 6th century. Our task is facilitated by material and cultural sources, especially by the clear stratigraphic connections between the finds from different cultures. We are also making efforts to support our results with radiocarbon datings carried out in the laboratory of the Institute. There are, though, major uncertainties in some datings probably caused by the condition of the materials.

At the moment we are examing excavated oak timbers dating back to the Slavonic settlement period of the 6th to 10th centuries. Here it is mainly a question of finding relative chronologic clues about the duration and sequence of settlement periods for a certain excavation complex with the help of dendrochronology. Our main aim is to elaborate a standard curve for the GDR and to date it exactly. For dendrochronology, we have to split for the time being the territory of the GDR into a central and a northern region, both lowlands.

RESULTS

Cathedral Island, Brandenburg

Excavations were carried out on Cathedral Island in Brandenburg (on the Havel), formerly territory of the Hevellian princes' castle and seat of the bishop of Brandenburg. Stratigraphic evidence places the more recent outer front of the castle in the 10th century.

Synchronization of samples produced a 135-year mean curve dated from the 9th till the middle of the 10th centuries with the help of the southwest German standard chronology. While the samples of the seventh fortification period dated to the middle of the 10th century, a plank of the preceding (sixth) period was dated to the end of the 9th century.

This is good evidence for the chronological classification of excavated materials and for the development of the fortification. The results confirm a major extension of the castle in the middle of the 10th century.

Slavonic rampart at Grodisch near Wiesenau

Wood samples from five sections excavated at this settlement (Eisenhüttenstadt district) were evaluated and a 200-year relative, i.e. floating, chronology was compiled.

The rather regular intervals of 30 to 50 years between the stages of construction in the sections attracted our attention. Within the various sections they often had the same synchronous position: this is probably not accidental.

Dendrochronological results do not confirm a clear succession of construction of the inner and outer sections. The bridge was first built simultaneously or shortly after the construction of the inner and the outer sections, about at the same time as the repair or extension of the outermost section. The reconstruction of the bridge was possibly partly connected with the repair or modifications of the rampart, but should probably be regarded as a continual process.

The results of this application of dendrochronology were of great value to the archaeological research. The excavations had been analyzed by the usual criteria and attempts made to find out the stages of construction of the rampart. Irregularities pointed to modifications, repairs and reconstruction but could not be determined in detail. The examination of very many oak samples has made it possible for the first time to establish a more exact dating of the progress of construction and modifications for a Slavonic 'burgwall'.

An important conclusion can be drawn from this research: even "one-phase" fortifications that appear to be homogeneous underwent repair, modification and changes more often than is generally supposed.

<u>Tornow and Vorberg</u>

Excavations at Tornow, Calau district (Lower Lusatia) have made it possible to examine an early Slavonic fortification. The results achieved were complementary to those from a similar stronghold located at Vorberg, also in the Calau district. This has brought new knowledge about early Slavonic history and culture. The fact that the excavated parts of the stronghold were in good condition allowed the reconstruction of a refuge (stronghold A) and a smaller residence (stronghold B) dating back to the 7th/8th century and 8th century, respectively.

Relative chronologies were compiled for the following phases of construction:

Lütjenberg for 169 years,
Borchelt, settlement phase A, for 186 years,
Borchelt, settlement phase B, for 154 years and
Vorberg, phase A/1, for 61 years.

The mean curves of Lütjenberg, Vorberg, and Borchelt, phase A were synchronized and gave a 231-year floating chronology. Even though the dating of this is not yet absolute because of the lack of a standard curve for Lusatia, the cross-linking of sequences for German and Slavonic settlements represents great progress.

At the moment we are trying to correlate the Slavonic strongholds of the 6th to the 10th centuries of the central region of the GDR with the help of dendrochronological evidence.

Relationship of Strongholds

The relative connections between the different strongholds make it possible for the archaeologist to come to conclusions about the development of strongholds in the Slavonic settlement period. We have a tentative dating for the various strongholds on the basis of the floating mean curve for Brandenburg from which the following picture emerges as a working basis:-

(a) The Lütjenberg fortifications would then date back to the 6th or the beginning of the 7th century;
(b) The Vorberg fortifications would date back to the first half of the 7th or the early second half of the 7th century;
(c) The Tornow, settlement phase A, fortifications would be dated to the second half of the 7th century or the beginning of the 8th century, while the phase B fortifications would be dated to the 8th century.
(d) The mean curve of Wiesenau would then date from the middle of the 7th to the middle of the 9th centuries.

This would give a relative chronology of 492 years, from the middle of the 5th to the middle of the 10th centuries for the central region of the GDR.

Rampart of the Mecklenburg fortifications

This fortification was one of the strongholds of the Obodritic princes in the 10th, 11th and 12th century. Excavations proved that the 'burgwall' dates back to the early Slavonic period and was used until late in the Middle Ages.

A 340-year mean curve, with a possible dating from the beginning of the sixth to the middle of the 9th century was derived for the excavation at the Mecklenburg rampart near Dorf Mecklenburg, Wismar district, by synchronizing the sequences of the oak samples. It has to be stressed though that this is a first hypothesis which needs support from further evidence. That is why dendrochronological examination of samples from the 9th/10th century Slavonic stronghold of Gross Raden, Sternberg district, will start this year (1977) to find parallels with the Mecklenburg rampart and to confirm their dating. It is possible that we will have to check further samples from earlier excavations at Teterow, Suckow and Behren-Lübchin (all in the Mecklenburg region). We hope that the hypothesis proposed here will be supported and complemented by dendrochronological evidence from the excavation site at Haithabu (Schleswig-Holstein).

When the examination of samples from Gross Raden is finished we hope to have a relative chronology of the Slavonic settlement period from the 6th to the 10th centuries for the northern region of the GDR as the second main investigation and thus be in a position in the near future, to present a standard curve for the northern and central regions of the GDR between the rivers Elbe and Oder on the basis of these results and others for historical buildings and from recent material.

References

1. Jährig, M. Zum bisherigen Stand der Eichenjahrringchronologie und zur Dendrochronologie in Norddeutschland, Ausgrabungen und Funde 13, 1968, 281-286.

2. Jährig, M. Zu einigen Grundsatzfragen der Dendrochronologie und ihrer Grenzen, Ethnographisch-Archäologische Zeitschrift 13, 1972, 39-67.

3. Jährig, M. Die Ergebnisse der dendrochronologischen Untersuchung von Holzproben aus Tornow (Lütjenberg und Borchelt) und Vorberg, in: Herrmann, J. Die germanischen und slawischen Siedlungen und das mittelalterliche Dorf von Tornow, Kr. Calau, Schriften zur Ur- und Frühgeschichte 26, 1973, 243-266, Berlin.

4. Jährig, M. Einige Bemerkungen zur Stellungnahme von A. Delorme, Zeitschrift für Archäologie 9, 1975, 135-142.

5. Jährig, M.: Zum Stand der dendrochronologischen Arbeiten, Ausgrabungen und Funde 21, 1976, 191-194.

6. Jährig, M. Die Ergebnisse der dendrochronologischen Untersuchung der Holzproben vom slawischen Burgwall "Grodisch" bei Wiesenau, Kr. Eisenhüttenstadt, Zeitschrift für Archäologie 11, 1977, 101-120.

5.7. DENDROCHRONOLOGICAL DATING OF THE MEDIEVAL SETTLEMENT OF HAITHABU (HEDEBY)

D. Eckstein, Hamburg

Summary

In the medieval settlement of Haithabu the stratigraphical conditions are very unfavourable for the archaeological evaluation of the findings. By dendrochronological analysis of thousands of wood samples from houses and other objects it has been possible first to get an absolute date for the beginning of the settlement; secondly, to establish the chronological relation both between the complex wooden constructions within the excavation, as well as with contemporary Viking and Slovanic settlements in Northern Germany and Denmark.

Introduction

The situation of the early medieval site of Haithabu (Hedeby) in relation to other sites between the North Sea and the Baltic is shown in Fig. 1.

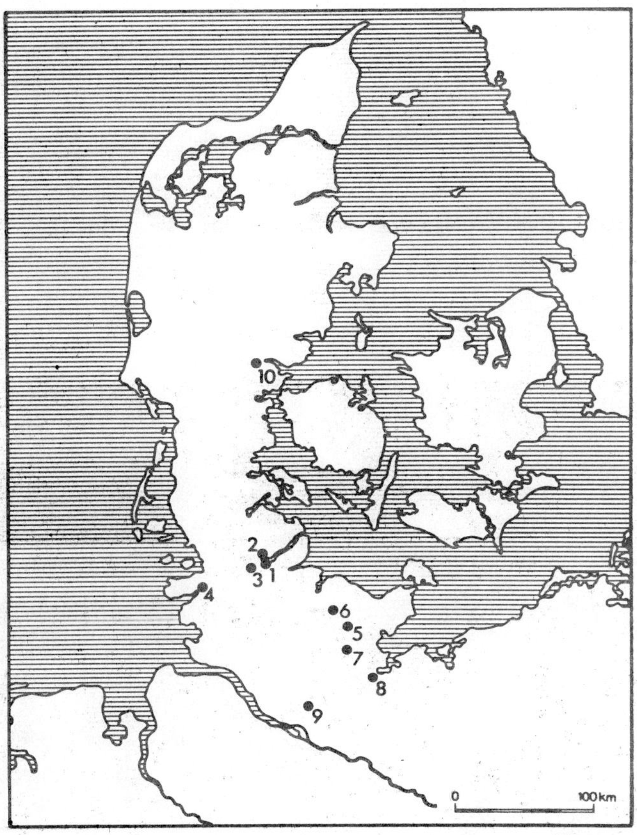

Fig. 1 Medieval Sites dated by Dendrochronology:

1. Haithabu: Trading Port of Viking times.
2. Schleswig: Subsequent settlement on the Schlei.
3. Danewerk: Early medieval fortification system.
4. Elsenhof: Rural settlement on a mound.
5,6,7,8. Bosau, Scharstorf, Warder and Lübeck: Slovanic fortresses.
9. Wittmor: wooden trackway.
10. Ravning: Danish bridge.

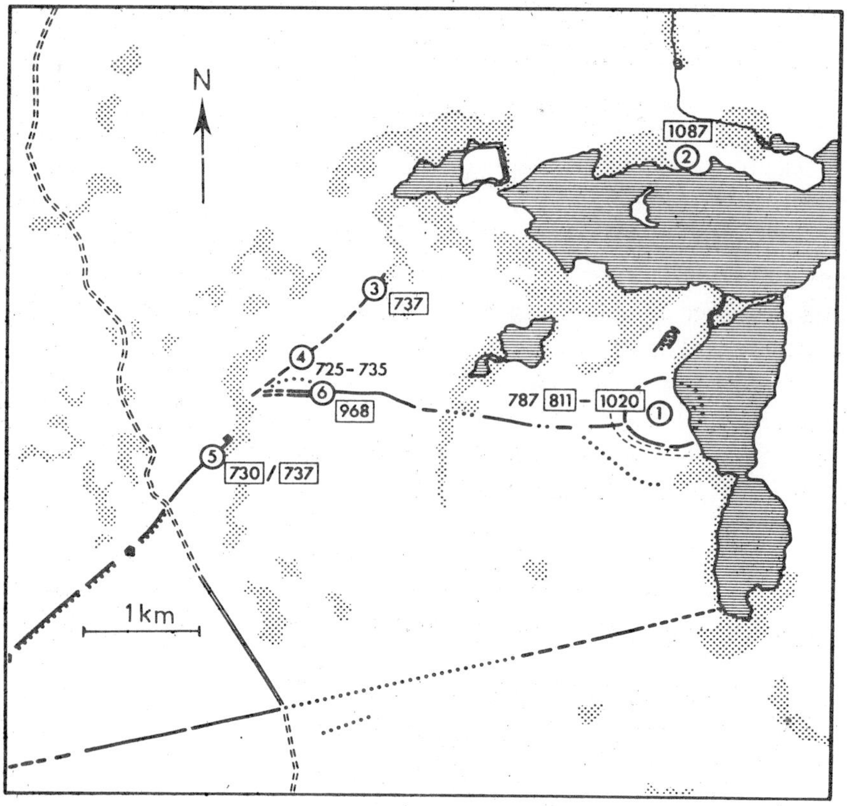

Fig. 2 Situation of Haithabu, 1, and its successor, Schleswig, 2, in relation to the Schlei and the chronology of Danewerk sites 3, 4, 5 on the lowland-pass across the peninsular.
Dates within frame, e.g. [737], imply exact dating from many samples with bark. Others refer to a single sample or to several without bark.

Plate 1. Haitabu. The settlement lay within the semi-circular area (cf. Fig. 2)

Haithabu (Fig. 2 and Plate 1) lay to the west of a bay of the Schlei, a long inlet from the Baltic (Jankuhn, 1976). Its rise to be one of the largest trading centres in Northern Europe during the early Middle Ages was determined by its favourite geographical situation at a junction of several trade routes and by political and economic conditions. At its culmination in the 9th and 10th century, its business connections extended in the west as far as Spain and in the east to Byzantium. Moreover, the settlement was an important place for trade from Scandinavia and, by manufacturing its own goods, was also of great local importance. Thus Haithabu attained an organised urban status (Schietzel 1975).

Haithabu is well suited to comprehensive excavations as Schleswig, the subsequent settlement, was founded on the other side of the Schlei. In the excavated areas (marked in white on Plate 1) the constructions were entirely of timber. They are well preserved due to the high ground-water level. Constructional details and the appearance of the houses can be evaluated even in the deepest layers.

The local peculiarities of the houses (Fig. 3) make it difficult, however, to follow consecutive phases from the archaeological evidence alone (Eckstein and Schietzel, 1977). Nearly all the buildings were erected in the form of halls without internal posts, so that the weight of the roof was carried exclusively by the outer walls. The supporting elements for these walls—particularly in the wattle-work houses ("Flechtwandhaus")—consisted only of thin planks stabilised by being inserted deep into the ground. For this a trench was normally made around the building site. This way of erecting houses caused comprehensive displacements of the soil (Fig. 3 shows three consecutive building activities in area A, and two in area B).

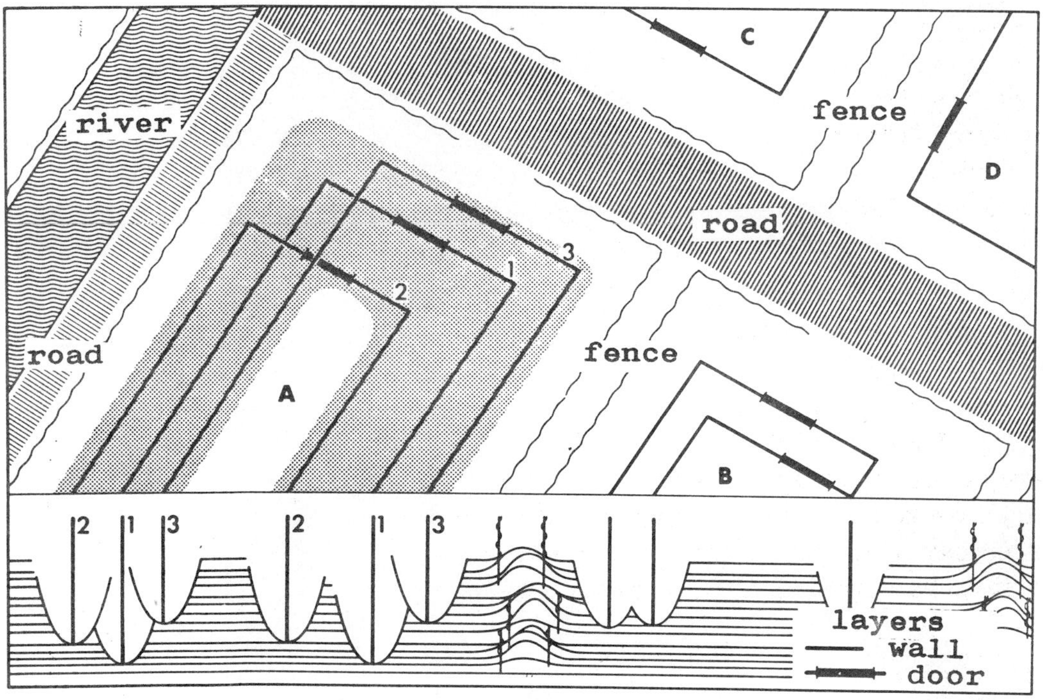

Fig. 3 Stratigraphical conditions in Haithabu

The narrowness of the sites did not allow a lateral displacement of consecutive houses and a zone of intensively rummaged layers was generated within every site; this made it impossible to stratify the constructions in terms of a time sequence. Added to that the limitations of the plots by parallel fence-lines make it difficult to determine contemporary horizons. The area between the fences was used to deposit wastes. This resulted in a dam-like increase of the layers of the settlement. On the other hand, the real building-plots are poor in archaeological findings. So extensive correlation of stratigraphical layers beyond the river or roads is almost impossible. This unsatisfactory condition was largely overcome by applying dendrochronological dating.

The excavated remnants of the constructions consist mainly of oak, for the most part split into planks and only seldom cut into halves or used in the round. A considerable number contain sapwood and even bark. Vertical timbers exist up to 1 m in length and at times the excavations reached 5 m below the present surface. When the riverlet, roads and fences around the building-plots, together with the various walls and entrances to the houses from the deepest layers of the settlement are converted into a single plan (Fig. 4), there emerges an impression of an extensively organised community. Its axes are determined by a canalized riverlet and by roads running at right angles or parallel to it. The continuance of these base-lines is obvious from the finding of constructions that penetrate each other intensively; continuous settlements took place on the same building sites.

More than 4000 oak wood samples were collected during the excavations from 1966 to 1969, in order to obtain chronological relationships between objects situated on top or adjacent to each other or even overlapping. On this basis, material from all over the settlement should be capable of being related in order to provide contemporary horizons.

The dating of Haithabu

The beginning of continuous building activities in Haithabu has been allocated by dendrochronology to the period 800-810 A.D. Only a few wood samples were found from the 8th century, the oldest being 783 A.D. These older timbers could not yet be related to any construction. Only after all the samples have been evaluated can it be decided whether they were in secondary use or of an original construction.

In the upper layers, most of the timbers have been destroyed by micro-organisms. Consequently, houses can only be recognised by their fireplaces. From the latest period some wooden wells are preserved because they had been dug deep into the ground-water. The oaks for the most recent well were cut in 1020 A.D. Thus, for the archaeologically expected period of 250 years for the settlement, about 220 years could be verified dendrochronologically. The 9th century is especially rich in preserved wooden remains. The local tree-ring chronology of Haithabu has prolonged the regional chronology for Schleswig-Holstein back to 436 A.D. (Eckstein 1976).

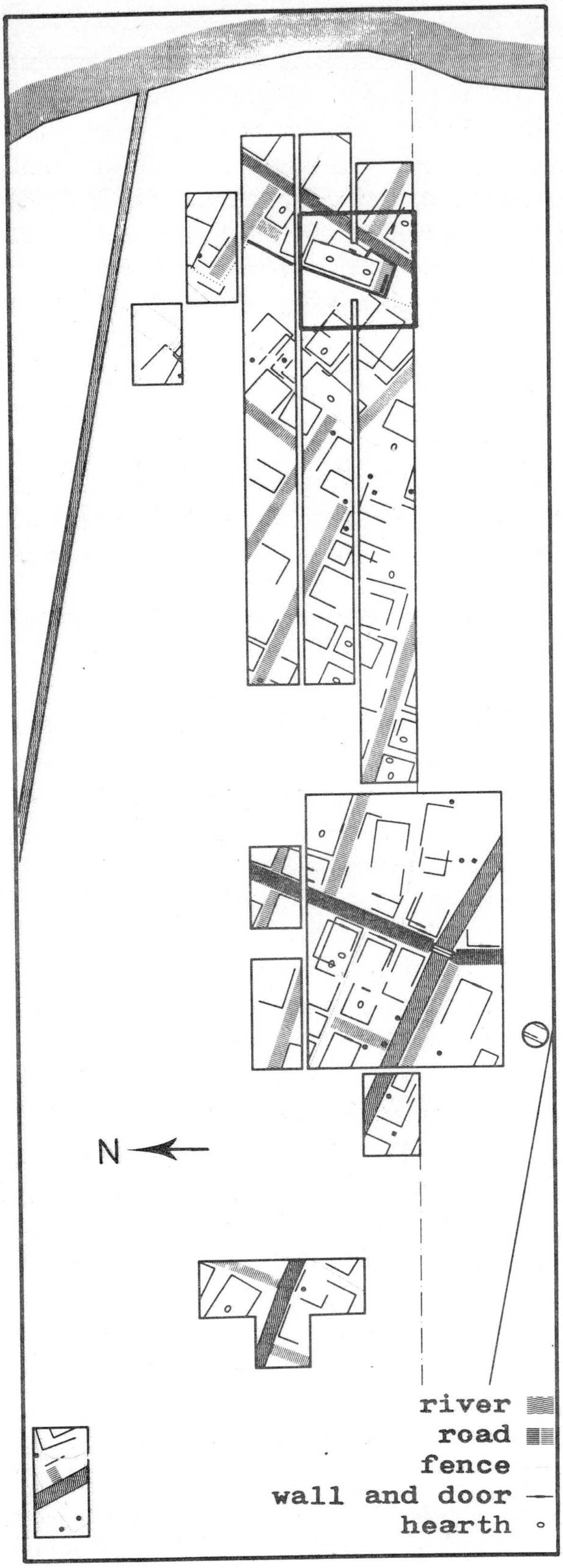

Fig. 4 The construction of the lowest layers of the excavated areas (marked white on Plate I). The waterfront of the Schlei is at the top.

Dendrochronological investigations of the construction timber of one building

The following example compares archaeological and dendrochronological results. From numerous horizontal and vertical planks, the whole ground-plan of a building, consisting of three-parts, was archaeologically determined. A rectangular stone fireplace indicates the middle part of the house as the real living room (Fig. 5).

The room with the two doors in its outside walls was used for working purposes, as judged by the remnants of a collapsed round-headed oven. The other end of the house, separated by a strong wall of planks, served presumably for some production purposes. The total building, including its oblique outside pillars, from which the constructional connections to the walls are no more preserved, was archaeologically determined as a chronologically homogeneous building.

This statement, as far as the vertical planks and horizontal doorsteps are concerned, was fully confirmed by dendrochronology. The building timbers were cut in 870 A.D.

In addition, there are three squared timbers near the northern gable, which on the basis of dendrochronological results belong to the same date of erection. They are cut out of the same tree from which also a plank of one corner of the house and another plank of the bulk-head were taken (Fig. 5 a + c). The trough in front of the oven was determined to be 12 years later than the building timber (Fig. 5b). Consequently, the house itself must have been in use at least that long. This statement is supported by two vertical planks beside the fire place which probably belonged to one construction — possibly to attach the boilers — and were used 12 years after the erection of the house; one of them originates from the same tree as the timber of the trough.

The oblique pillars, which archaeologically belong to the house, give felling dates which are different from the main building period. Those from the front of the house and the foundation timbers were felled in 845 A.D. (Fig. 5c). Since a storage of wood over 25 years can be excluded, they must be re-used material, presumably from a former roof. Two pillars from the back of the house together with a sample from the fence running parallel to it, belong to two individual trees. Both oaks were felled in about 873 A.D. This indicates simultaneous building of the house and the fence.

Chronological relationships between consecutive and adjacent findings

By tree-ring evaluation of all the samples (some 1500) from one and the same area (see Fig. 4) much information has been derived about the temporal sequence of the finds and their relationship to adjacent building sites, for example in the period 831 to 882. Thus a detailed impression of the developments at Haithabu during half a century, that would have been impossible to obtain by other methods, has been obtained (Eckstein and Schietzel 1977).

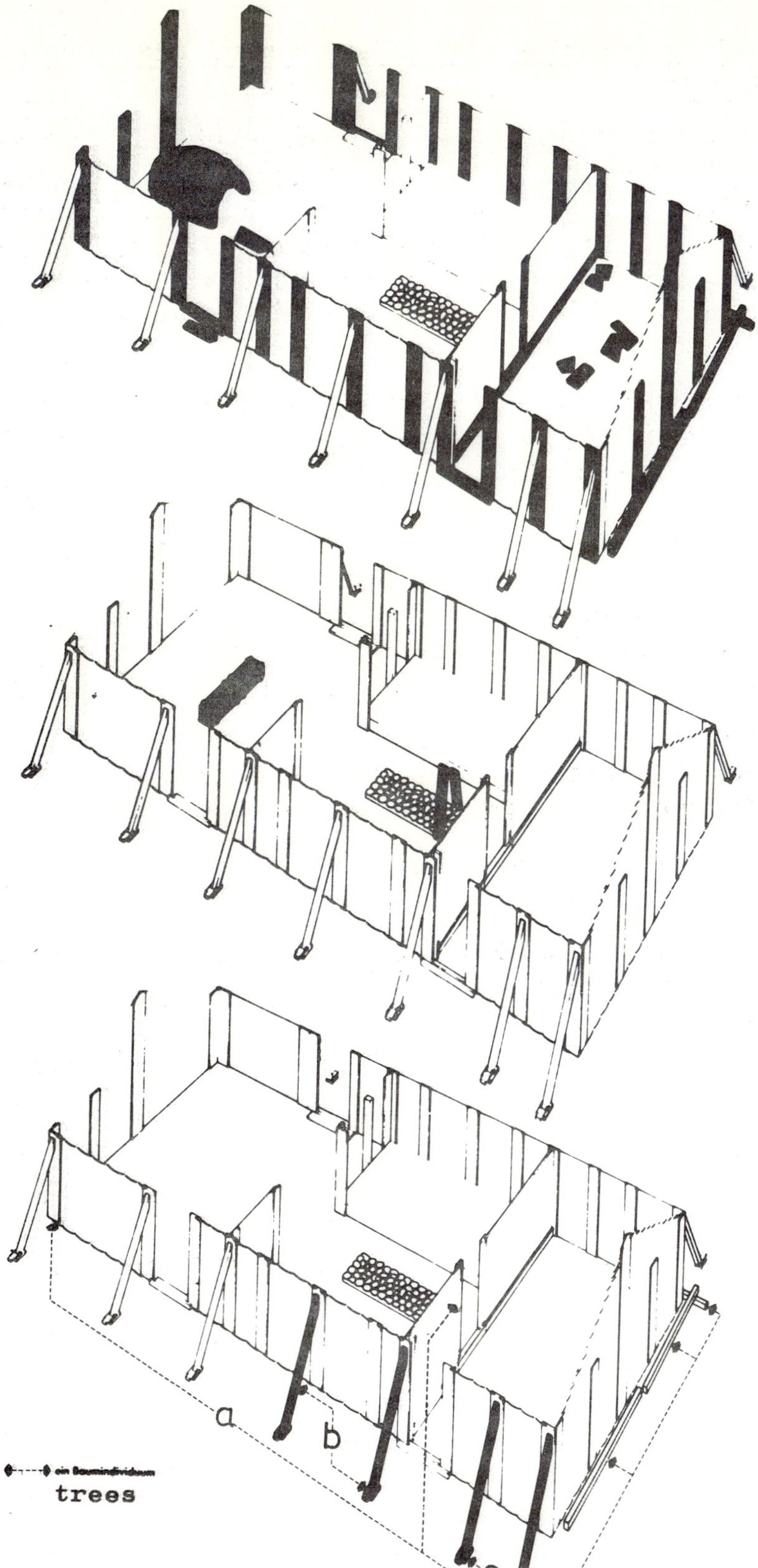

Fig. 5 A timber-framed wattle house (position indicated by the heavy frame of Fig. 4) showing timbers dated by dendrochronology:
Top: The original arrangement, timbers of 870 A.D.
Middle: Situation 12 years later.
Bottom: Timbers from individual trees, a. b. c.

Regional chronological relationships between Haithabu and other medieval sites in Northern Germany

By means of the regional chronology for Schleswig-Holstein other excavations, some shown in the map (Fig. 1), have been dated. The main results for the area around Haithabu are included in Fig. 2. Between the youngest remnants of Haithabu (1020 A.D.) and the oldest wooden findings of the settlement of Schleswig (1087 A.D.) — the successor of Haithabu — there is a gap of more than half a century, although this period seems to be covered by archaeological findings.

The oldest part of the comprehensive system of fortifications of the so-called Danewerk are dated to 730/737 A.D. Thus the defences of the lowland-pass near Schleswig were erected some 75 years earlier than the excavated areas of Haithabu.

References

Eckstein, D., 1976, Die absolute Datierung der wikingerzeitlichen Siedlung Haithabu/Schleswig mit Hilfe der Dendrochronologie. Naturwiss. Rundschau 29, 81-84.

Eckstein, D. and Schietzel, K., 1977, Zur dendrochronologischen Gliederung und Datierung der Baubefunde von Haithabu. Berichte über die Ausgrabungen in Haithabu Nr. 11, 141-164.

Jankuhn, H., 1976, Haithabu. Ein Handelsplatz der Wikingerzeit. Neumünster.

Schietzel, K., 1975, Haithabu — Ein Beitrag zur Entwicklung frühstädtischer Siedlungsformen Nordeuropas. Ausgrabungen in Deutschland 1, Part III. Mainz, 57-71.

Acknowledgement

The figures are contributed by "Ausgrabungen Haithabu".

5.8. THE DATING OF COFFINS AND STAVES IN MEDIEVAL LUND

T. Bartholin, Lund

Summary

Samples from archaeological excavations in medieval Lund show the town was founded about 1000 A.D. Coffins from two of its cemeteries show the older was in use from 1000-1050, the younger from c. 1050-1100. The construction date of a stave church has been found by tree-ring analysis to be about 1050 and various secular buildings have been dated.

The main work in the construction of a chronology for southern Sweden has been concentrated on samples from archaeological excavations of the early medieval town (Mårtensson 1976). Dendrochronological dating (Bartholin, this volume, page 125) has made it possible to derive many historical facts.

The discovery of wooden coffins in two cemeteries gave a good base from which, with other finds, a 561-year chronology was constructed. At first it was floating but thanks to dates kindly provided by Dr. Eckstein for his Haitabu chronology (Eckstein 1976) it is now known to cover the years 578-1138. Parts of the respective mean curves for the stave church and for the coffins are shown in Fig. 1. Notice the sensitivity of the mean curve for the wider-ringed oaks from the staves (cf. Mayer 1957) compared to the narrower-ringed ones from the coffins. The curve has been extended to 1293 by samples from churches, but more samples in the period 1250-1350 are required to link with the chronology that stretches from the present time back to 1274.

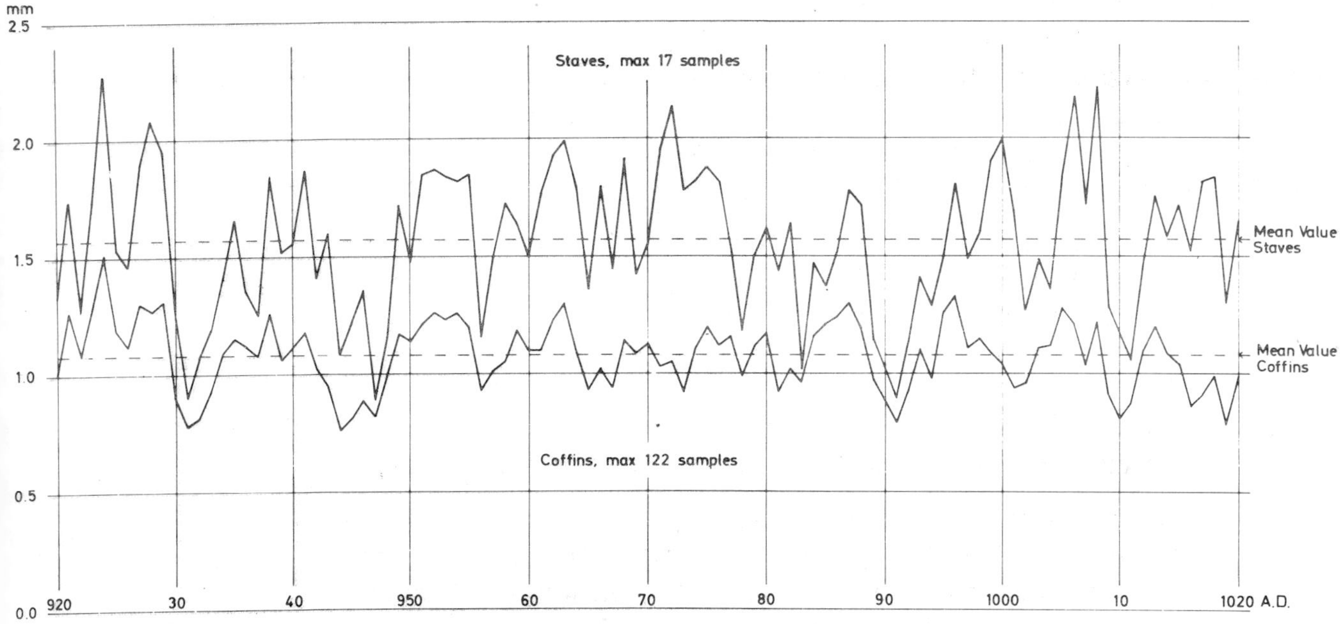

Fig. 1 Part of oak mean curves, Lund.

Coffins

There is a reused coffin plank on which the oldest ring is for the year 994. 26 coffins from cemeteries 1 and 2 have all their sapwood rings present and four of these even have bark. The years covered by their sequences and others with some sapwood are shown in Fig. 2. The terminal dates shown in the figure and for others in this paper are based on a correction for missing sapwood rings of 20 $\pm$5 years. When sufficient samples to make a statistical examination of the variation of tree rings in oaks from south Sweden become available, the results may be changed slightly. The dates at which the trees were felled are evenly distributed in time, and in view of the size of the cemeteries, we accept them as indicating the respective periods of use of the two cemeteries.

A layer of sandy clay, deposited by a flood over a short period, was formed after the closing of the earlier cemetery and before the consecration of the later one. The best estimate for the date of this deposit (and consequently for the closing of the earlier and the consecration of the later cemetery) is obtained by prolonging the sapwood from the youngest coffin in the older cemetery, and for the oldest coffin in the younger cemetery; this gives the year 1051.

From Fig. 2 we can see that the consecration of the older cemetery must have been around the year 1000, and that the closing of the younger cemetery must have been close to the year 1100, even though there is no sapwood on the youngest coffin.

Two coffins from the third churchyard (cemetery 3) belonging to a 12th century church, have also been dated.

Stave Church

Part of this, situated in cemetery 2, has been excavated. Due to lack of sapwood all the enormous staves could not be dated exactly; for many, the earliest year for felling the tree corresponded in date with the oldest coffin in the cemetery. Archaeological observations about repairs to the church are confirmed by dendrochronology, which also indicates the occurrence of reused staves with the earliest year for felling, corresponding with the oldest coffin in the old cemetery (Fig. 3).

Remains from secular constructions have also been dated; the main interest concerns a Viking house of the so-called "Trelleborg Type", dated to c. 1016, with repairs six years later.

Conclusions

Before the dendrochronological datings, the prevailing opinion of the early history of Lund (Blomquist and Martensson 1963) was as follows:

> "The town was founded by Canute the Great in 1020. The stave church, the biggest in Scandinavia, 25.7 m long, was built as a bishop's church for the first Bishop of Lund about 1060 and was used, together with its cemetery, for nearly 100 years to c. 1160".

We can now see that there was activity in the town earlier than 1020 and so theories about coin fabrication around 970-980 have once more become of

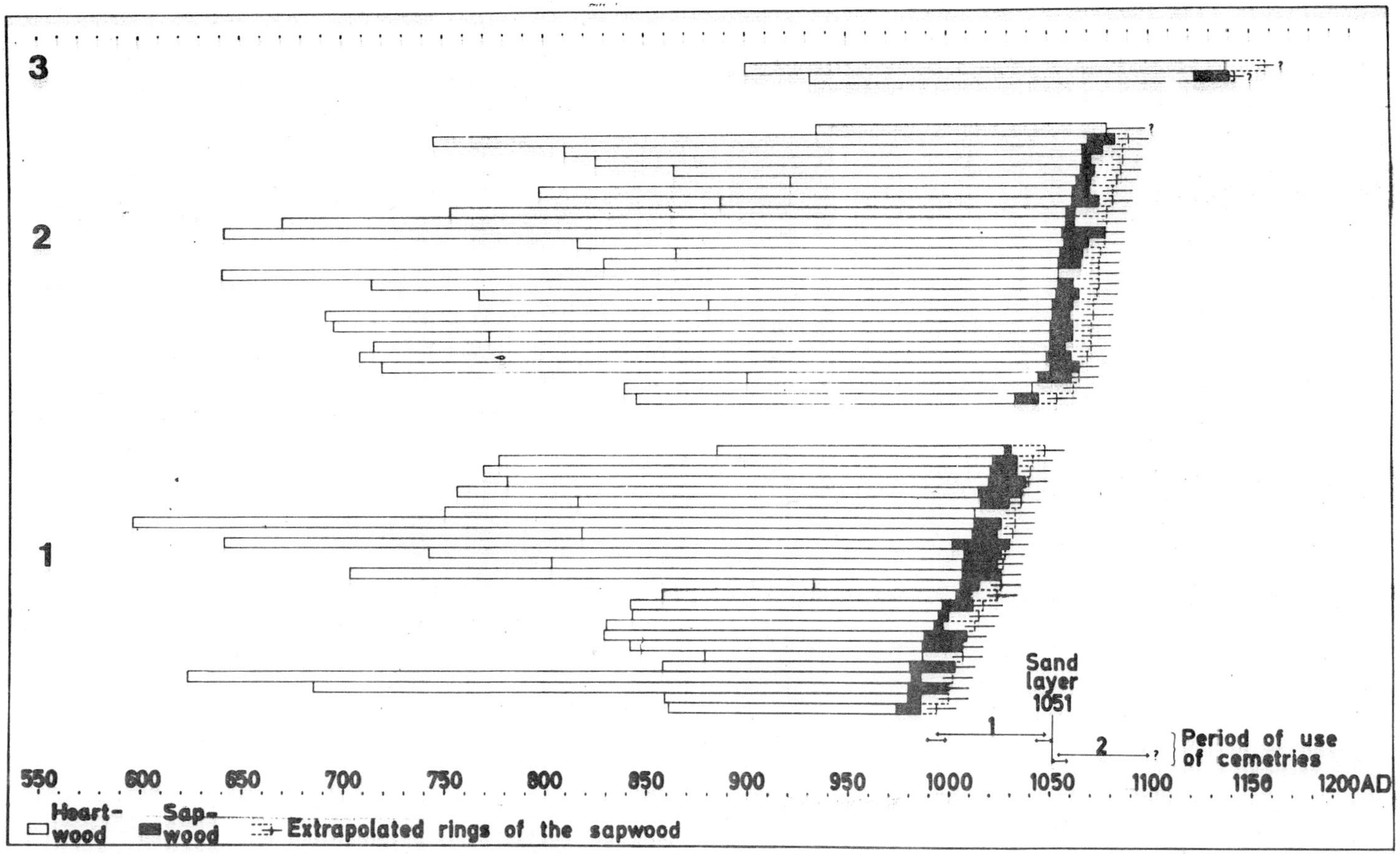

Fig. 2. Tree-ring sequences for oak coffins, Lund.
1,2,3 refer to Cemeteries (see text).

of immediate interest (Holmberg 1977). The younger cemetery (No. 2) and presumably also the stave church in it existed for only about 50 years, from c. 1050 to 1100 and the year 1060 is still reasonable as the year of their consecration.

References

Blomqvist, R., and Mårtensson, A.W., 1963: Thulegrävningen 1961. Archaeologica Lundensia, 2, 1-300.

Eckstein, D., 1976: Absolute Datierung der wikingerzeitlichen Siedlung Haithabu/Schleswig mit Hilfe der Dendrochronologie. Naturwissenschaftl. Rundschau, 29, 81-84.

Holmberg, R., 1977: Den skånska Öresundskustens medeltid. Acta Archaeologica Lundensia, No. 11, 1-250.

Mårtensson, A. W., ed., 1976: Uppgrävt forflutet for PKbanken i Lund. En investering i arkeologi. Archaeologica Lundensia, 7, 1-411.

Mayer, R., 1957: Untersuchungen über Kronengrösse und Zuwachsleistung der Traubeneiche auf süddeutschen Standorten. Allg. Forst- u. Jagdzeit., 129, 105-114, 151-163 and 191-201.

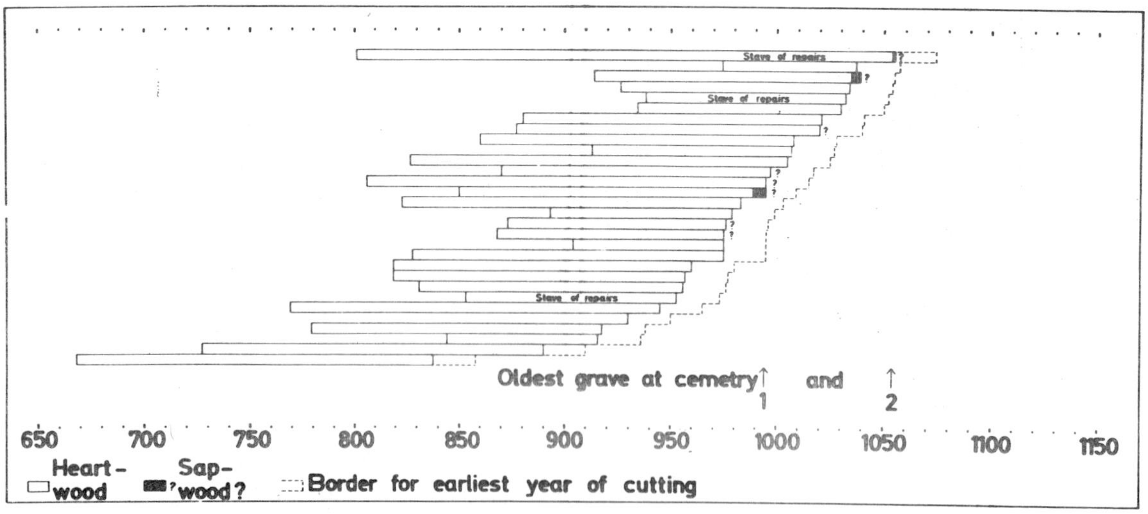

Fig. 3 Tree ring sequences for oak staves from church, Lund

5.9. SUBMERGED FORESTS AROUND THE BRITISH ISLES: THEIR DATING AND RELEVANCE AS INDICATORS OF POST-GLACIAL LAND AND SEA LEVEL CHANGES

Alan Heyworth, Aberystwyth

Summary

The situation of submerged forests around the coasts of the British Isles is shown. They are visible on the foreshore as stumps or fallen trunks. They were formed by the drowning of coastal woodlands by the post-glacial rise in sea-level, Fig. 1.

In age they range from more than 8000 years to less than 1000 years. Due to the superimposition of the isostatic rise of the land on the eustatic rise of the sea, the whole age-range is accessible in spite of the rise in sea-level during the last 8000 years of about 20 m. From their ring-width patterns in a particular locality at different elevations the rate at which the sea-level rose can be determined.

Trees far apart (in Bridgwater, Cardigan and Morecambe Bays) and now at slightly different elevations are found to have died, as a result of marine transgressions, at almost exactly the same time (about 4500 bp). This has enabled the relative vertical increments of the land to be calculated for the three sites.

Around the coasts of the British Isles are many sites where 'submerged forests' are exposed on the beach. These forest beds consist of stumps, still in the position of growth, and fallen trunks; often with associated peats. The trees may be very large and are usually extremely well-preserved. Plate 1 shows the submerged forest at Borth, in Cardigan Bay, Wales. The trees shown here date from about 5300 bp (all dates in radiocarbon years). These submerged forests are the result of the drowning, by the Post-Glacial rise in sea-level, of coastal woodland, Fig. 1.

Dendrochronological dating of the submerged forest trees can establish the patterns of local forest development in great detail, and is also invaluable in determining past sea-levels and vertical movements of the land. As the ages of the trees can be fairly accurately predicted, they are also a very suitable source of wood for radiocarbon calibration determinations.

THE FORMATION AND PRESERVATION OF SUBMERGED FORESTS

The trees in these forests were usually killed by a rising water-table, or by the ingress of salt-water rather than by catastrophic and permanent marine inundation. It was thought, at one time, that the submerged forests were all formed during the same episode, the Submerged Forest Period. It has since been shown, however, by pollen analysis and radiocarbon dating, that they were formed throughout most of the Post-Glacial period, the earliest probably as soon as trees reappeared with climatic amelioration.

Fig. 2 shows the rate and extent of eustatic sea-level rise in the last 8500 years. The differences between the three curves shown will be dealt with later, but all three are of the same general form. Submerged forests occur at all heights in the vertical range shown, the practical limit for dendrochronological purposes being set by the lowest tides. In the Bristol Channel, which has a very large tidal range, (13 m), Low Water Spring Tides is at about -6 m OD. In other places it is higher, normally about -2.5 m OD.

The trees in the submerged forests grew at about contemporaneous Mean High Water Mark, usually behind a coastal barrier of some kind. The earliest trees accessible are therefore those which grew when MHWM was from -6 m OD to -2.5 OD, depending on the tidal range. This would put the age limit at about 7000 bp in the Bristol Channel, and later, c. 6000 bp, elsewhere.

Another factor is, however, important. As well as the eustatic world-wide rise in sea-level there has, during the Post-Glacial, been an isostatic uplift of the land, as a recovery from the depression due to ice-loading. The amount of this recovery varies, depending on the thickness of the original ice cover. In places where ice was not present, or present only sporadically, no isostatic uplift may be manifested.

The degree of isostatic uplift must be subtracted from the eustatic sea-level to give the relative sea-level rise in any one area. To complicate matters further there may be other, longer term, tectonic vertical movements of the land. In Britain, the net result of isostatic uplift and long-term subsidence of the land is that coastal areas in the north-west have risen during the Post-Glacial, while coastlines in the south and east have sunk. Between these two areas, much of the coastline of Wales and the south-west of England appears to have undergone little or no vertical movement. This is also the case in the vicinity of the Humber, on the east coast. It is not possible to draw contours for this vertical movement, as we can only guess at its extent in inland areas, where there are no sea-level deposits.

The further north the site, the greater the degree of uplift, and when the Solway Firth is reached the amount of uplift during the last 8,000 years is approximately equal to the amount of eustatic sea-level rise, so that submerged forests formed at about mean high water mark (MHWM) 8,000 years ago, are at MHWM now, even though there has been a rise in sea-level of about 20 m during this period. Fig. 1 shows the extent of vertical movements of the land during the last 8,000 years, and the ages of submerged forests at present exposed in the intertidal areas.

It will be apparent that, as a result of these movements of land and sea, submerged forest trees of all ages from a few hundred years to more than 8,000 years are easily accessible. Furthermore, it can be seen that the altitude of the trees is closely related to their age, which can therefore be predicted quite accurately by reference to established sea-level curves. This is a very useful feature for dendrochronological purposes.

The most common age of submerged forests is between about 4,000 and 5,500 bp, since this was a period of a moderate rate of sea-level rise which allowed the growth of extensive coastal forests which were progressively inundated. Before this period sea-level rise was more rapid, so that there is only a narrow belt of submerged forest trees of any particular age, whilst after this period sea-level rise has been so slow that the sedimentary environment has been comparatively static.

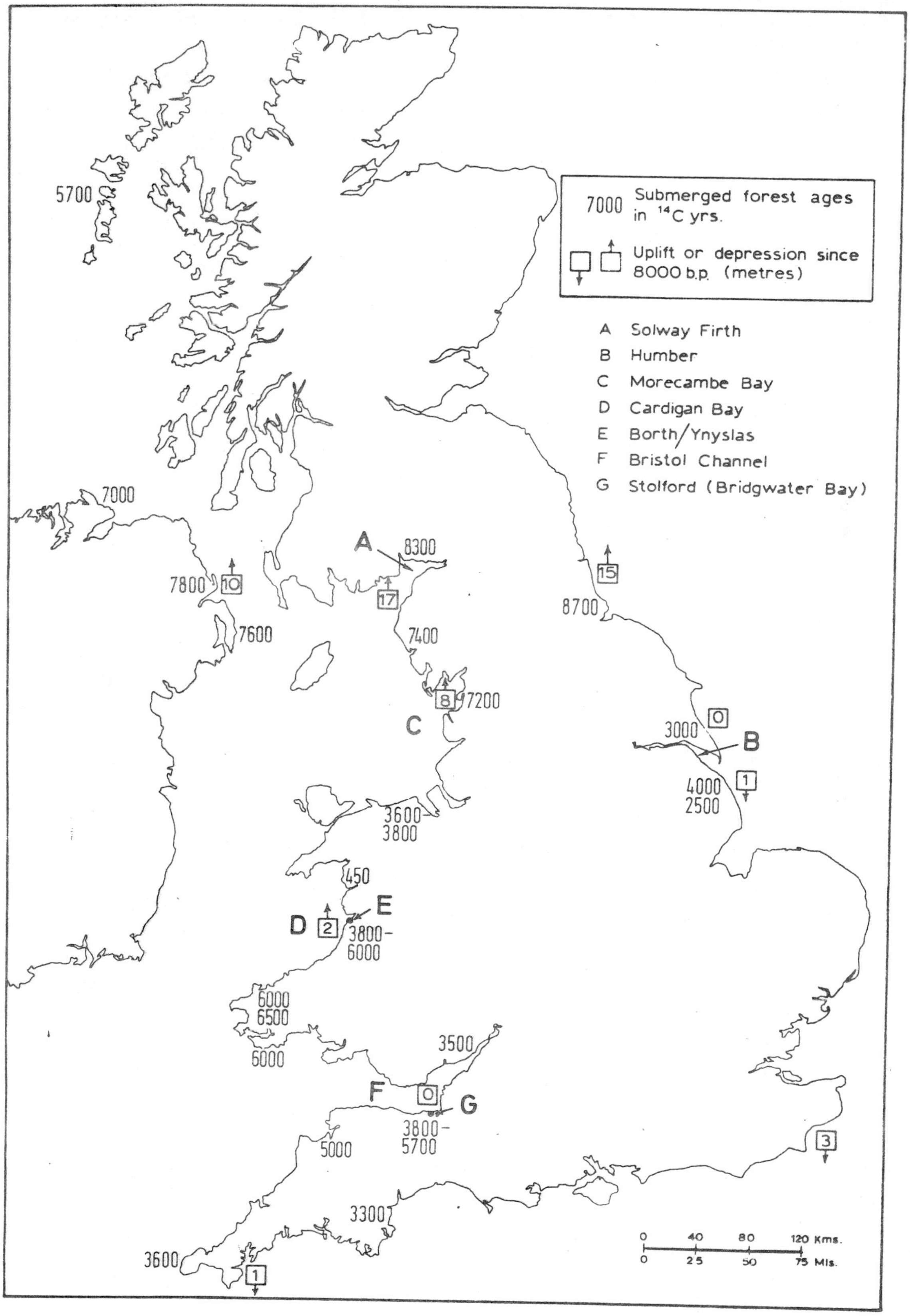

Fig. 1. SUBMERGED FORESTS. SITES A TO G WITH THEIR AGE, AND UPLIFT OR DEPRESSION OF THE LAND SINCE 8000 bp.

Almost all the submerged forests now found owe their preservation to the fact that, following the death of the trees, they were buried by later growths of peat, or by clays, silts or sands, from which they have only recently been exhumed. This pattern of sedimentation is largely determined by, and synchronous with, the rise in sea-level. Throughout most of the Post-Glacial, period, sedimentation and peat growth have kept pace with, or outstripped this rise, and only in the period before 8000 bp did the sea make any significant inroads into areas of land.

The rarity of submerged forests from the last 3000 years is thus explained, since the negligible rate of sea-level rise has meant, firstly, that few coastal woodlands have been drowned, and secondly, that if trees have been killed, they have not been subsequently buried and have, therefore, rotted away or been carried away by the sea. Most submerged forests from the last 3000 years are, therefore, from the last few hundred years where the preserved trees were killed by the proximity of the sea, e.g. the 450 year old one in the north of Cardigan Bay (Fig. 1).

It may seem paradoxical that a rising sea-level should result in greater accumulation of coastal sediments and less destruction of these sediments than a static sea-level. It should, however, be apparent that sea-level is only the indirect cause of coastal processes, in that a rising sea-level produces a mobile coastal environment and conditions in which sedimentation is encouraged, whereas a static sea-level means a virtual halt to the accumulation of coastal alluvium, whilst its sharply defined seaward edge is constantly removed by wave-attack.

It should be noted here that the best preserved submerged forests are those which are seldom exposed, being normally buried by beach sand or shingle and thus protected from wave-action. A good exposure of a well-preserved submerged forest is, therefore, a rare event.

Most of the trees are oak, though at some sites pine is dominant. It is hoped, eventually to produce a continuous oak ring-width curve for the last 8,000 years and a programme of tree-ring calibration of radiocarbon dates from this source is under way, in collaboration with the Radiocarbon Dating Laboratory at the University of Glasgow.

The other impetus for these tree-ring studies was the dating of sea-level rise with a precision greater than that achievable by radio-carbon dating.

Theories for Sea-Level Rise

A controversy has arisen between those who hold that sea-level rise up to the present may be represented by a smooth curve and those who hold that oscillations can be observed, some, perhaps, reaching heights above that of today's sea. These alternatives are shown in Fig. 2.

Two smooth curves are shown: a composite one from the Bristol Channel (Kidson & Heyworth, 1973 & 1976) and Cardigan Bay, and one from the Thames estuary (Devoy, 1977). The former is regarded as representing the eustatic sea-level rise, whilst the latter shows tectonic subsidence in the S.E. A third, oscillating curve is also shown, from Tooley (1974), derived from the

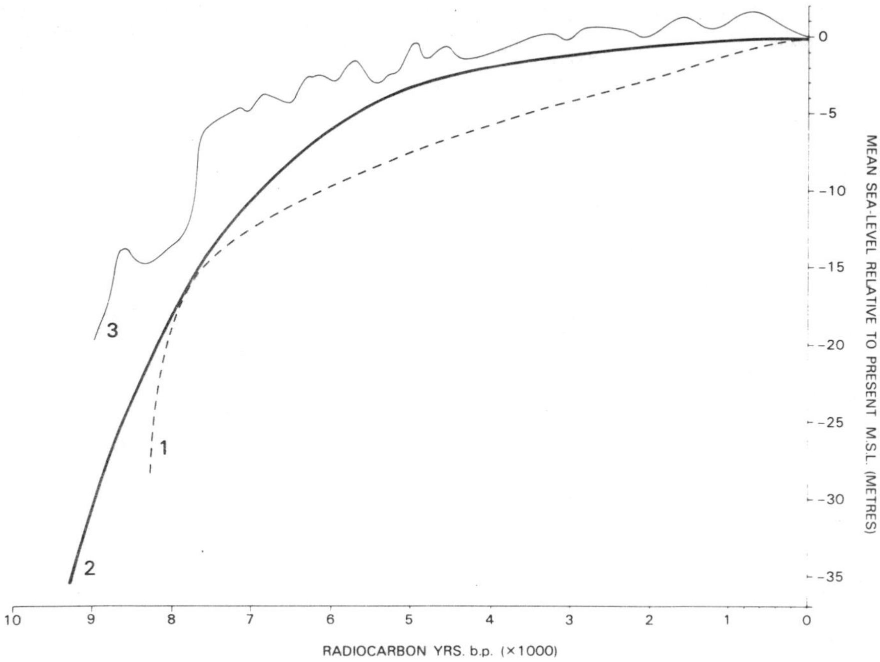

FIG. 2. SEA-LEVEL CURVES
1. Thames Estuary (Devoy 1977)
2. Bristol Channel (Cardigan Bay)
3. Lancashire (Tooley 1974)

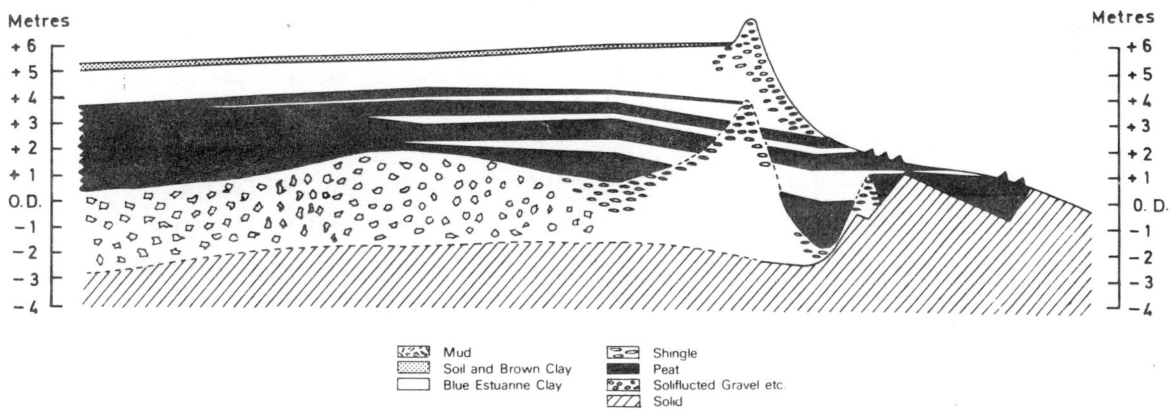

FIG. 3. STOLFORD, BRIDGWATER BAY. Section showing origin of submerged forest.

coastline of south-west Lancashire and Morecambe Bay. This is above the eustatic curve, and no doubt represents a degree of isostatic uplift in this area.

The supporters of the oscillating rise claim that transgressions and regressions can be recognised in widely separated localities and can be correlated with changes in eustatic sea-level and with various palaeoclimatic records. Supporters of the smooth, exponential curve argue that the rapid rise and fall of sea-level suggested is, because of the inertia of the glacio-eustatic system, inherently improbable and that other explanations are far more likely. These are:

1. A failure to take into account the differences in tidal range at different localities. This can produce differences in the surface heights of coastal areas, of as much as 4 m, far bigger than the size of the 'wiggles' on the curve. In estuaries, particularly those with tidal bores, this can be significant even over short distances. Unless allowance is made for these differences, any study of sea-level using dates from more than one locality is liable to produce misleading results.

2. Systematic errors in radiocarbon dating as a result of atmospheric radiocarbon variation. This would tend to produce 'bunching' of dates giving the illusion that the terrestrial deposits were concentrated in particular age bands, the intervening periods appearing as 'transgressions'. The question of whether these variations in atmospheric radiocarbon have occurred awaits a final and agreed dendrochronological calibration. When this is available, the forests will, of course, be dateable entirely by the tree-rings, and the radiocarbon dates will be largely superfluous.

3. The occurrence of exceptional storms and storm surges. These may be so infrequent that, in many areas, their effects have not been documented. The results of storm surges in the North Sea in recent years are however, well known, with high tides at least 2 m above normal, and areas many miles from the sea inundated. Such surges may occur only once in 50,100 or more years but their effects, killing trees well inland and at heights above High Water Spring Tide level, without any change in the secular sea-level, are very similar to those which would result from actual oscillations in sea-level. The probable frequency of storm surges is, significantly, of the same order as that suggested for eustatic variations and the errors inherent in radiocarbon dating make it difficult, if not impossible, to distinguish the real cause.

4. Consolidation of sediments. Some forest beds lie more or less directly on solid rock, whilst others may be above considerable thicknesses of clays and peats. In the former case, the underlying beds will not have changed in thickness whilst in the latter, consolidation of the sediments as a result of gravitational compaction may have been considerable. Errors of up to 2 m can easily be introduced in this way.

5. Changes in the coastline. These may occur as the result of normal coastal processes, without any change in sea-level, although they may be misinterpreted as the results of such change. The growth and destruction of spits, shingle ridges, sand-dunes, etc., can bring about dramatic

changes in the coastline, with large areas being flooded or building up above the reach of the sea, and with trees buried by advancing sand or shingle.

SUBMERGED FORESTS AS INDICATORS OF THE SEA-LEVEL

Many of the above uncertainties can be resolved by tree-ring studies and it is not essential to have a continuous, absolutely-dated sequence from the present, to produce useful results. If one of the supposed oscillations is studied in detail, the time span which it represents and the rate of sea-level change during that time can be determined.

Bridgwater Bay, Somerset: In the submerged forest at Stolford, large oak stumps and fallen trunks lie at heights between zero and +2 m OD. Associated peat beds continue the sequence down to extreme low water (-6.5 m OD) and below. The submerged forest is an extension of the buried peats inland (Fig. 3). The trees are now being exhumed by the sea, from their coverings of peat, clay, or shingle.

The distinction between the peat beds and the submerged forest seems, to a large extent, greater than it really is. In many places the buried peats contain large trees, but it is only on the beach, where the peat matrix is washed away, leaving the firmly rooted stumps, that they become so distinctive.

Most submerged forests are rooted in fresh water or brackish clays and it is the altitude of surface of this clay which is regarded as their altitude. This surface represents the upper limit of sedimentation and is, therefore, a measure of contemporaneous sea-level. At first sight, the presence of such a clay band, with trees on its surface, might be interpreted as evidence of a fall in sea-level, allowing trees to grow on a previously submerged area, and it is this kind of reasoning which has led some workers to introduce numerous oscillations into the sea-level curve. However, if these clays are examined in detail, it is clear that they were deposited very rapidly. Where the clays occur as thin layers, c. 0.5 m thick, intercalated with peats, radiocarbon dates from above and below the clays are very close, sometimes indistinguishable.

Furthermore, individual stems of Phragmites, Typha, Sparganium, etc. are often seen to be continuous throughout the layer. The conditions indicated by the Foraminifera in the clays are often too saline for the growth of these species and it is clear that, in many cases, the stems of a reed-swamp were buried by the clays before they could collapse. In other environments, of lower salinity, the clays probably accumulated around the living stems. In either case, such deposition could only be due to the breaching or overtopping of a coastal barrier or river bank, by storms or floods, and represent a purely temporary episode. Contrary to first impressions, such an episode resulted not in wetter, more marine conditions, but in a drier environment. This followed from the rapid deposition, by abnormally high water levels, in a reed-swamp, of a layer of clay which raised the surface above the normal water-table and produced a habitat suitable for the rapid establishment of forest cover.

Even in localities, e.g. Borth, where the submerged forest trees are rooted in thick estuarine clays, which have accumulated over thousands of years, the top part of the clays was deposited very quickly and by an exceptionally high water level which at a time of only gradual sea-level rise was not reached again for some hundreds of years, allowing forest growth on a clay surface raised above the water-table.

Rings have been measured in many of the trees, such as those at Stolford, where trees with radiocarbon ages of 4800 to 5400 bp lie at +0.1 to +0.7 m OD.

At this point, Tooley's curve (Fig. 2) shows very marked oscillations, the rate of rise and fall being, of course, much greater than the rate shown by the smooth eustatic curve at this point. However, the ring width matchings of the trees show that adjacent trees did not die at about the same time, as would be expected if the rapid rise of sea-level was responsible for their deaths. For example, in the case of two adjacent oaks, although both are at the same height, the former did not die until 160 years after the latter.

The bark is present in both cases, as in most of the submerged trees, so there is no error due to missing outer rings. The combined age range of the two trees covers a period of 331 years and suggests an environment which remained stable for at least this time, in which oaks could grow reasonably well, but which was sufficiently wet and acid to preserve the stumps and fallen trunks when the trees died. At this site, the topography is such that waterlogging could only have been brought about by ponding-up of fresh-water, more or less directly related to sea-level.

Other, nearby trees extend the period of stability to over 500 years (a period which is inconsistent with marked oscillations at 4,500-5,500 bp). The relative ages of the trees, and the fact that some have long sequences of narrow rings at the outside, whilst others have not, suggests long periods of little change, with a stable coastal barrier, interspersed with short episodes of marine incursion when the barrier was destroyed or overtopped. Stratigraphical and pollen analytical evidence from boreholes confirms that such a barrier was in existence at this period, the situation being very similar to that of today.

The shingle ridge is now occasionally overtopped or breached by storms coincident with high tides, and when this happens considerable areas behind it are flooded and remain so for weeks. This would be much more marked were it not for artificial embanking and drainage. Any trees in this area would be killed; in fact, there are none. Similar events must have given rise to the submerged forest, although at the date of its formation the rate of sea-level rise was still significant. It was, however, small in comparison with storm waves which were the agents responsible for the form and position of the storm beach. To an observer the flooding of areas behind the storm beach would seem to be entirely the results of storm waves, and sea-level rise would have been imperceptible.

It is certain that in that period the whole of the transgression due to rise in sea-level was accomplished in storms. The intervening periods would be ones of stability, perhaps even of apparent regression of the sea. In these periods, however, the height of the storm beach would gradually lag behind sea-level, so that each very exceptional storm, occurring at intervals of, perhaps, 100 or more years, would flood land at higher levels and reach farther inland than the preceding one.

The belt of trees killed by each such event would be interspersed by trees which had died in the previous stable period, as a result of waterlogging, or from other causes, so that adjacent trees are found to have died at different dates. The stable intervals may have been long enough for trees to recolonise the previously flooded areas, so further lengthening the period represented by trees at any one date.

Radiocarbon dating of the resulting submerged forest can, therefore, give the impression of discrete eustatic transgressions, the frequency of which will depend entirely on which trees are dated. Only by tree-ring dating can the true picture of the origin of the submerged forests and the pattern of sea-level rise be determined.

Vertical Movement of the Land

Similarly, in the resolution of isostatic movement of the land, the precision of dendrochronological dating can remove uncertainties. For example, trees at Stolford, Borth (Cardigan Bay) and Alt Mouth (South Lancashire), all with radiocarbon dates of about 4,500 bp, have been matched and thus shown to have died as a result of marine transgression, within a period of less than 50 years. Their altitudes are:- Stolford, +2.0 m, Borth, 0.1 m, Alt Mouth +3.3 m OD. Allowing for the differences in tidal range, the equivalent heights relative to a zero datum at Stolford are Borth, -0.1 m, Alt Mouth, +2.5 m. This implies that since the trees died there has been an isostatic uplift relative to Stolford of zero at Borth and 2.5 m at Alt Mouth.

The very close agreement between the sites in the Bristol Channel and those in Cardigan Bay makes it very probable that these areas both lie in a region of isostatic and tectonic stability. It is very unlikely that a block of this size would move vertically without any tilting. It seems, therefore, that the relative sea-level curve for these areas is a close approach to a true eustatic one. In that case, taking Stolford as an example, present MHWM is at about +4.5 m OD. At 4500 bp it was +2.0 m, and so eustatic sea-level rise in the last 4500 years is 2.5 m. This fact, incidentally, makes it very difficult to accept the idea, reiterated by Mitchell (1977) that, at 4500 bp sea-level was almost 4 m above that of today.

It is intended, eventually, to link all the known submerged forests dendrochronologically in this way and so provide a measure of absolute and relative sea-level rise of far greater precision and reliability than hitherto.

References

Devoy, R. J. N. (1977). Flandrian Sea-level changes in the Thames Estuary. Nature, Vol. 270, 712-715.

Kidson, C. and Heyworth, A. (1973). The Flandrian sea-level rise in the Bristol Channel. Proc. Usher Soc., 2, 565-84.

Kidson, C. and Heyworth, A. (1976). The Quaternary Deposits of the Somerset Levels. Q.Jl. Engng. Geol. 1976, Vol. 9, 217-235.

Mitchell, G. F., (1977). Raised Beaches and Sea-Levels. In British Quaternary Studies, ed. F. W. Shotton, 169-186.

Tooley, M. J. (1974). Sea-level changes during the last 9000 years in north-west England. Geog. J., 140, 18-42.

Wilks, P. (1977). The Flandrian sea-level change in the Cardigan Bay area. Ph. D. Thesis. University of Wales.

PLATE I. THE SUBMERGED FOREST BETWEEN BORTH AND YNYSLAS, CARDIGAN BAY.
Most of the eroded stumps are pine.

5.10. DENDROCHRONOLOGICAL EVIDENCE FOR HOLOCENE CHANGES IN THE DRAINAGE SYSTEM OF SOUTHERN CENTRAL EUROPE

B. Becker, Stuttgart-Hohenheim

Tree-ring analyses of more than 2000 subfossil oak cross-sections provide datings of Holocene valley fillings and details of their genesis. The dates are based on 1) conventional ^{14}C-dates, 2) dendrochronologically calibrated ^{14}C-dates of the La Jolla Laboratory, and 3) absolute tree-ring dates, as indicated by the numerals below.

Early Holocene

A first phase of fluvial activity during Preboreal times, 9600 to 9200 BP, deposited pine and birch trees. At that time the river channels must have been lowered due to the position of these trunk layers at the very basis of the Holocene gravels. Numerous crossdated oak trunk layers of Boreal times date from 8700 to 6500 BP (1).

Their slow growth rates as well as the occurrence of peat bogs in the valley plains give evidence for wet sites caused by a high groundwater level.

In contrast, the tree rings of oak layers, which have been accumulated in Early Atlantic times, 5900 to 5300 BC (2), indicate an improvement of the floodplains, probably due to a better drainage caused by a downcutting of the rivers. Between 5300 to 4300 BC (2) the fluvial activity was decreasing.

Late Holocene

The drainage system changed drastically in Late Atlantic and Subboreal times. A renewed phase of trunk accumulation took place from 4300 to 3100 BC (2). Overflooding reached its first maximum from 2800 to 1500 BC (2), where extensive areas of the valleys have been redeposited. Crossdated trunk layers of this phase occur in a wide distribution along the Danube, Rhein, Main, Regnitz, Lahn, Werra and Iller valleys ("first main horizon").

In Late Subboreal times, 1500 to 277 BC (2,3), the activity of the rivers decreased remarkably, whereas in Subatlantic times, beginning 277 BC (3) the rivers repeatedly overflooded the forests. The deposition of this "second main horizon" can be followed in many exposures. This phase ended 236 AD (3). The following 80 years are characterised by a complete lack of trunk deposition within the whole area studied.

In Early Medieval times a "third main horizon" was accumulated from 311 to 780 AD (3). Along the Rhein, Main, Danube, Isar and Inn valleys crossdated trees could be found. Findings of even younger trunks (Medieval times to Present) are rare.

Considerations

The subfossil oak horizons of Bronze Age times, but especially of the two following phases in Iron Age/Roman Period and Early Mediaeval times give, according to their broad tree rings, evidence for deep, fertile floodplain soils with a balanced water supply. The significant change in the growth rates from the Early to the Late Holocene trunk layers is in our opinion explained only by increasing floodloam sedimentation in the valleys since Subboreal times. On the other hand, the deep alluvial loam soils of our recent valley plains are caused by an increasing soil erosion due to forest clearance in Prehistoric and Historic times. It seems that the tree-ring analyses of the trunk horizons of Late Holocene provide evidence not only for the activity of the rivers, but also for the influence of man to the Central European hydrosystem.

Dr. Bernd Becker cutting a sample from a sub-fossil oak from the banks of the Danube near Ulm. The oak died in 524 A.D.

DISCUSSION

Section 5. Applications to Archaeology and Geomorphology

Mr. BIDDLE introduced a discussion on the needs of archaeologists by referring to their desire to have firm calendar dates based on links with the present day or with well-established historic dating from standing buildings. He himself believed that for many archaeological finds there should be radiocarbon analysis as well as dendrochronological analysis (see this volume page 170). If giving more than one possible match for a timber, the dendrochronologist should show why he does so. He wondered whether dendrochrologists might have been pressured by archaeologists to give absolute dates when this was not yet possible.

Baillie said that he agreed with Biddle that floating chronologies should be clearly identified as such. He was however amazed at the view that every piece of wood dated by dendrochronology should also be analysed for radiocarbon. If you believe in dendrochronology, he said, you believe it dates with the precision of its method. If each piece of wood is to be dated by two methods then radiocarbon is being vested with a precision equivalent to that of dendrochronology. He considered this would be opening flood-gates to all sorts of problems.

Becker agreed in principle with Baillie. The system of using statistics as a guide but with the ultimate decision on dating based on visual comparison over a relatively large number of years had been used in German laboratories for many years and was accepted as providing dates. He agreed, however, that the dendrochronologist should be prepared to show the archaeologist the visual match on which the result was based. He himself wanted at least 100 years in a single sample before he would try to date it.

Fletcher pointed out that there were two ways of dating chronologies: by overlapping each from the present day; or by cross-links with dated reference curves for adjacent regions. He had chosen for the area covered by his samples to use the latter method as being more practical and because chronologies so dated were likely to be more reliable than overlapping with relatively few samples from the present day. Biddle asked him on what grounds dates based on overlaps from the present are less reliable than those based on long sequences using teleconnections. The answer was that the statistical reliability could be higher (see page 291).

Dr. McGrail said that, as the archaeologist concerned with samples, for example from the Graveney boat, he wanted to be given the date of the latest growth-ring in addition to any deductions made about felling date. He asked Fletcher: are you doing this, or not? The answer given was in the affirmative: in addition to giving the date of the latest ring it was also usual to aid the customer (sometimes, due to his historic and archaeological interests, he was himself the customer) by assessing either the earliest date at which the tree concerned was felled, <u>or</u> the likely period in which felling occurred.

The dendrochronologist's examination would show if any sapwood was present and if the waney edge—the sapwood/bark edge—existed (see page 34). There were three situations as follows for oak samples for a ship or archaeological find in which the latest ring from several samples was found to be, say, 1000 A.D.:

(a) Sapwood/bark edge identified at the latest ring. Tree felled in 1000 or early in 1001.

(b) Some, but not all, the sapwood rings present. Number of absent rings assessed on the basis of the age of the trees and the average width of their latest rings. Period in which felling was likely, e.g. 1005-1015, recorded or shown visually in a diagram.

(c) No sapwood and no evidence for heartwood/sapwood boundary. Likely minimum allowance for absent sapwood rings estimated. Earliest date for felling of trees given, e.g. 1023.

The dendrochronologist might also have evidence about the interval between felling and use of the timber (see page 37).. On other matters such as the building of the ships, delays before use etc., the archaeologist must decide for himself.

Finally, the pressure which archaeologists put on dendrochronologists was stressed by Dr. Baillie as being an important matter of which they had heard too little at the conference. They had to realise that dendrochronology is not a universally applicable method. Unlike radiocarbon, you cannot pay your money and start demanding an answer in three months. As a young dendrochronologist, he had suffered from that attitude. People would give him bits of wood and the next thing he knew they were round, hounding him as to why he had not given them an answer. He would say he could not date it reliably and would never put his name to a date for that piece of wood. By and large, if a sample had a short ring-width sequence, it was his advice, and most other dendrochronologists, not even to attempt to date it.

Miss ORME (Paper 5.2) was asked by Bauch about the criteria used in the Somerset Levels Project to identify timber from individual trees. He was referred to Ruth Jones' publication of her methods in Coles, J. M. *et al.*, Somerset Level Papers No. 3, 1977. M. Hughes asked whether she had taken into account work by Prof. Newbold of University College, London, on the coppice cycle in Kent which suggested that coppicing might produce the same tree-ring pattern again and again. Miss Orme replied that this possibility had been considered. They now had coppiced material from sites spread over *c.* 2000 years and only two had identical patterns and they were believed to be contemporary. Conversely, coppiced material from two sites known to be contemporary had produced different ring patterns.

With reference to the Roman timber from the London Waterfront (Paper 5.3) Schove said he had tried to relate the sequences for these to a skeleton plot for the Walbrook site in London derived from Lowhter's curves that were obtained some years ago. Fletcher said that not only is there no southern English chronology back to that time but it may be many years before one can be formed. Cross-dating by a continental curve offers the best possibility, but so far only parts of a reference curve for that period have been published by Hollstein. He asked how matters stood.

Becker replied that problems still exist with the B.C. part of Hollstein's chronology because a check of its continuity and hence of the dates of part of it are lacking. On the other hand the validity of the tree-ring pattern for the Roman and pre-Roman parts published by Hollstein are confirmed by their cross-dating with central German mean curves derived at Stuttgart. He regretted that Herr Hollstein had been unable to attend the symposium and discuss this matter. Editorial Note: In view of the interest in the status of the Central German oak chronology, Dr. Becker subsequently prepared, in association with Dr. Delorme, the paper on pages 61 to 66 of this volume.

In reply to a question, Dr. BAILLIE (Paper 5.5) stated that there was poor correlation between the Dublin and Belfast regional curves for the period 1000 to 1300 A.D., but good correlation between 1350 and 1650 A.D., perhaps as a result of some climatic shift.

Schove asked whether the differences between the Belfast chronology on the one hand and the Dublin-Scottish one on the other, might be due to site differences giving rise to contrasting sensitivity to drought. Baillie replied that material from all three regions had narrow ring-widths: however, in the early part of the Belfast chronology the material was sub-fossil and thus might be atypical.

Dr. Becker asked if, in addition to the correlation with Huber's southern German chronology, the GDR archaeological samples from before the 13th century (Paper 5.6) matched Eckstein's north German chronology. Dr. JAHRIG replied that although his samples came from the northern part of GDR, i.e. more from the Baltic Sea area than the southern part, they gave a better match (about 70%) with Huber's chronology than with Eckstein's.

Prof. ECKSTEIN, in answer to a question, said that there was no gap in the chronologies for Haithabu (Paper 5.7) and for the Schleswig region. The gap to which he had referred occurred between the felling dates found for Haithabu and the nearby settlement at Schleswig. This indicated a possible break between the end of the settlement at Haithabu and the beginning of the town of Schleswig.

In reply to another question, Eckstein said that the dates in his paper given with a precision of one year were from samples with bark. His published dates were felling dates without any allowances for storage etc. There were many indications however, that timber at Haithabu was used immediately after it was cut.

Dr. McGrail asked whether any other material besides trees from submerged forests (Paper 5.9) had been used for dating sea-level changes. Mr. HEYWORTH replied that the vast majority were from submerged forests.

Baillie questioned whether the trees were giving an accurate estimate of the rise of sea-level. In his experience, trees encased in peat which look to be at a similar level can often be 500-600 years apart in age. Heyworth said that they had to eliminate many trees obviously formed in bogs which had no connection with sea-level. Unless pollen and diatoms provided evidence for salt water conditions, trees were excluded.

Becker asked about the two trees mentioned as being close together physically, but separated by 150 years in time. If one tree died, why had not both died? Heyworth replied that they looked at associated pollen and diatom evidence, and concluded that the sea-level would only rise a few centimetres in the period under investigation. The first tree died for some unknown reason and was preserved; the second tree carried on after the first one died, and was then overtaken by the sea and buried in shingle.

6.1 THE BARN AT GREAT COXWELL, BERKSHIRE

V. Siebenlist-Kerner, D. J. Schove and J. Fletcher

Summary

Samples from four of the twelve pads under the massive aisle posts of this historic building (Plate 1) have been dated (i) by the weather history of the late 12th and of the 13th century and (ii) by reference chronologies constructed at Oxford for oaks of southern or eastern England with comparable average ring-widths.

Sections cut from two of the pads had 165 and 198 annual rings respectively but cores from another two had many fewer. The mean curve, probably for a single tree, consists of 225 rings, and is dated to the years 1043 to 1267. The heartwood-sapwood boundary is absent in all four samples. The tree would not have been felled before 1280, perhaps considerably later.

Five other cores and one section from different timbers have relatively few rings (42 to 81) and can only be dated tentatively. None of them has sapwood and so no firm date can be deduced for the felling of the trees that were used for constructing the barn. While the present evidence suggests that the barn was built within a decade of 1300 A.D., further sections for tree-ring analysis are desirable.

Plate 1. Barn at Great Coxwell. From N.W. (Photo. W. Horn)

Introduction

The success achieved by dendrochronology in dating buildings in central and southern Germany, in Novgorod and Schleswig, in Ireland and Sweden, is well recognised. So far, however, the method has made little impact on the dating of buildings in southern Britain in spite of the area being rich in medieval and post-medieval timber-framing. The failure is not for lack of samples and effort. It is due, as already noted by Berger, Giertz and Horn (1971) to the prevalence of oak that was fast grown and usually had fewer than a hundred rings as timber for houses and barns. Slow grown oak, usually radially split or sawn boards, if present at all, occurs in special items such as doors or panelling. Up to the present, dendrochronology has been of little help to building history in southern Britain, but the situation is improving with the formation of oak chronologies that apply to the wider rings of fast grown trees (Fletcher, 1977).

Oaks growing under certain stresses associated with wet and hilly areas in Wales or the west Midlands enabled the tree-ring sequences (from 8 buildings) to be matched by the Huber-Giertz master curve for central and southern Germany (Charles, 1971 and Siebenlist-Kerner, this volume, p. 156). However, the sequences of many other samples collected by the Munich team to obtain information for the study by Prof. Walter Horn of barns in southern England, could not be matched by that master curve. Those samples included some from the Great Coxwell barn, near Faringdon, a Cistercian barn famous as an historic building of great merit (Plate 1) and the subject of a scholarly book by Prof. Horn (Horn and Born, 1965).

The oaks for the barn are likely to have been felled in the hilly country that exists within two or three miles of the village of Great Coxwell and extends to the crest of the Cotswolds. There is (at present) a woodland about a mile to the west at an elevation of 300-350 feet and the wooded Badbury Hill at 400-500 feet lies about a mile to the NNW, while Furze Hill near Fernham to the south is noted for its fine trees. These woodlands are at an elevation and sufficiently far to the west of England to have an average annual precipitation about 25% greater than that around Oxford (about 650 mm, 25 inches) which lies some 30 km (20 miles) to the north-east (Map I).

The date of the construction of the Coxwell barn has remained a matter of speculation for well over a century. Estimates have varied by as much as 150 years: from the first half of the 13th century (William Morris, Horn) to the second half of the 14th (Smith, 1960, well-known for 'late' assessments). Rigold (1966) gave c. 1275 as the consensus of the Vernacular Architecture Group on the evidence available. The barn was, however, usually attributed to the 14th century by Turner and Parker (1853) and by eminent architectural historians of Berkshire buildings, such as Keyser (1907), writing at the beginning of this century.

Two samples from a post, one reported as having sapwood and a waney edge, the other heartwood, were analysed for their radiocarbon content by Berger and gave, after calibration and adjustment for absent rings, the date c. 1250 A.D. for the construction of the barn, the uncertainty being ± 30 to 40 years (Berger, 1970).

This paper deals with dating the ring-widths of samples from the barn by weather history and by dendrochronology and ends with a discussion in which the reliability of the date derived from radiocarbon analysis is questioned.

Samples and The Floating Mean Curve from them (by V. Siebenlist-Kerner)

At Munich the chronology of Huber and Giertz-Siebenlist had been successful in dating barns in Normandy and the Ile de France (Horn, 1970). To extend this research to England, we drilled cores or cut sections from timbers in several English barns. Details of the ten samples are given in Table 1.

Table 1 Samples and Ring-Width Details

Sample	Origin	No. of Rings	Width, mm Average	Width, mm Terminal	Mean Curves
b	N Pad, Truss 6	198	2.0		MC 4 Pads 225 rings
b	N Pad, Truss 6	Last 179*	2.15	1.5	
c	S Pad, Truss 5	165	1.9		
6	N Pad, Truss 5	102			
3	S Pad, Truss 2	113	2.0	2.5	
2	Post, Truss 2	74	2.9	2.5	MC 2/5 89 rings
5	Curved support between Trusses 3 & 4, N side	81	1.6	2	
7	?	33	2.9	4	MC 7/a 49 rings
a	Beam, N side of Nave	42	2.7	2.5	
4	?	61	2.2	1.5	MC 4/1 61 rings
1	?	53		c.1	
b,c,6,3	MC 4 Pads	225	2.0		
b,c,6,3	MC 4 Pads	Last 206*	2.1		

Notes

a, b and c are sections; 1-7 are cores.

The Mean Fluctuation of the Ring Widths is 18-19% for the pad samples, and 20-24% for the others.

* As rings 3 to 17 are very narrow (c. 0.5 mm or less), the first 19 rings are here excluded.

The samples with more than 100 rings were those from the square pads (about 40 cm x 40 cm and 15 cm high) on which each massive aisle post rests (Plate 2). The rings cover 225 years (Fig. 1). The 237 ring mean curve, MK 5, which was formed from them and from the 42 wider rings for a disc cut from beam 'a', could not be dated by the Munich master curve nor by any of the English sequences from samples we had taken there. The ring-widths of the mean curve were accordingly sent to the other authors of this paper and,

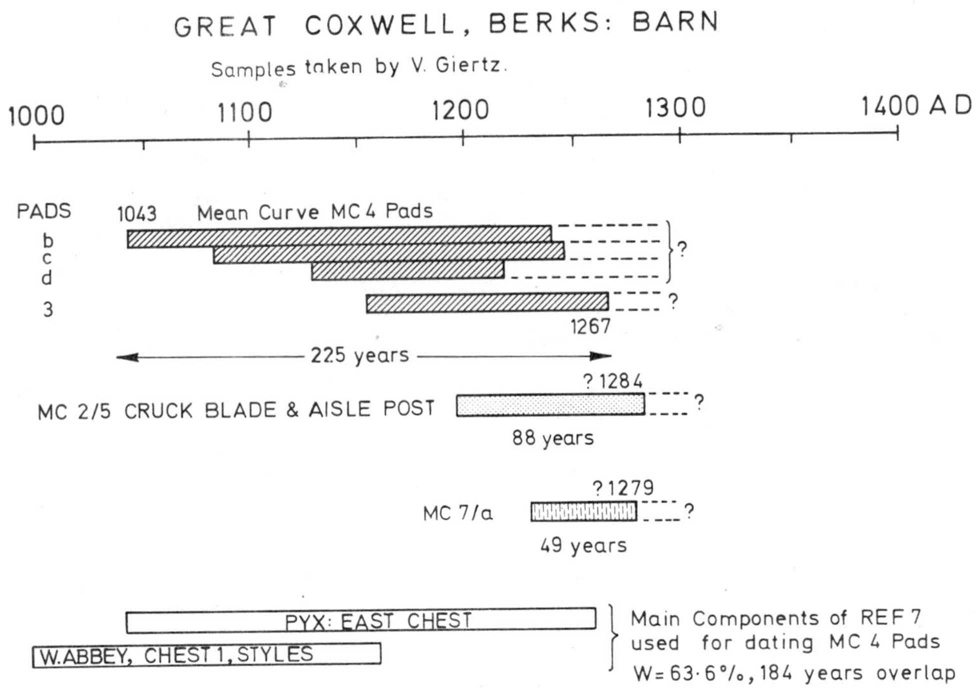

Fig. 1. See text.

Dating the Ring-Width Sequence by Weather History (D. Schove)

In Europe, specimens of archaeological timber are often difficult to cross-date when there is no master tree-ring curve for the period in question (Schove, 1974b). The matching of skeleton plots (Schove and Lowther, 1957) either with one another or with droughts, has long been used successfully in the semi-arid parts of the U.S.A. In Europe the results of such matching by eye are often ambiguous: the writer tried to turn the art into a science by using cross-correlations to express quantitatively the matching of a ring-width plot with climatic data. Outstanding seasons were given weights of 1, 2, 3 or 4, corresponding to multiples of the standard deviation; and outstanding ring-width maxima and minima were listed and weighted independently in a similar manner. For any particular position of matching (Schove, 1971), the scores were multiplied. If we assume that dry springs and summers (and, say, good harvests) correspond with narrow rings, each product has a sign.

The driest summers in N.W. Europe (Schove, 1974a) in the period 1175-1275 were considered to be 1177, 1188 (and 1189), 1205, 1211, 1222 (and 21), 1236, 1241/2, 1247/8, 1250, 1252, 1263 and 1272. By assuming that the final ring for the Coxwell Barn lay in the 13th century, MK 5 was found to fit a summer drought pattern with its final ring at 1279 (the corresponding final ring for MC 4 Pads is 1267). This fit gave the highest score when all seasons were included, the adjacent scores being:

Date of Last Ring	1277	1278	1279	1280	1281 A.D.
Net	-12	-13	+72	-37	+21

No other date for the final ring gave a score greater than + 25.

Dating by Ring-Width Reference Curves (J. Fletcher)

By early 1977, the mean curve for the Coxwell barn offered (even more than those from the strakes of the Graveney ship) a challenge to the work of the Oxford laboratory. Would they be as suitable to date as the oak boards of Tudor paintings on English panels? For the latter, over 90% (out of over 140 of those with more than 100 rings) had been matched and dated by the reference curves built up (this volume, page 139).

Those reference curves are certainly not specific to panel paintings of the 15th and 16th centuries; many boards with up to 250 rings from chests in Westminster and Oxford form part of them and also extend them back to Norman times. There was the possibility, however, that though the reference curves applies in the lowlands of eastern England and south-eastward to Flanders for distances up to over 200 km from London, their usefulness might be limited westward. There the terrain is more hilly and the precipitation increases substantially westwards from Oxford to the crest of the Cotswolds and the Severn valley. At least, however, the Coxwell barn series was likely to be derived from local trees and so of inland origin.

A systematic comparison of the sequence for each of the separate timbers was undertaken starting with those from the four pads. Values of W and t were obtained from about 40 runs with the Hamburg or Belfast computer programmes. The W values between three of the pads lay in the range 76-80% (most if not all of the twelve pads appear to be from one tree-trunk divided centrally). In forming mean curves we avoid combining sequences of different average widths and so limited our mean curve (MC 4 Pads) to the four pads by reason of the similarity of their average ring-widths (1.9 to 2.15 mm) and because the relative position of the short sequences (Table 1) seemed too tentative to warrant their inclusion.

The pattern of our wider (Type B) set of oak reference curves, the one of average width c. 2 mm (Fletcher, 1977) appeared to offer the best chance for matching and dating MC 4 Pads. Within the likely period our relevant chronologies are:

Ref 5, average width 1.8 mm. Years spanned, 1147-1368, and
Ref 7, average width c. 2 mm. Years spanned, 1018-1257.

Both at present contain relatively few (6 and 5 respectively) single curves. The former is built up mainly from the wood of panel paintings, the latter from large chests at Westminster. The decisive computer results for dating MC 4 Pads to the years 1043-1267 are given in Table 2. This result is equivalent to Schove's for MK 5. In the selected position, visual comparisons showed the coincidence of a number of signatures on the reference curve with prominent maximum, minima or group of rings on the Coxwell curve.

Cores with Shorter Sequences: Visual comparisons suggested these could be tentatively grouped into the pairs shown in Table 1. In the relative position proposed by V. Siebenlist-Kerner for MC 2/5, 88 rings long, the agreement with MC 4 Pads was W = 70%, t = 4.0 and the date of its last ring 1284 (Fig. 1).

Table 2 Agreement Values for MC 4 Pads, 1043-1267 A.D.

Rings Compared	Agreement With	W	Overlap
All (225)	REF 5	65%	120 years
All	REF 7	62	214
The last 206	REF 7	64	184

Note

For comparison, W between the 231 year single-tree curve (from 1370 to 1600) for the Bailiff's house, Bewdley and the S. German curve is 61.1% (Berger, Giertz, Horn, 1971): important additional evidence was provided by the presence of the SAW from 1529 to 1541.

DISCUSSION AND CONCLUSIONS (J. Fletcher)

The results confirm the difficulty of dating by tree-rings the timber buildings in southern England. Only the presence of pads, an unusual feature, has made reliable dating of any of the samples possible and had the samples been limited to the two cores, dating would have been uncertain on account of their relatively small number of rings. Inspection of the timbers in the barn shows no obvious example of sapwood and no heartwood-sapwood boundary could be readily recognised.

In assessing the allowance for sapwood rings of oaks in southern Britain it must be remembered that the wider their rings, the fewer in number. In our experience the overall width of sapwood on the trunks of mature trees remains relatively constant (a width of 1 to 1½ in., 25 to 35 cm, is usual).

Applying this principle to the samples from the Coxwell barn we can assess the number of absent sapwood rings by assuming their width to be that of the respective terminal rings listed in Table 1. For Sample 3, the pad sample with the latest ring (for 1267), this gives removal of 12-16 sapwood rings and an indeterminate number of heartwood rings. Felling of the tree before 1280 is excluded. If they are positioned correctly in Fig. 1, samples 5 and 'a' would extend this terminus post quem to about 1297.

In the light of these tree-ring results it seems possible that the carefully measured and calibrated radiocarbon date of c. 1250 derived from two samples from the same post by Berger (1970) would be early by as much as 30-50 years. A substantial part of this may be explained by the allowance for absent rings being incorrect. What was thought to be the waney-edge, i.e. the sapwood-bark boundary, may in fact be the heartwood-sapwood boundary (the external appearance of which is often similar to the waney-edge) or even within the heartwood, such is the erosion of the timber to be seen on many of the posts. The uncertainty from sampling may be as much or exceed that from the analysis by radiocarbon.

Finally, various matters relating to dendrochronology in southern Britain may be mentioned:

(a) Whereas exact dates were derived by Hollstein (p. 36) for buildings in Germany, there is some uncertainty for the date of the Coxwell barn due to the absence of sapwood and the absence (or doubt) about the heartwood-sapwood boundary.

(b) The need for many samples to date a building may increase the effort on sampling and on analysis considerably. About 1/4 of a man-year has been spent on analysing the results at Oxford on these samples.

(c) The usefulness over a large area of our lowland (or London) reference curves has already been well established for the period from 1250 to 1600. This application is somewhat earlier and in an area some 60 miles (95 km) to the west of London (Map, p. 148). Still further to the west (and north) some trees likely to have grown in woodlands above the Severn valley or in other parts of the West Midlands (cf. Morgan, 1977, for Sheffield) have the West Midlands/Welsh pattern (p. 156), i.e. the one that is matched by that of central-southern Germany. It and our lowland pattern may therefore intermingle in the higher ground to the north of the Upper Thames and in the west Midlands.

Acknowledgements

Jennifer Hillam and Margaret Tapper are thanked for helping with the computer runs and analysis of the ring-width sequences and Prof. Walter Horn for providing Plates 1 and 2.

References

Berger, R., 1970. In Scientific Methods in Medieval Archaeology, ed. R. Berger, University of California Press, 109-110.

Berger, R., Giertz, V. and Horn, W., 1971. Vernacular Architecture, 2, 3-6.

Charles, F. W. B., 1971. Timber Trades Journal. 30 Oct.: supplement 14-19. Quoted extensively in Berger, Giertz and Horn, 1971.

Fletcher, J. M., 1977. J. Archaeol. Science, 4, 335-352.

Horn, W. and Born, E., 1965. The Barns of the Abbey of Beaulieu at its Granges of Great Coxwell and Beaulieu—St. Leonards. University of California Press.

Horn, W., 1970. In Scientific Methods in Medieval Archaeology, 23-87.

Keyser, C. E., 1907. Berks. Arch. J., 13, 67.

Morgan, R., 1977. Vernacular Architecture, 8, 809-814.

Rigold, S. E., 1966. Archaeologia Cantiana, 81, 1-30.

Schove, D. J., 1971. Geografiska Annaler, 53, Series A, 214-234.

Schove, D. J., 1974a. Proc. XXIII Internat. Cong. Hist. Medicine, 1265-1272.

Schove, D. J., 1974b. Medieval Archaeology, 18, 165-172.

Schove, D. J. and Lowther, A. W. G., 1957. Ibid., 1, 78-95.

Turner, H. and Parker, J. H., 1853. Domestic Architecture of the Middle Ages, 2, 275.

Plate 2. Interior of Great Coxwell barn. Note wooden pad between each stone support and the wooden aisle posts. (Photo. W. Horn)

6.2 TREE-RING ANALYSIS OF PANEL PAINTINGS

J. Fletcher, Oxford University

Summary

Ring-width measurements have been made on two hundred oak panels used for paintings of late medieval, Tudor and Jacobean times and over 80% of the boards dated. Examples of the interval between the date of the latest ring and the known date of a painting are given. These intervals form the basis for estimating the likely period of paintings of unknown date. Historic evidence can reduce the length of this period and indicate the purpose of a painting or group of paintings.

There are innumerable oak panels in churches, mansions and more humble domestic dwellings in England but almost all are firmly held in grooves and their edges inaccessible to tree-ring study. Research on panels has therefore been limited to those that are the support for paintings. They fall into three categories:

(a) Panels formerly in ecclesiastical buildings now no longer in situ; they include a 15th century tester in Westminster Abbey and two panels formerly associated with the shrine of St. Etheldreda in Ely Cathedral.

(b) Small panels of single boards, used for paintings in the workshops which flourished in the decades around 1500 in centres such as Brussels; a number of these came over to England early in the 16th century before portrait painting became well-established here; one or two were specially brought by a Flemish artist with a particular commission, e.g. by Sittow for his painting of Henry VII in 1505.

(c) Portraits that originated in England from the reign of Henry VIII (1509-1547) until the Commonwealth (mid-17th century).

Access to these works of art has only been possible with the consent kindly granted by the owners: first and foremost, I acknowledge the gracious permission of Her Majesty the Queen to work on paintings in her collections at Windsor Castle and Hampton Court.

Information from Tree-Ring Analysis about the Date of a Painting

As a result of experience gained in this research over the course of six years the relation between the date of the latest of the tree-rings and the likely date for the use of the panel has been refined.

Clearly a board or panel could not have been used when the tree from which it came was still growing. The tree had to be felled, divided (while still green to avoid unnecessary labour) and converted to the radial sections used for panels (Fig. 1). By itself, tree-ring analysis can only estimate the approximate date when the tree was felled, this being the earliest date at which a board from it could have been first used. Reuse at a later date, a rare occurrence, can be detected by X-radiography.

Allowance for Absent Rings: To estimate the date of felling requires (i) allowance for all the sapwood rings removed (about 5% of the paintings examined have sapwood); and (ii) allowance for heartwood rings removed. The width of sapwood on mature oaks is relatively constant (about an inch) and so if the final rings are narrow, there are more in the sapwood than when the final rings are wide. As the boards of panel paintings were usually derived from aged oaks, the number of sapwood rings tends to be somewhat greater (20-30) than the number (15-20) in building timber that is fast grown.

Information about the number of heartwood rings removed in preparing boards has come from two sources:

(a) The date, from tree-ring measurements, of the latest ring together with the recorded date for several portraits of the mid-16th century; the interval between the two gives the number of years that elapsed up to the time the board was used by the artist. Table 1 gives results that apply to trees about 200-250 years old when felled. The last column gives the likely number of years that are accounted for by heartwood removed plus seasoning or storage. It appears that for both items only a few years were involved at that time (cf. Bauch, this volume, page 309).

(b) Evidence produced when measurements are made on boards from the same tree: this happens not only with a multi-board panel but with separate panels, perhaps now in different collections. The three royal portraits on panels of identical size (two now interpreted as portraits of Henry VIII early in his reign) that are in the exhibition this summer (1977) on Dendrochronology at the National Portrait Gallery in London provide an example. The respective dates of their latest ring (1487, 1488 and 1493) are close to one another so it would seem that the panel maker was trimming the boards close to the heartwood-sapwood boundary. There is confirmation for this: one of the panels (with the latest ring, 1488) has at some time been reduced in width by about 1 cm, on either side, to remove holes due to insect attack on sapwood that must have existed when the panel was used. This modification, which removed about 5-7 rings, enables the year of the heartwood-sapwood boundary to be located almost precisely.

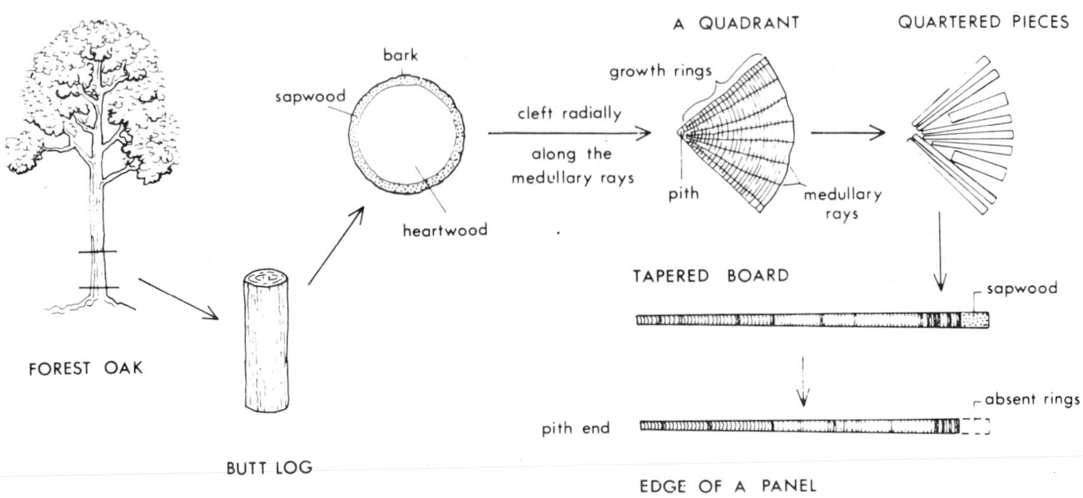

Fig. 1 Derivation of tapered quarter-sawn panels, free from sapwood, from the trunk of a forest oak.

For these three royal portraits, the felling of the tree can be estimated as about 1512 or soom after. The panels could have been made and used within a year but in general a somewhat longer period elapsed. The likely date of the portraits was therefore given (Fletcher, 1976) as 1513-20.

With royal portraits there is often evidence from aspects of fashion, costume and jewellry which indicates the date more exactly, once the period has been identified by tree-ring analysis. This applies to the group of three to which reference has been made above. A date as late as 1519 is excluded as an X-radiograph shows one portrait was overpainted about then to reconcile the appearance of Henry VIII with his newly grown beard and a change in the fashion of his hat.

Boards from very old trees, aged 300–400 years when felled, and from 15th century paintings indicate that large numbers of heartwood rings were sometimes removed. Hans Holbein during his first visit to England (1526-1529) used panels each made up of boards from an exceptional tree that was at least 400 years old. Two such boards about 26 cm wide, with a narrower (15 cm wide) one between them, make up the panels both of his portrait of Archbishop Warham and that of Kratzer, the astronomer—paintings now in the Louvre.

The weekday when Mr. Walker and I worked on those paintings was chosen to coincide with the day the gallery was closed to the public. Even when darkness put an end to our labours, we had not finished marking and measuring the many rings on the boards from that tree. The rings of the two completed were dated 1168 to 1414 and 1183 to 1421, while the rings on a third board began at 1091, were measured as far as 1250 and thereafter counted, the last c. 1440.

Table 1. Information from Panel Portraits with Recorded Dates

Portrait	Artist	No. of Boards	A. Date of Latest Ring (Heartwood)	B. Date of the Painting	Interval B-A years	No. of sapwood rings (estimate)	Difference. years
Henry VII	Sittow	1	1476	1505	29	25	4
Robert Cheseman	Holbein	2	1509	1533	24	20	4
Jane Seymour	"	3	1506	1536/7	30	18	12
Thomas Howard, Duke of Norfolk	"	2	1509	1539	30	19	11
Man in a Black Hat	Bettes	3	1521	1545	24	21	3
Thomas Cranmer	Flicke	3	1525	1546	21	19	2
John Luttrell	Eworth	4	1525	1550	25	20	5
Thomas Chaloner	Unknown	2	1521	1559	28	23	5
Nicholas Heath	Eworth	2	1546	1566	20	17	3

Note: The estimates for sapwood rings vary with the average width of the latest rings present.

After our research had laid down guide lines for deriving the likely period of time at which a panel was used, the method was applied to nineteen paintings belonging to the Society of Antiquaries of London, of which I am Fellow, and to twenty-one brought together at the National Portrait Gallery for the Richard III Exhibition in 1973 (Fletcher, 1974).

Influence of Tree-Ring Dates on Iconography

My interests in this research have extended beyond deriving an approximate date. One object has been to link the latter with historical information in order to discover the purpose and perhaps even the artist for certain Tudor portraits that have presented problems.

The date and purpose of the portraits in the Royal Collection of Henry V, Henry VI and Richard III (Sir Oliver Millar, 1963) provide an example. They are companion portraits of the same size and each panel has a board from the same tree. Ring-width measurements on those boards indicate about 1520 or soon after as the likely date for the use of the panels, the workmanship of which suggest Flemish origin. Quite early in his reign (1509-1547) Henry VIII seems to have wished to have portraits of his ancestors but the actual occasion for these three being painted would have been the opportunity to use them to adorn a panelled wall in a new royal palace. Bridewell Palace, which was completed c. 1522, seems appropriate for the date derived from tree-ring analysis.

The purpose of copies, apart from obvious examples of those for Long Gallery sets, is also emerging. Some are nearly contemporary, some provide illustrious family ancestors or revered bishops or archbishops. The portrait of Jane Seymour at Woburn Abbey, p. 138, has been regarded as a copy of Holbein's portrait now in Vienna. Tree-ring evidence shows that it was most unlikely to have been painted before Holbein's death (1543) but rather in the subsequent decade. So it was obvious in the light of the succession in 1547 from Henry VIII to Edward VI, Jane Seymour's son, that sometime after 1547 was the actual date. This can even be narrowed to 1547-50 as control of the Council and of the young King by the Seymours was abruptly terminated by the rise of the Dudleys in 1550. Furthermore it can be suggested that the artist was one John who was commissioned by Princess Mary in 1544 to paint her. That portrait happened to be adjacent in the Dendrochronology Exhibition to the Woburn portrait of Jane Seymour and the similarity of the hand then became apparent for the first time.

References

J. Fletcher, 1974. Burlington Magazine, CXVI, 250-58.

J. Fletcher, 1976. Studies in Conservation, 21, 171-78.

Sir Oliver Millar, 1963. Tudor, Stuart and Early Georgian Pictures in the Collection of her Majesty the Queen, London.

6.3. DENDROCHRONOLOGY APPLIED TO THE DATING OF DUTCH, FLEMISH AND GERMAN PAINTINGS

J. Bauch, University of Hamburg

Summary

Dating of the oak panels of Dutch, Flemish and German paintings of the 14th to the 19th century has been demonstrated. The felling date of the tree derived from the panel gives the art historian a "terminus post" for the creation of the painting. Further information can be obtained from this analysis, in particular when the results for numerous panels used by a single artist are compared. More than 440 oak panels used by about 75 artists have been analysed in 26 galleries: approximately 70% have been dated.

INTRODUCTION

In the last twenty years more and more techniques based on natural science have been used in art-historical research. Dating methods relative to the various materials used are of increasing importance.

In 1965 the Institute of Wood Biology, University of Hamburg began to apply dendrochronology to the dating of oak-panels of paintings, especially those of Dutch, Flemish and Old German artists who worked in the 14th to 19th century (Bauch, 1968). A particular problem with such paintings is that the wood and even the art objects themselves are transportable and their origin is not always obvious. In order to date the paintings it therefore seemed necessary to have local master-chronologies. During recent years such chronologies have been established in several laboratories (this vol. page 117; Fletcher, Tapper and Walker, 1974).

A statistical analysis on more than 2000 paintings showed that during the 15th to 19th century approximately 50 percent of the paintings were on oak panels (Eckstein, Brongers and Bauch 1975). Others were on canvas, copper and even, to a limited extent, on tropical woods.

The oak panels are most usually sawn from the tree trunk in an approximately radial orientation and therefore contain a large number of growth rings. The smaller panels were made of one solid board; the larger from several boards glued together (Bauch, Eckstein, Meier-Siem, 1972; Fletcher, 1976).

Synchronization of a growth ring sequence with a master-chart can give the date of the most recent growth ring on the panel and so give a "terminus post" for the felling of the tree. Thus it can be estimated when the panel was made and painted. If there is evidence of sapwood rings then the felling date can be assessed relatively precisely. The method employed has already been described by Bauch, Eckstein and Brauner (1974). The following illustrates with a few examples the possible contributions of dendrochronology to art-historical dating problems as well as the limitation of the technique.

TREE-RING DATING OF THOSE DUTCH AND FLEMISH PANELS OF THE 17th CENTURY WITH AN INSCRIBED DATE

Our first example compares dendrochronological dates with the dates on signed panel paintings of Dutch and Flemish artists of the 17th century. The

Fig. 1. *Dendrochronological dating of paintings with a time signature*

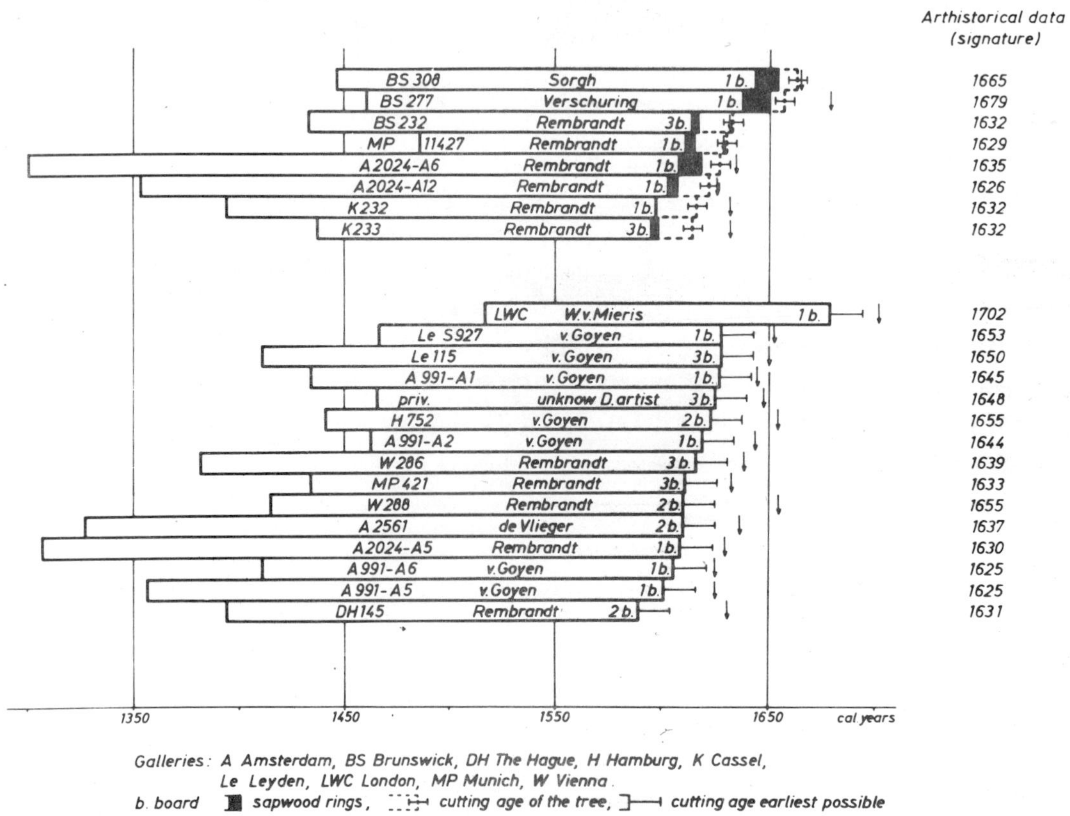

Fig. 2. Dendrochronological dating of the wood of Rubens paintings Maria de Medici series, ordered 1622

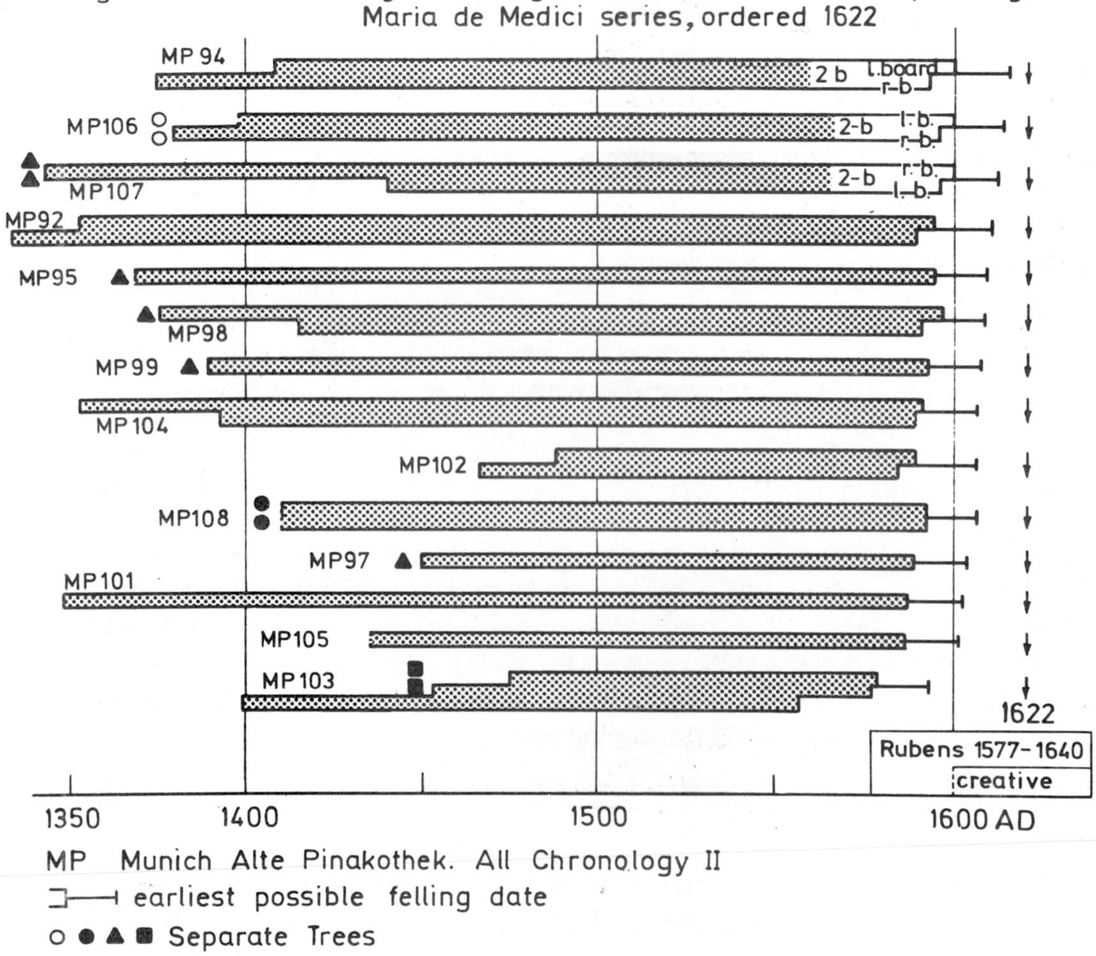

first group of eight panels listed in Fig. 1 contained sapwood rings, marked in black. In this case it is usually possible to determine the felling-date of the tree used for the panel with an accuracy of $\pm$ 5 years. From the figure, where an allowance is made for 20 $\pm$ sapwood rings, it can be seen that only a few years elapsed between the felling-date and the painting of the pictures—indicated by vertical arrows. For BS 232 by Rembrandt for instance, the earliest felling date likely is 1628 and the painting is signed 1632. So drying and seasoning the wood, making the panel and any storage before use occupied no longer than 4 years, perhaps less.

The second group of 15 paintings were also signed and dated but there was no sapwood. Therefore only the earliest possible felling-date can be derived. A comparison of the respective intervals revealed that frequently only the less durable sapwood had been cut off and almost no heartwood.

These examples indicate that dendrochronology can provide some help when we try to attribute a painting with no signature and stylistically of doubtful date to a particular period.

IDENTIFICATION OF THE ORIGIN OF WOOD USED FOR PANELS

In addition, the origin of the wood used for panels can frequently be suggested. This possibility was shown by 41 paintings by Ph. Wouwerman (this vol. page 133). Because the interval between felling date and creation of the painting was short, an approximate territorial classification of the paintings is possible. This is of special importance, because the paintings are not signed or dated and little study has been made of style-criticism for this artist. With dendrochronological dating, it is possible to classify the 41 paintings into three groups; early, medium and late. This classification is greatly helped by the fact that up to about 1650 only wood from provenance II—a coastal one—was used; later, wood from provenance I—more inland—was the only type available. This is of special interest, because Wouwerman (1619-1668) worked exclusively in Amsterdam throughout his life.

IDENTIFICATION OF WOOD FROM THE SAME TREE IN SEVERAL PANELS USED BY AN ARTIST

The analysis on 42 Rubens paintings supports the results on Wouwerman and shows what additional information can be expected from series analysis for paintings of one artist. Fig. 2 shows the result of tree-ring analysis of 14 Rubens paintings which belong to the Medici series, ordered in 1622 in Antwerp.

It is very striking that between the earliest felling date and 1622 there is an interval of only a few years. Furthermore it could be proved that the boards with identical symbols came from one and the same tree. From this it can be concluded that the logs were sawn into boards while still "green", and the boards then dried.

The next figure (Fig. 3) shows the years spanned by the panels of another 28 paintings by Rubens. At the right side is the art historical information. It is notable that only the wood of painting Ca. 1942 belongs to the inland provenance; all the other Rubens panels are of the coastal wood type. A re-examination of the attribution of Ca. 1942 to Rubens would be of interest.

Fig. 3. Dendrochronology Applied to the Dating of Rubens Paintings

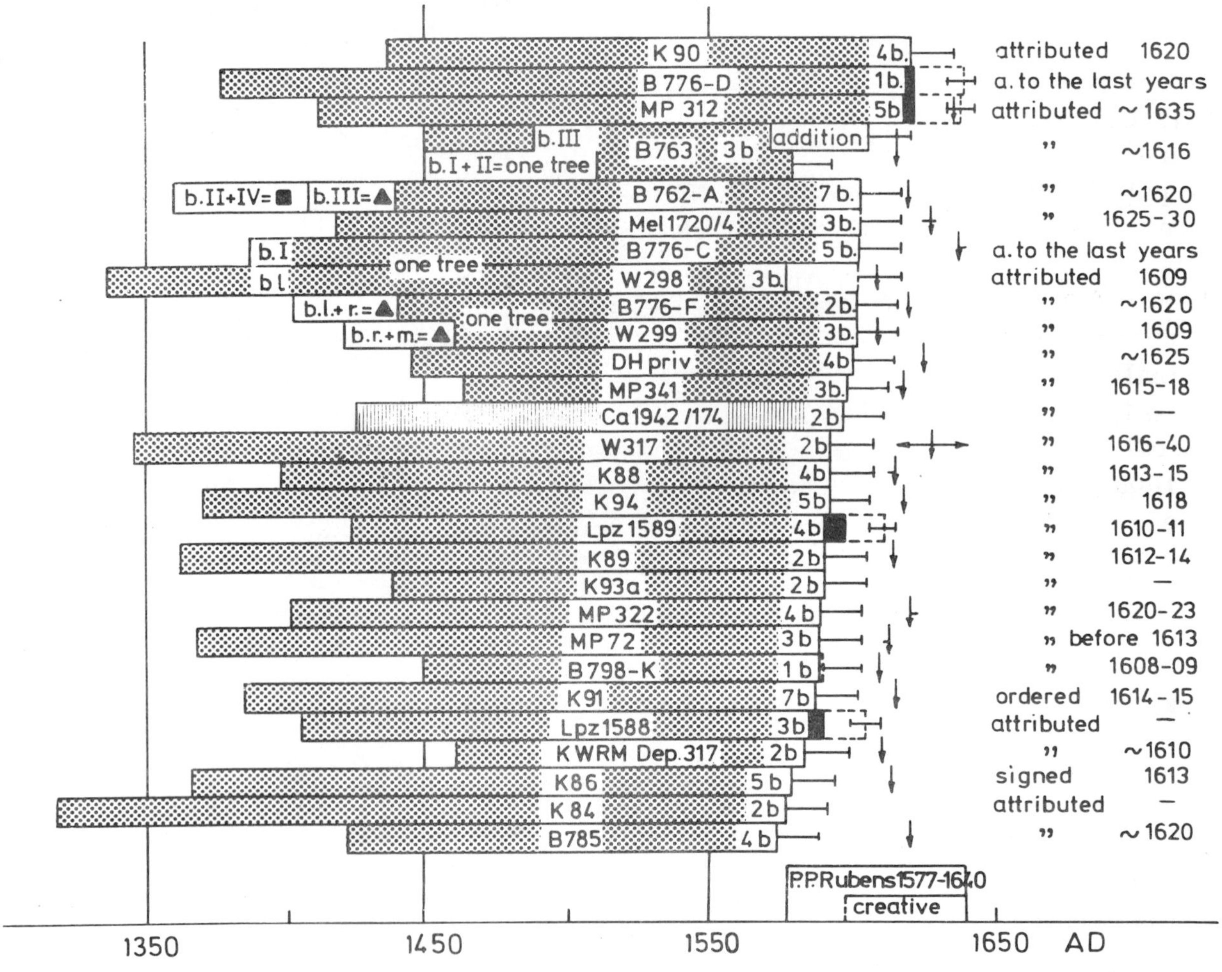

B Berlin, Ca Cambridge/Mass., DH The Hague, K Cassel, KWRM Cologne, Lpz Leipzig,
M Melbourne, MP Munich, W Vienna, b.: board
■ one tree with MP103 (b. I+III), ▲ one tree with MP107 (b. r. + l) and others

Fig. 4. 18th Century Paintings by Adriaen van der Werff in Alte Pinakothek Munich (MP)

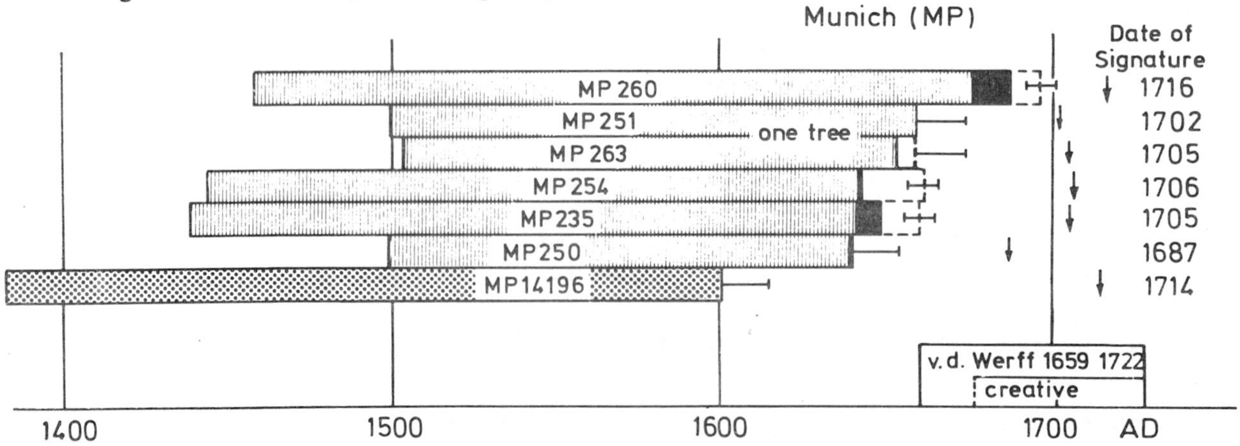

earliest possible felling date of the tree, felling date, sapwood rings
chronology I chronology II

The companion paintings W 298 and W 299 represent Archduke Albert and Infanta Isabella respectively. Both are definitely attributed by documentation to 1609 as originals. But for W 299 (Infanta) we find from the growth-rings that the painting could not have been made before about 1620. This is confirmed by the growth-ring pattern for the panel B 776-F, which contains board from the same tree as W 299 and is appropriately attributed to about 1620. The growth-ring sequence for the companion painting W 298 (the male portrait), has its youngest heartwood ring as early as 1577. From this sequence alone we originally assumed that the portrait of Archduke Albert could have been painted in 1609. However later we found that the growth ring pattern of W 298 is identical with that of one board of the painting B 776-C; its latest annual ring is 1603, implying that both paintings (W 298 and W 299) could only have been executed from about 1620 onwards. Thus this companion pair seem not to be the originals but copies, very probably executed by Rubens himself.

DATING OF PANELS OF ARTISTS OF THE 18th CENTURY

For paintings of the 18th century the relationship between the date of the wood and the creation of a painting is, in contrast to that found for the 17th century, often remote (Bauch, Eckstein and Brauner, 1974). This is demonstrated by our work on the paintings of A. van d. Werff (Fig. 4). The painting MP 14196 is a copy from the original MP 251. MP 251 and its wood is of the coastal provenance that was only available up to about 1650. It could be assumed therefore both from the years of its sequence and the provenance of its wood that MP 14196 was reused in the early 18th century. Subsequent X-ray analysis revealed the original painting from the first half of the 17th century underneath.

It is known that panels became scarce in the second half of the 17th century and that the reuse of earlier ones was not uncommon. Indeed, because of the scarcity of local wood, A. v. Werff frequently used tropical wood.

DATING OF PANELS BY EARLY GERMAN ARTISTS, 14th to 16th CENTURY

Fig. 5 summarises our work on the dating of early German paintings. A series of paintings of the Cologne School serves as an example.

The oldest panels we were able to date are the companions (WRM 4 and 5). The wood for both panels came from the same tree and the latest heartwood ring was laid down in 1222. By adding a minimum of 15 sapwood rings, we conclude the painting could have been made from about 1240 onwards. However, the art historical attribution to the period 1310-1340 is considerably later. It is not possible to estimate how many heartwood rings were cut off; but it would appear from these and other examples that the storage of wood, in particular of sawn timber, was sometimes of longer duration in Germany in the 14th and 15th centuries than in later centuries. For example, WRM 65 has boards from trees that were felled 15 years apart, and WRM 165 boards from trees felled 14 years apart. In each case, storage of the older tree must have lasted at least 14 years.

Another result of interest is the conclusion that some of these German paintings are on wood that had come from the Netherlands. An example is

Fig. 5. Dating of Oak Panels of Paintings of the Cologne School, 14th and 15th century.

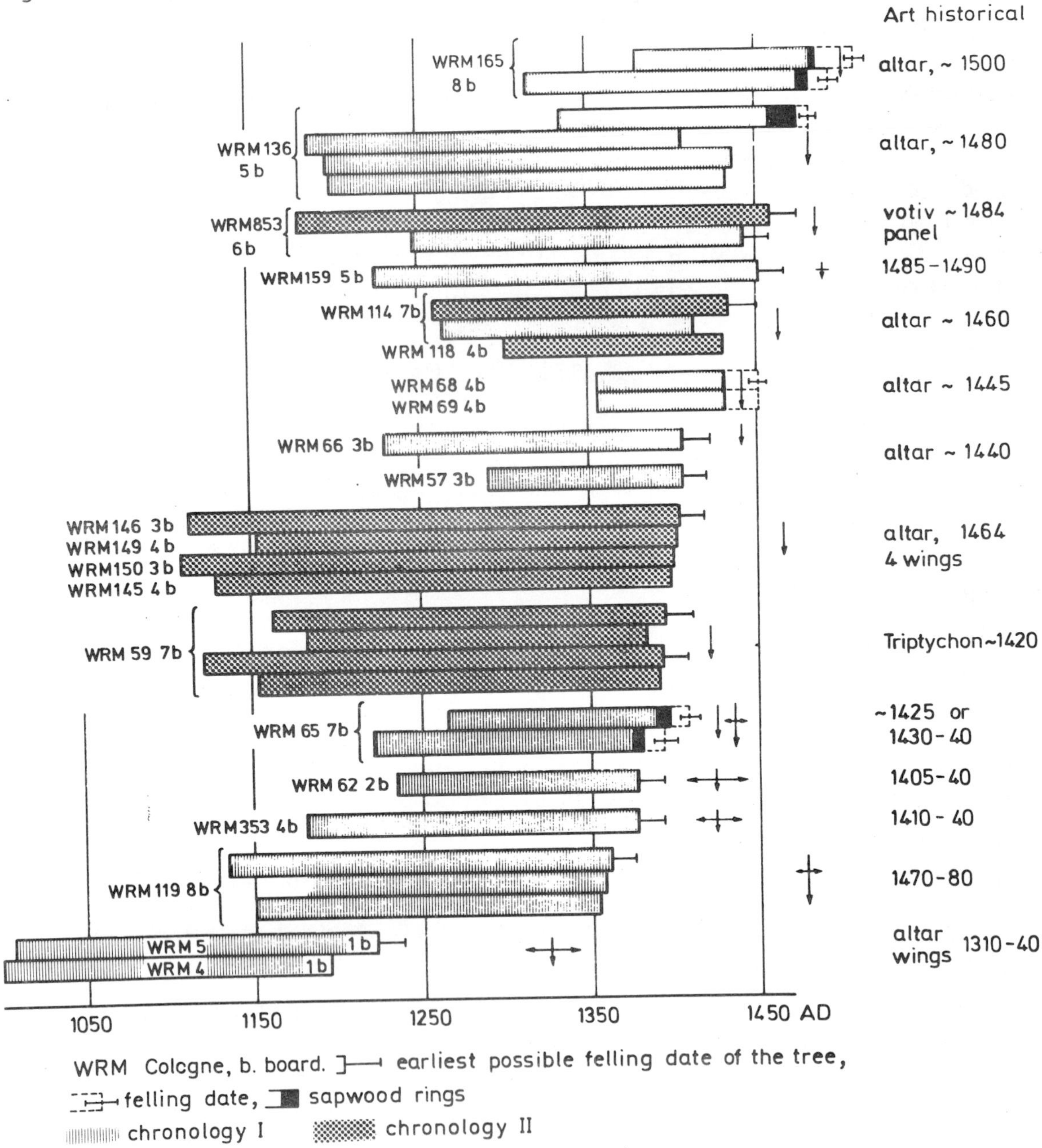

the Triptych at Cologne, WRM 59, for which boards of the central panel and of the wings were derived from the same tree. The panels themselves were probably made in the Netherlands and transported to Cologne. For instance we have indications that the artist Funhof used panels which came from the Netherlands to Hamburg. However, another Cologne panel (WRM 114) from an altar of 1460 is composed of some boards of the coastal provenance, others are of Cologne origin.

CONCLUSIONS

We find it more difficult to deduce the approximate date for a painting of the 15th or 14th century than for those of later centuries. In the Netherlands and Germany the interval of time between the felling of a tree and its use for a panel painting was somewhat greater in the 16th than in the 17th century. By then it was usual for the interval between felling and the date of the painting to be only a few years. By the 18th century, when oak wood became scarce, increasing reuse of panels adds a complicating factor.

Very often the "terminus post" for a painting is an essential requirement. This can be decided by tree-ring analysis of the panel and frequently other details, e.g. the provenance of the wood of the panel can be obtained. Of particular importance is the detection of the use of wood from the same tree for several panels. This was found often during the examination of about 100 Rembrandt and more than 50 Rubens paintings situated in 26 galleries in many countries.

Dendrochronological analysis of panels can provide important information for the art historian for use in combination with other methods.

Acknowledgements

These investigations were carried out in cooperation with Professor Dr. D. Eckstein and Mrs. G. Brauner. We are indebted to the following institutions and galleries for their support: Rembrandt Research Group, Amsterdam; Rijksmuseum, Amsterdam; Gemäldegalerie Berlin; Museum of Fine Arts, Boston; Herzog-Anton-Ulrich Museum, Braunschweig; Fogg Art Museum, Cambridge; Staatliche Kunstsammlungen, Dresden; Städtische Galerie, Frankfurt; Mauritshius and Museum Bredius, The Hague; Kunsthalle, Hamburg; Hessisches Landesmuseum, Kassel; Walraf-Richartz-Museum, Cologne; Museum der bildenden Künste, Leipzig; Hermitage, Leningrad; National Gallery and the Wallace Collection, London; National Gallery, Melbourne; Alte Pinakothek, München; Metropolitan Museum of Art, New York; National Gallery of Art, Washington; Kunsthistorisches Museum, Vienna.

References

Bauch, J., 1968, Kunstchronik, 144-145.

Bauch, J., Eckstein, D. and Brauner, G., 1974, Maltechnik 80, 32.

Bauch, J., Eckstein, D. and Meier-Siem, M., 1972, Nederlands Kunsthistorisch Jaarboek, 23, 485.

Eckstein, D., Brongers, J. A. and Bauch, J., 1975, Tree-Ring Bull. 35, 1.

Fletcher, J. M., 1976, Studies in Conservation 21, 171.

Fletcher, J. M., Tapper, M. C. and Walker, F. S., 1974, Archaeometry 16, 31.

DISCUSSION—Section 6

Prof. BAUCH, asked by Mr. Lewis to describe briefly his technique for making measurements on panel paintings, replied "The panel is taken out of its frame in the gallery or museum. We clean with a razor blade the part of the edge away from the painted side. This usually gives us an area large enough for measurements on the panel itself. Within a few hours the painting is back in its frame and in its place. We make sketches and record details to supplement what we know about the painting". Photographs or micro-photographs he said, were not taken but use was made of any available from the gallery.

In answer to a question from Becker, Bauch said that Chronologies I and II related not to a few but to many panels, including several used in a workshop in the Cologne area and taken there by Netherlandish artists. Mr. Rigold enquired if a Baltic source was feasible for one of the Chronologies. In his reply, Bauch said that regional chronologies for the Baltic as well as for Scandinavia and England have been tested for synchronisation without success and so they had finally decided an indigenous origin must be correct. Eckstein added that it was difficult to believe that timber had been imported into the Netherlands over a period of 250 years exclusively from the same country.

Miss Geddes asked if the English panels mentioned were from English or Flemish timber. Dr. FLETCHER said in reply that the tree-ring sequences of slow-grown oaks used at that time in south-east England and in Flanders were so similar that one could not decide this point on tree-ring evidence alone. The construction and finish of the panels usually indicated the provenance and thus it appears that almost all those used in England after about 1525 were from English trees. Flemish panels of the 15th and early 16th century that had been examined in Paris (the Louvre), Brussels and Vienna show that Flanders was a generation or so ahead of England in panel-making—as in many other crafts. For larger panels, such as those of two or three boards used in England for portraits in the mid-16th century, importation was unlikely on grounds of bulk alone.

Mr. Farrell asked if the panel on which the portrait of Henry VII was painted by Sittow might have been in store for some time as, according to Salzman, timber was getting scarce at that time. Fletcher replied that its latest ring was laid down in 1476. With no sapwood present, the interval of 29 years up to 1505—the date of the painting—implied that seasoning and any storage before use was unlikely to have occupied more than about 5 years. Timber scarcity is a relative term: the prevalence in England of household panelling after 1505, even more than before 1505, suggests no particular shortage of oaks for panels.

7.1. THE NEW DENDROCHRONOLOGY LABORATORIES AT NEUCHATEL AND ZURICH

O. U. Bräker, Birmensdorf and G. Lambert, Neuchâtel

Summary

New dendrochronological laboratories at Zurich and Neuchatel have become important as their concept and structure incorporates recent developments. In the Zurich laboratory, methods that use the latest techniques of microcomputers have been developed.

The pioneer research in Switzerland of Huber and Merz encouraged the archaeologist of the city of Zurich, Dr. Ruoff, to establish in 1969 a laboratory for dendrochronological studies (Braker, this volume, page 203). Initially a binocular microscope Zeiss/Jena was used; later the measurements were made with the classic mechanical equipment from Addo-Eklund. All curves were drawn by hand. The calculations of the means and synchronizations by the procedure of Eckstein (1969) were processed at the computer centre of the Zurich Institute of Technology.

Later, a series of excavations of prehistoric sites rich in wood remains in the Bay of Auvernier (Lambert and Orcel, this volume page 208) led to the development of a new dendrochronological laboratory at Neuchatel, western Switzerland, in 1973. The Swiss National Science Foundation sponsored the purchase of Addo-Eklund equipment (Fig. 1). The large Addo tape-punch was replaced by a smaller Facit puncher branched to a keyboard for alphanumeric input.

This digital system is more versatile and can be used for other archaeological purposes such as input of post maps, archaeological maps, files of archaeological materials as ceramics, axes, artefacts, etc.

The processing of the paper tape was done by the computer centre at the University of Neuchâtel. Methods used for the calculation of synchronization and mean curves were similar to those used in Zurich. The curve plotting was drawn with a computer programme.

Since 1974, excavations of neolithic sites also occurred at Lake Zurich. While excavations frequently unearth more than 2000 posts in a couple of months, only 400-500 samples could be measured and synchronized per year; thus new measuring equipment was acquired for the Zurich laboratory (Fig. 1). With the knowledge acquired at Neuchâtel, low efficiency processing media such as paper tape or cards were avoided. With the new microcomputer on the market, we opted for a system which integrated the calculations and plotting in the measuring equipment. The existing mechanical measuring platform was branched to an encoder converting the measuring values from the spin to digital values. These values are accepted directly into the microcomputer and stored on magnetic discs (Diskette).

The computer programmes give prints and plots of all measured samples, calculate mean curves, and synchronize values. All computations and drawings required by a dendrochronologist for his work can be done. Access to the computer centre of the Zurich Institute of Technology is not necessary.

The advantages of this working concept are the following:

a) elimination of punching or drawing errors,
b) no displacement between measuring laboratory and computer centre, no waiting time for prints and plots (turnout time for plot 1 day, for print ½ hour),
c) continuity of the work after the measuring, immediate control of the fit of two radial readings, possibility of immediate correction with the wood sample as reference
d) synchronizations to already measured samples can be made and the mean curve can be calculated, later synchronizations can be made with the most recent mean curve
e) preparation of all drawings and figures for final publication about a studied site.

The evaluation of different microcomputers started in 1976. Encoder and interface were installed in early 1977. The equipment and necessary programmes are now operating. The equipment is also usable for the other archaeological purposes mentioned above for Neuchâtel: mapping, files for ceramic and medieval material etc.

In comparison with the magnetic tapes (cassettes) used by other computer systems, the disc permits the reading of single values from the storage memories rather than processing the entire file. This saves working storage space for the programmes.

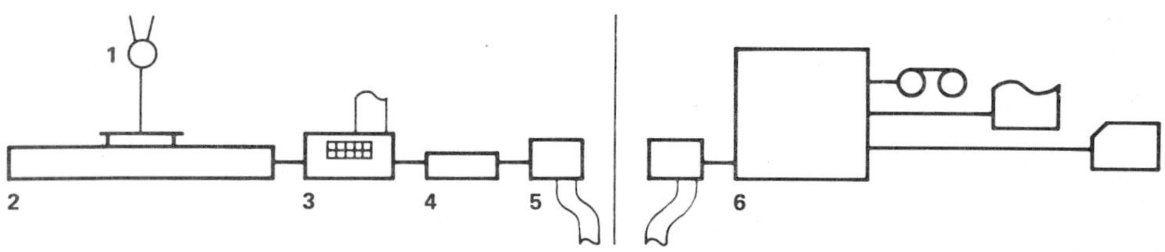

Neuchâtel: **1** microscopic optic, **2** electronic measuring unit with small platform, **3** calculator with numerical print, **4** alphanumeric keyboard and interface, **5** paper-tape puncher, **6** data processing center of the University of Neuchâtel: input by paper-tape, conversion to cards, output by printer and plotter

Approximate prices in 1976			
1+2	sFr.	32000.–	£ 6400.–
3		4300.–	860.–
4		7700.–	1540.–
5		6600.–	1320.–
Total	sFr.	50600.–	£ 10120.–

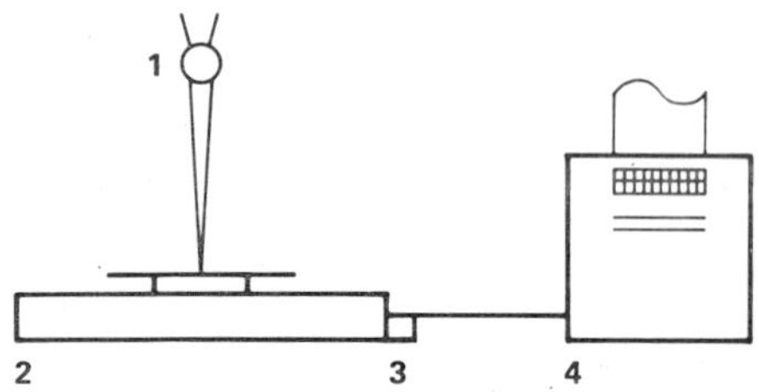

Zurich: **1** stereoscopic optic, **2** mechanical measuring unit with big platform, **3** A/D converter-encoder and interface, **4** microcomputer with 16 K-bytes memory, keyboard, thermic printer with plot capabilities, 2 disc stations, each with 1/4 M-bytes memory

Approximate prices in 1976			
1	sFr.	2000.–	£ 400.–
2		8000.–	1600.–
3		7000.–	1400.–
4		36000.–	7200.–
Total	sFr.	53000.–	£ 10600.–

Figure 1 Dendrochronological equipment used at the laboratories in Neuchâtel and Zurich.

7.2. X-RAY DENSITOMETRY AS A DENDROCHRONOLOGICAL TECHNIQUE

S. J. Milsom and M. K. Hughes (Liverpool Polytechnic)

Summary

X-ray densitometry has been set up. Wood samples are exposed over film by tracking under a collimated X-ray beam. This technique produces very small parallax. Results for slow-grown sessile oak (Quercus petraea) from Maentwrog, North Wales, show cross-dating between maximum density curves both within and between trees. The 227-year maximum density chronology derived from these trees is similar to their ring-width chronology.

Introduction

In moist, temperate regions, the production of annual growth rings by trees is less dominated by simple environmental factors than in semi-arid regions. The radial width of the annual ring is, however, only one measure of growth in in a year. Measurement of other properties of the annual ring may enable a better understanding of the tree ring/environment relationship than is possible using ring width alone.

Polge (1965), Parker and Meleskie (1970), Fletcher and Hughes (1970) and Hughes and Sardinha (1975), have given accounts of the principles of X-ray densitometry of wood, while a number of workers have used the method, principally with softwoods Polge (1966), Parker and Henoch (1971), Schweingruber, Bräker and Schär, this volume, (page 89). This paper reports the establishment of an X-ray densitometry laboratory and gives results of an attempt to apply the method to the study of oak, the principal material for European dendrochronology. In one technique, a closely collimated beam with a relatively short source-film distance is used to expose the wood sample, tracked beneath the beam (Echols 1970, Parker and Meleskie 1970). Our use of this technique is the first in Britain. It is capable of reducing parallax for a 20 cm long sample to that equivalent to a source-film distance of 27 metres in the static technique. Techniques have been developed for sample preparation and standardisation. The samples are sawn to standard thickness using a twin blade saw unit and oven-dried prior to radiography. The radiographs are developed using standard, repeatable methods, and analysed using a Joyce-Loebl microdensitometer with output to a chart recorder. Calibration of each radiograph is achieved by including a synthetic step wedge of known density with each sample.

Sample collection and preparation

Samples are collected from the field as 5 mm increment cores, billets or complete discs. Cores are dried, then glued with fibres vertically aligned between grooved supporting sticks making a final "sandwich" with a 2 x 4 cm cross-section.

Larger samples are seasoned for several months in a well ventilated, unheated room. 2 x 4 cm samples are then cut using a band-saw, extreme care being taken in arriving at vertical alignment of the vessels. Lathe centring has been used to align the samples correctly before cutting. Poor ring boundary alignment results in truncation of peak density values in the radiograph (Lenz, Schär and Schweingruber 1976).

The 2 x 4 cm billets or "sandwiches" are then cut to an accurate and uniform thickness of 3.18 mm (coefficient of variation of 48 measurements = 1.5%) using a twin-blade 15.2 cm circular saw unit. The unit incorporates a pair of hollow ground, 2 mm thick, combination blades set on the shaft of a 1 h.p. motor. The sample is held in position relative to the spaced blades by means of a line bar and sprung roller assembly.

The 3.18 mm thick sawn samples are clamped flat in pairs in order to reduce warping; then oven dried to constant weight at 105°C, and stored in flat "Perspex" dessicator trays until needed for X-radiography.

Radiography

The equipment for making radiographs consists of a water-cooled "Pantak" X-ray tube with a 50 kV and 17 mA rating. This is enclosed above a film carriage mechanism within a lead-lined cabinet with a drawer (Fig. 1). The tube incorporates a 1 mm beryllium window behind a collimator with either a 2.0 or 8.4 mm slit respectively 0.5 and 2.9 cm above the film. The focal point to film distance is 9.4 cm. The film carrier mechanism consists of a screw-driven trolley with a film carrier plate 60 cm long by 20 cm wide. The total traverse distance is 60 cm, the carriage speed 16.6 cm/minute (c.v. 1.0% n = 10). Kodak Type M, fine grain film is used in 35 mm format.

The wood sample and calibration chip are held in clamps on top of the naked film during exposure. A peak kilovoltage of 10 is used. A current of 3 mA with the 8.4 mm slit, and 12 mA with the 2.0 mm slit gives an effective exposure of 18 mA seconds for each collimator. After exposure the film is processed in developer agitated by a one second burst of nitrogen every 10 seconds, then stopped, fixed, washed and air-dried. All stages of the photographic process are carried out at fixed times and temperatures.

Fig. 1
X-RAY UNIT cabinet and carriage assembly

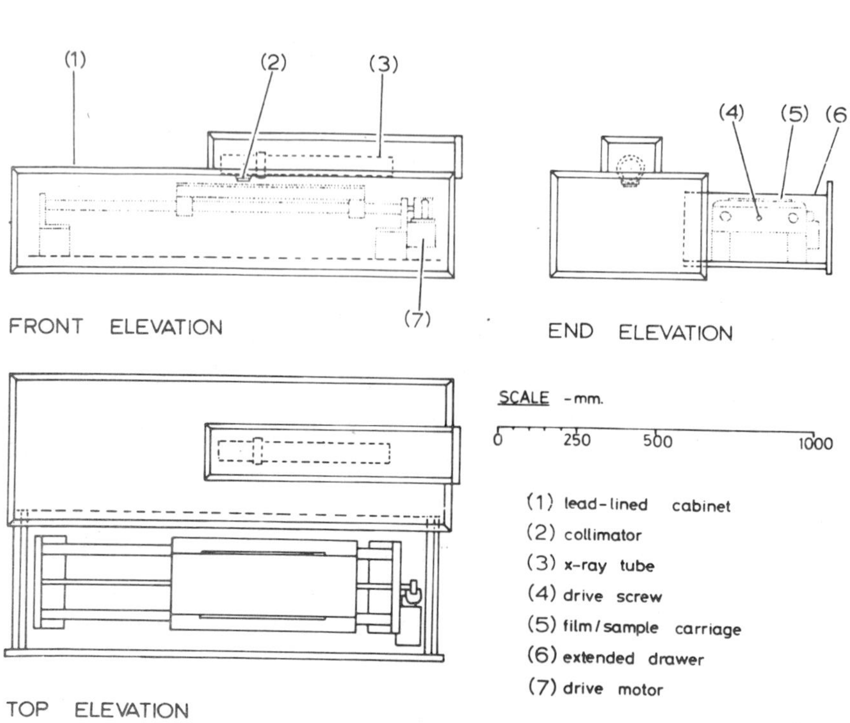

Densitometry and Calibration

The radiographs are analysed using a Joyce-Loebl microdensitometer. Currently a 1.5 D linear reference wedge is used, but a non-linear wedge of the type used by Fletcher and Hughes (1970) is being manufactured. The radiograph is scanned by an effective slit of 20 x 850 microns. The use of arm ratios of 1:1, 2:1 and 5:1 and output to a chart recorder with variable input and paper speed allows considerable flexibility.

Wood is moderately transparent to X-rays so that the beam should consist of poorly penetrating, long wavelength radiation to give sufficient contrast to the film. Unfortunately it is this very type of radiation that results in a non-linear relationship between the optical density of the radiograph and the physical density of the wood. Pending the availability of the compensating optical wedge for the microdensitometer mentioned above, calibration is achieved for every radiograph by fitting a regression line to the values for radiographic optical density and the superficial density of each of the five steps of a synthetic step-wedge. The linearity is achieved by transforming both variables to their natural logarithms. The regression equation is then used in the interpolation of physical densities from the known optical densities read off the densitometer chart output.

The synthetic step-wedge is made from the polyacetal, "Glacetal", a plastic similar to the kematal used by Fletcher and Hughes (1970). The wedge is 4.5 cm long by 1.5 cm wide with thicknesses from 0.80 to 3.08 mm. These correspond to superficial densities of 0.35, 0.50, 0.67, 1.02 and 1.34 (c.v. $< 2\%$ at each level $n = 10$) in a 3.18 mm wood sample.

As each radiograph is calibrated individually, minor variations in, for example, developer condition are accounted for and may be checked by comparing the regression statistics. At the moment optical density values have to be read off the densitometer chart output and then linearised by use of a calibration equation as described above.

Variability in X-ray dosage

Variation in X-ray dosage along the length of a sample may arise from (i) a variable rate of carriage movement and (ii) a variable output of the X-ray tube. These will be dealt with in turn:-

(i) variable rate of carriage movement. The use of a stepping-motor supplied by a good power supply to drive a high-quality screw is vital to the minimisation of this problem. It is also important that the screw be used to pull the carriage towards the motor and that the screw be kept absolutely clean. Extensive tests showed that overall end-to-end carriage speed variability was smallest, as was variation along the traverse, when a carriage speed of 16.6 cm/min was used. At this speed the run-to-run coefficient of variation was 1%, and the variation in carriage speed among five separate 10 cm stretches of the traverse was less than 2%.

(ii) variable output of the X-ray tube. The use of a stabilised power supply and careful adherence to the manufacturer's working instructions helps reduce this problem. Before each exposure the tube was allowed to equilibrate at the working voltage and current, and throughout the run the current is closely monitored. If there is any detectable fluctuation the run is abandoned.

The effective variation in X-ray dosage along the traverse has been tested by exposing 50 cm strips of "Glacetal" under the normal procedure applied to wood samples. Two thicknesses of "Glacetal" were used, corresponding to superficial densities of 1.34 and 0.67. The coefficients of variation for optical density measurement taken every 0.5 cm along the resulting radiograph were 1.65% (n = 30) and 3.86% (n = 54) respectively. No systematic drift in effective dosage along the traverse is detectable, providing that the mechanism is properly maintained.

Radiographic consistency

The repeatability of radiographic results obtained from a single sample has been tested by examining maximum late wood density for a 20-year period from oak sample LS 102B. This was a radial section viewed tangentially, taken from the top of the butt log. Ten radiographs of this sample were made and maximum latewood optical density taken for each year on each radiograph. Coefficients of variation for the ten estimates of each of the twenty rings ranged from 0.71 to 3.65%. Variability within a single radiograph was examined by measuring ten slightly different tracks within the central centimetre of the negative. In this case the coefficients of variation for the twenty years' maximum latewood densities ranged from 1.79 to 3.68%. Variability arising from the densitometer-recorder part of the system was measured by scanning the same track of a single radiograph ten times. In this case coefficients of variation for the maximum latewood densities of the twenty rings were from 0.08 to 0.47%.

The checks on the radiographic and densitometric procedures described above are repeated on a routine basis so as to guarantee consistency of application of these techniques. It should be emphasised that the term optical density used in this paper applies to a relative measure produced by the densitometer, rather than to a calibrated measure. Similarly, the term physical density used here refers to a relative, not an absolute, measure of wood density. It is for this reason that no units have been given.

Fig. 2

X-RAY DENSITY PROFILES. - OAK ANNUAL RINGS.

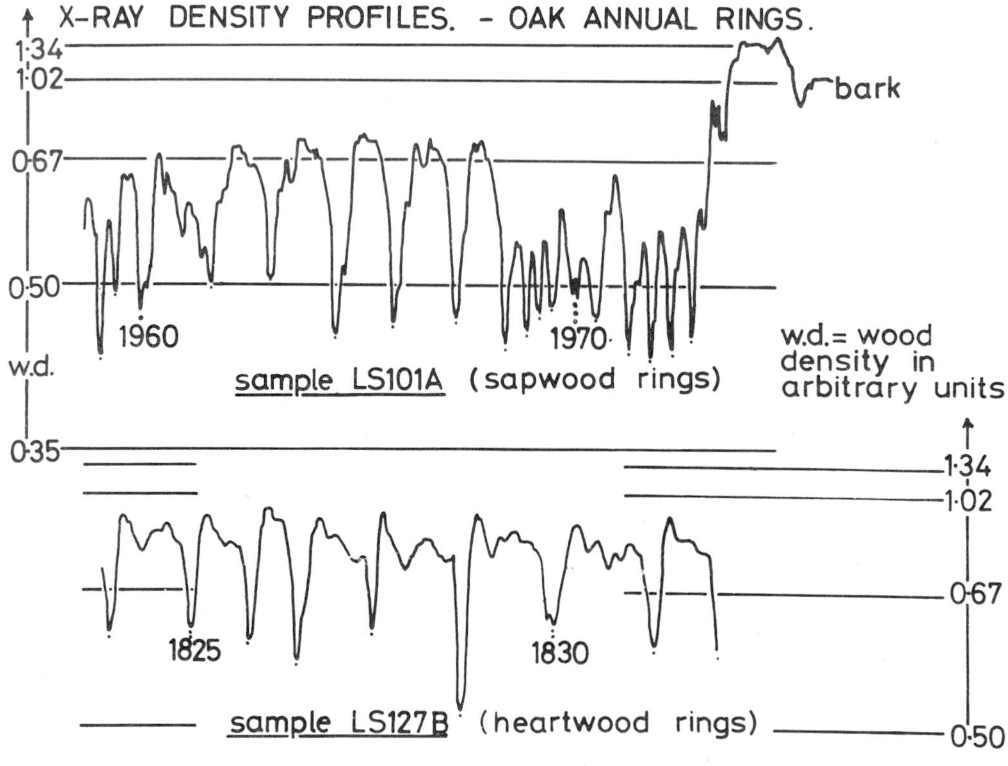

X-ray densitometry of oak

Severe problems exist in the densitometric examination of oak. First among these is the irregular orientation of the various elements of the wood structure when viewed in any direction. As shown by Lenz, Schär and Schweingruber (1976) poor alignment will produce errors in densitometric analysis. Secondly, a considerable proportion of the wood volume is taken up by medullary rays. When the transverse surface of oak is examined these rays characteristically take a sinuous route from the centre of the tree towards the cambium. Thirdly, early wood vessel lumina may be both wide and long enough to extend completely through a transverse section and to allow unhindered or almost unhindered transmission of X-rays. Fourthly, oak does not produce a density trace characterised by a single sharp latewood density peak, but rather it has a complex form as shown for samples LS 101A and LS 127B in Fig. 2 often with a peak in the inner latewood.

Fletcher and Hughes (1970) worked with radial sections of about 3 mm thickness viewed tangentially. This was done "to avoid complications from medullary rays". It also avoids the problems produced by earlywood vessels. In the present work this was the first approach used. It was satisfactory when dealing with short samples containing wide rings, such as LS 127B (Fig. 2). In work on material from the Maentwrog oaks (Leggett et al., this volume) which have a mean ring-width of 1.58 mm, severe difficulties were encountered, particularly in the many narrow rings. A close examination of the material showed variation in vertical alignment of vessels to be less of a problem than orientation of ring-boundary and presence of medullary rays. As a result, transverse sections are now in use, viewed vertically. These are of 3.18 mm thickness, as described earlier in this paper. Radiographs of such sections can be compared directly with the wood sample and its dated ring-width curve for absolute cross-dating of each ring. Problems would clearly arise in examination of early wood, but that is not a priority at the present time.

In order to avoid large medullary rays and to maintain normality between the ring boundaries and the movement of the radiograph relative to the densitometer light beam it is necessary to analyse several, overlapping tracks on each radiograph. This procedure results in clear densitometer traces, reduces problems of ring identification and minimises errors arising from misalignment.

Preliminary Results

Fletcher and Hughes (1970) reported X-ray densitometry of medieval oak beams. They concluded that "microdensity records obtained by X-radiography of oak grown in the past will be useful (i) for indicating climatic conditions both within a growing season and between one growing season and another: (ii) for establishing tree-ring chronologies and hence for dating material of interest".

In order to determine how great a potential exists for pursuing these two lines of work we have undertaken a detailed analysis of ring-width and density at various heights in thirteen oaks (Quercus petraea (Mattuschka) Liebl) from the Maentwrog site in North Wales for which a ring-width chronology has been established (Leggett, Hughes and Hibbert, this volume, page 187). It was decided to work first with modern trees from a site with a local chronology so that the advantage of absolute dating and of extensive replication could be enjoyed. The decision to examine ring-width and density at varying heights should, through the examination of oblique, vertical and horizontal series as modified from Duff & Nolan (1953), allow an examination of the relationships between cambial age, height, position relative to canopy, the environmental conditions in particular calendar years and the ring measurements.

As this extensive programme of measurement and analysis is at the time of writing not completed, preliminary results will be presented here. It is believed that these will be of interest to dendrochronologists.

Maximum density is the first variable to be examined, principally because it is the simplest to derive manually from densitometer chart output. Maximum density for a 145 year period in two radii at breast height in a single tree is shown in Fig. 3a. The similarity of these two curves is striking and shows extremely good cross-dating between these two radii. This pattern is repeated in the other trees examined. Fig. 3b shows the tree mean curves for maximum density (reduced to a mean of unity) for seven trees. Again, the similarity between these curves is striking and demonstrates clear cross-dating between the trees. Table 1 gives the results of statistical analysis of the density measurements for the fourteen radii, compared with those for ring-width indices taken from the same trees. The ring-width indices were derived as described by Leggett et al. (this volume, page 187). The 51-year period analysed is that in which both ring-width and maximum density measurements were available for two radii in all seven trees. It can be seen that mean serial correlation is lower in the density series than in the ring-width indices, whilst mean sensitivity, average correlation between trees, and variance held by the common chronology (% variance, years) are higher in the ring-width indices than in the density chronology. All of these differences indicate that a chronology derived from the ring-width indices would be of more use in cross-dating than one derived from maximum densities, at least in material taken from the lower part of the stem.

Fig. 3c shows chronologies derived from maximum density and from ring-width indices for these seven trees. The density chronology is the mean of the seven curves shown in Fig. 3b. The two chronologies are similar, having a correlation coefficient for the period 1860-1910 of 0.561. This period has been taken because both chronologies are based on fourteen radii between these dates.

Table 1 Comparison of statistics

	Maximum density	Ring-width
Period of analyses	1860-1910	1860-1910
Number of trees	7	7
Number of radii per tree	2	2
Mean serial correlation[1]	0.438	0.445
Mean sensitivity[2]	0.163	0.241
Average correlation between trees[3]	0.347	0.558
Analysis of variance:-		
% variance, years	23.46	39.87
% variance, years x trees	22.54	11.78
% variance, years x trees x cores	54.00	48.35
	100.00	100.00

Fig. 3

X-RAY DENSITOMETRY OF MAENTWROG OAKS

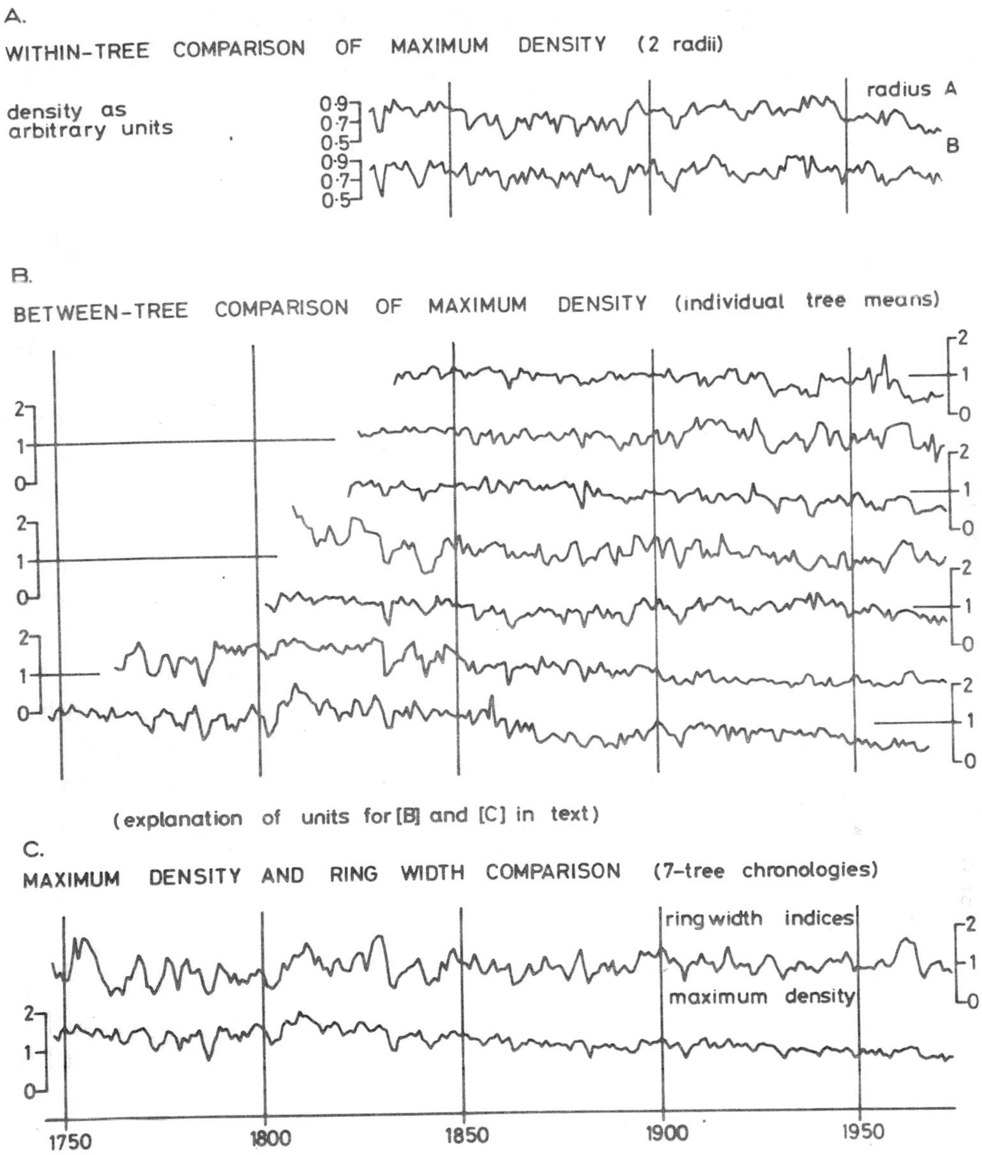

1. Mean serial correlation is a measure of the correlation between, say, the ring-width in one year and that in the previous year. Where this figure is large and positive a direct relationship will exist between successive ring-widths.

2. Mean sensitivity is a measure of the fluctuation of a set of measurements. The greater year-to-year differences are relative to their mean, the greater this figure will be. It is the mean fractional change from each measured ring value to the next.

3. Average correlation between trees is the mean of all the possible pairs of correlations between the trees measured. In this case there were 7 trees and hence 21 correlations. The average correlation increases as the similarity between the seven sets of measurements increases.

Conclusions

The very limited results presented here support the conclusion that further investigation of X-ray densitometry of oak for dendrochronological purposes would be of interest, if density chronologies contain information not recoverable from ring-widths. The questions yet to be answered may be summarised:-

1) will the cross-dating demonstrated here be repeated with larger numbers of trees and with trees from other sites?
2) to what extent do density variables change independently of ring-width in a systematic manner that can be related to factors external to the tree?
3) how are density variables and ring-width related in faster-grown timber than that sampled at Maentwrog?

Work currently in progress in our laboratory, particularly the extensive within-tree analyses referred to above, should give answers to these questions.

Acknowledgements

The research is supported by N.E.R.C. grant GR3/2244. Thanks are due to John McHarrie, Ken Sharples, and Walter Neal for help in the design and construction of apparatus, and to Pat Leggett for help in cross-dating. The Snowdonia National Park Administration allowed sampling at Maentwrog.

References

Duff, G. H. and Nolan, N. J. 1953. Can.J.Bot.31, 471-513.

Echols, R. M. 1970 in Tree Ring analysis with special reference to N.W. America. Univ.B.C.Fac.For. Bulletin 7, 34-36.

Fletcher, J. M. and Hughes, J. F. 1970 in Tree Ring analysis with special reference to N.W. America. Univ.B.C.Fac.For. Bulletin 7, 41-54.

Hughes, J. F. and Sardinha R. M. de A. 1975. J. Microsp. 104, 91-103.

Kleuters W. 1964. For. Prod. J. 14 (9) 414-420.

Lenz, O., Schär, E. and Schweingruber F. H. 1976. Holzforschung 30, 114-123.

Parker, M. L. and Meleskie, K. R. 1970. Tree Ring Bull. 30, 11-21.

Parker, M. L. and Henoch, W. E. S. 1971. Can.J.for Res. 1 (90), 90-98.

Polge, H. 1965. Proc.IUFRO, Sec. 41, 2, 1-30.

Polge, H. 1966. Ann.Sci.For. 23 (1).

7.3. TECHNIQUES FOR TREE RING DATING

J. F. Hughes, Oxford University

At present tree ring dating depends largely on the measurement and comparison of sequences of ring widths. With the development of techniques for the measurement of within-ring variation in density and in cell dimensions, it would now seem worthwhile to investigate the possibility of using features other than ring width for comparing series of rings. This paper describes some of the automated and semi-automated methods that are available to examine ring structure and some procedures for data sorting that may be useful to dendrochronologists.

DENSITY

There is considerable evidence that the proportion of low density earlywood and of high density latewood within tree growth rings is influenced by the particular pattern of seasonal climatic variation in the year of formation. It therefore seems likely that the mean density, the proportion of low density wood and high density wood, and possibly maximum and minimum values for density can be used to identify individual growth rings or sequences of growth rings that could not be distinguished from examination of ring widths only.

The density of individual annual rings can be measured accurately and rapidly by gravimetric methods, provided the specimens are not too small. It is, however, very difficult to set up a satisfactory technique and strategy, based on gravimetric methods, for sampling within ring variations. For example, it is often not possible to distinguish clearly and consistently between earlywood and latewood, and as samples of smaller size are taken, the accuracy with which they can be measured and weighed decreases.

Of the other methods available, the radiography and densitronomy technique, first used by Polge (1966), seems to be the most promising for general application, including dendrochronology, because it provides a permanent photographic record of within-ring variation in density which can be analysed in detail and with accuracy.

The radiography and densitometry technique, as used at the Commonwealth Forestry Institute, Oxford, (Fletcher and Hughes, 1970; Hughes and Sardinha, 1975) is probably very suitable for dendrochronological investigations. Wood specimens are machined to 4 or 5 mm in thickness, by passing them between a pair of tungsten carbide-tipped saws mounted on a light-duty high-speed spindle moulder. Specimens are then conditioned to a known moisture content in a constant temperature oven, in which the humidity is controlled by a saturated solution of a mineral salt (sodium dichromate is frequently used). Radiographs of the conditioned specimens are prepared by exposure techniques similar to those developed by Polge and intended to represent the best compromise between uniformity of exposure, contrast and time required for photography (50 kV X-ray tube, specimens at 2.5 m from the source, peak kV 15-20, at 25 milliamps for about 4 minutes, with Kodak Industrial C film).

Films are very carefully processed in an automatically operated developing tank, constructed here with modifications from designs provided by Polge, Nicholls and others, in which film background variation is reduced virtually to the effect only of film grain. Every film, however, is checked from end

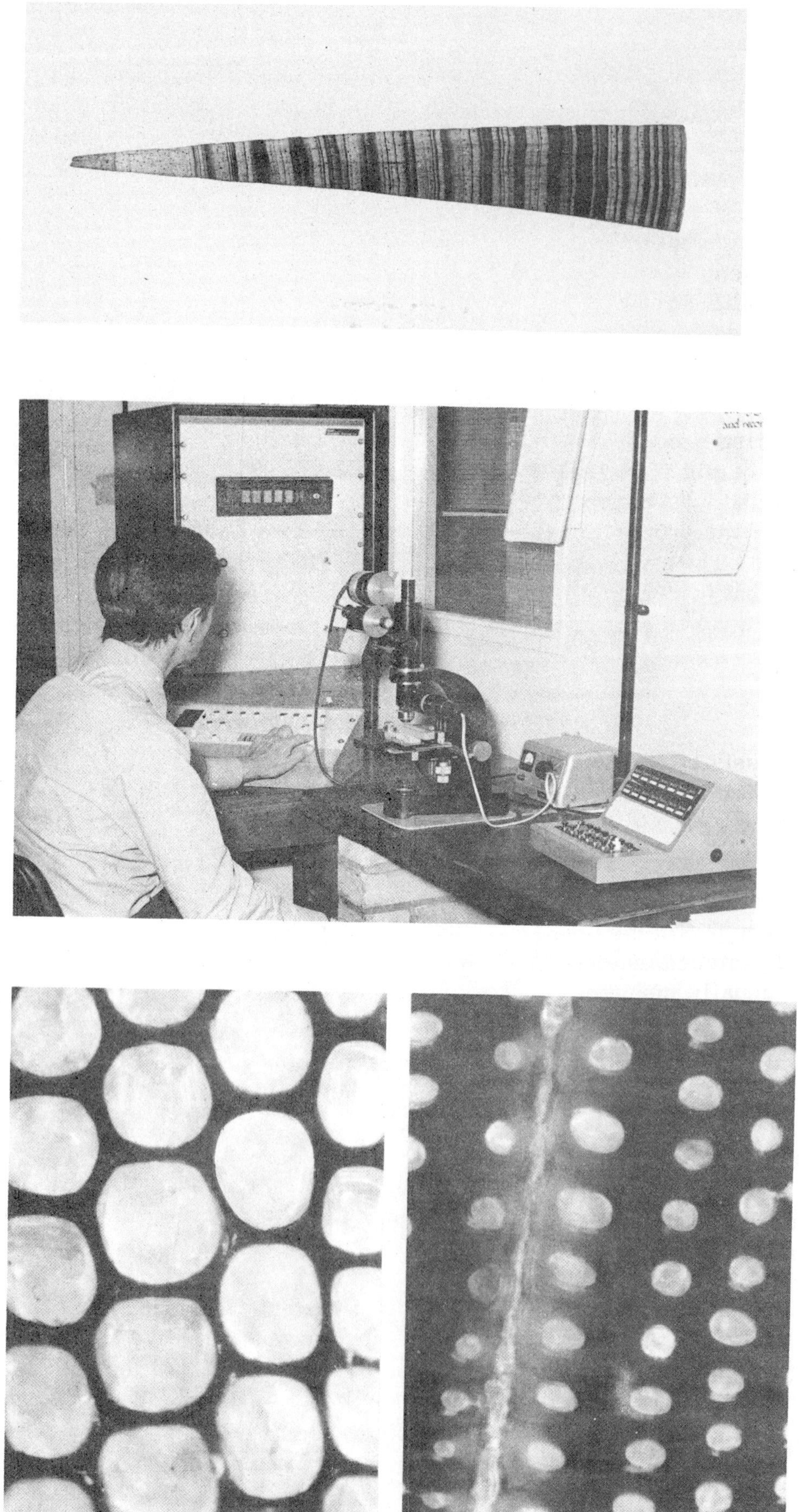

Fig.1 Cell cross-sectional dimensions by incident light microscope.
Top: Polished specimen Centre: Equipment
Below: Cells (<u>Pinus caribaea</u>)

to end by placing on it three strips of a polyacetal plastic, 1mm, 2mm, and 3mm in thickness. These are in fact used for instrument calibration, but when their images on the radiographs are scanned for this purpose, they also provide a check on the uniformity of development of the film.

Films of wood specimens are scanned with a Joyce Loebl microdensitometer, which records the variation in density in a line from the pith outwards and presents this information on a chart recorder and also by a punch tape output. The instrument is fitted with optical wedges that have been developed at the Commonwealth Forestry Institute to compensate for the non-linear relationship between the density of wood specimens and the optical density of their images on radiographs. The non-linear effect occurs because of the low peak kV conditions that are necessary in film exposure to obtain adequate contrast.

Computing procedures have been developed to present the data from the tape punch output in various forms. These include presentation of mean density, maximum and minimum densities for each growth ring separately, and presentation of the proportion of each ring in density class at .1 gm/cm^3 intervals separately. The tape output can also be shown in graphical form on a reduced scale, so that for example the patterns of variation from the pith outwards and within each growth ring at various levels in a tree can be easily examined together.

CELL DIMENSIONS

It is possible that the pattern of within-ring variation in cell dimensions and the proportion of different tissues may differ appreciably from year to year, depending on climatic conditions, and to an extent which would distinguish individual growth rings and sequences of rings.

Cell cross-sectional dimensions can be measured on macerated specimens, but there are problems with this method in distinguishing between radial and tangential dimensions and it is slow. At the Commonwealth Forestry Institute, cells are usually measured on the transverse surfaces of wood specimens that have been prepared by polishing on a belt sander, using a fine grade of silicon carbide grit (No. 400).

A few minutes only is required to produce a smooth and evenly polished surface (Fig. 1) with no apparent distortion and suitable for examination with an incident light microscope at reasonably high magnifications. The specimens require no special treatment after polishing, but it was found that the application of a small amount of paraffin wax on the polished surfaces just before microscopic examination enhances the images of the cells, especially when they are viewed with a combination of yellow and green light filters.

The cross-sectional cell dimensions of wood specimens prepared in this way are measured with equipment developed in conjunction with Faul-Coradi (Scotland) Ltd. This consists of a screw micrometer eyepiece, which is used to traverse the specimens, geared to an optical digitiser, where it registers the movement of the micrometer and is interfaced to a counter unit and paper tape punch to record successive measurements (Fig. 1).

A simpler and less expensive system is now being developed in which the digitiser will be interfaced to a commercially available data logger. To facilitate the selection of cells for measurement, a Swift point counter-moving stage is attached to the stage of the incident light microscope. This can be set to move at fixed intervals, so that a random or systematic selection of within-ring sampling positions can be easily and quickly located.

Techniques under development

A technique is now being developed at the Commonwealth Forestry Institute for measurement of cell cross-sectional dimension, in connection with a study of pulping properties of wood, but it may be useful in dendrochronology.

Microphotographs of polished cross-sectional surfaces of wood are taken with an incident light microscope. Enlarged black and white positives of the cross-sectional images of the wood cells will be scanned, using a Quantimet Image Analysing Computer which will count the number of cells per sq. mm and classify them by shape and size. With this equipment it should be possible to measure within-ring variations in cell cross-sectional dimensions with speed and accuracy approaching that of the densitometrical measurement of density.

Data processing

There would be no difficulty in developing computer programmes to sort and classify measurements of cell cross-sectional dimensions. Some of the data on within-ring variations should be suitable for treatment by smoothing techniques — for example, the proportion of latewood or maximum and minimum values.

References

Fletcher, J. M., and Hughes, J. F., (1970). Uses of X-rays for density determinations and dendrochronology. Bulletin, Faculty of Forestry, University of British Columbia, No. 7, 41-54.

Hughes, J. F., and Sardinha, R. M. A., (1975). The application of optical densitometry in the study of wood structure and Properties. Journal of Microscopy, 104/1, 91-103.

Polge, H., (1966). Annales des Sciences Forestières, 23/1.

7.4. PEDUNCULATE AND SESSILE OAKS: SPECIES DETERMINATION FROM DIFFERENCES BETWEEN THEIR WOOD

F. S. Walker, Oxford

Summary

Results from the study of wood specimens from adjoining plantations in Bagley Wood, near Oxford, are mostly in accordance with those reached by Huber, Holdheide and Raack ((1941) and extra characteristics useful for diagnosis have been observed.

For the two species of European oak, differences in wood features are often regarded as insufficient to distinguish the species either from timber presently grown (Forest Products Research 1960) or from the prehistoric past (Godwin and Deacon 1964). Four characteristics, however, were found by Huber, Holdheide and Raack (1941), working at Tharandt, three of which when found together in a specimen gave positive identification with very limited exceptions (Table 1). Their samples were taken from 52 pedunculate and 41 sessile trees identified in Tharandt, Saxony or in forests outside Germany. The preference of pedunculate oak for moist lowlands and heavy soils and of sessile oak for warm dry hillside positions was noted by them. Individual trees were no doubt taxonomically assessed on the best standards of the time which could not have anticipated, however, the future importance to be attached to the phenomenon of introgression.

TABLE 1 Characteristics for Species Determination (after Huber, Holdheide and Raack)

Feature (see Fig. 1)	Pedunculate	Sessile
1. Earlywood vessel shape	CLEARLY OVAL	Approximately circular
2. Number of rows of earlywood vessels in rings over 2 mm wide	4 or more	1 or 2, EXCEPTIONALLY AS MANY AS 3
3. In such rings, the ratio of the earlywood vessel zone to the total ring width	Over 0.25	UNDER 0.25
4. Appearance of latewood vessel groups, called "flames"	Clear and wide, particularly in youth. When older, often faded.	Narrow and clear

Note: A matrix composed of these characteristics showed that the known species was highly likely to be diagnosed correctly provided (i) the one feature in CAPITALS for pedunculate, or the two in CAPITALS for sessile were present; and (ii) in all, at least three of the appropriate four characteristics applied. Applied to their samples of known species this rule gave the correct result for 36 out of 40 pedunculate and 33 out of 36 sessile samples.

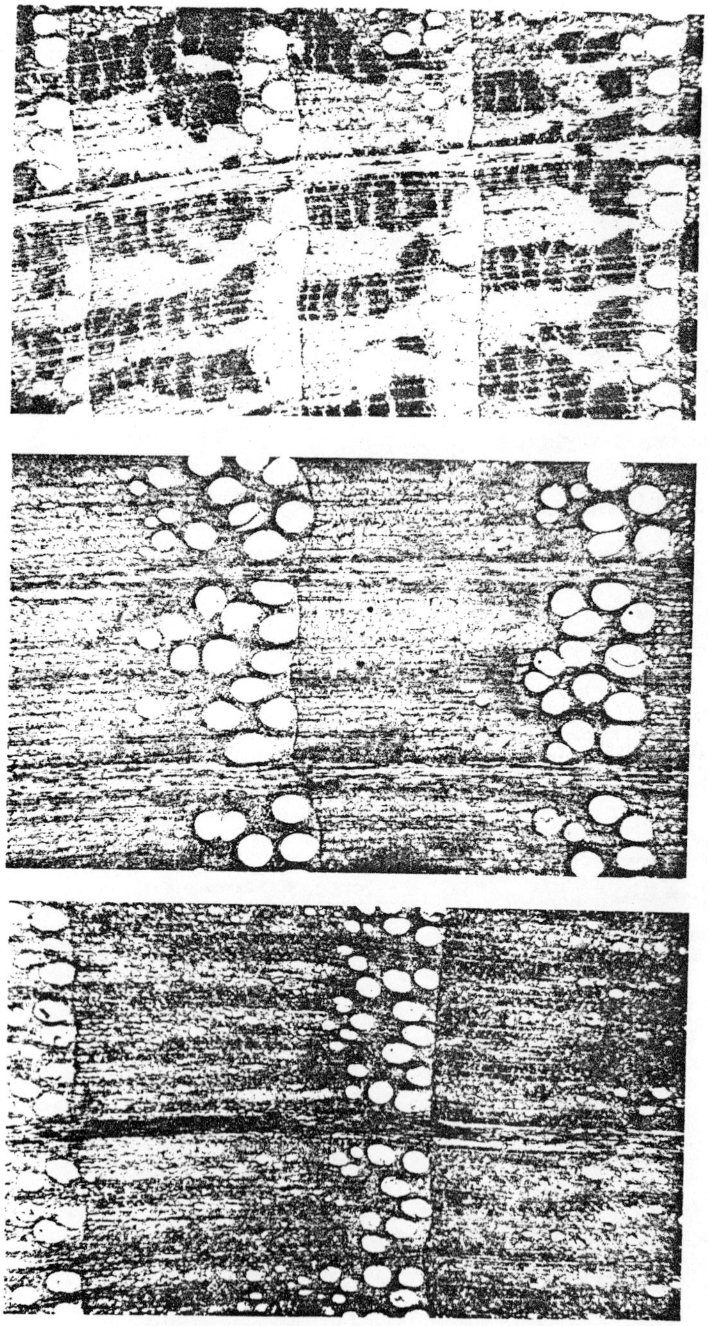

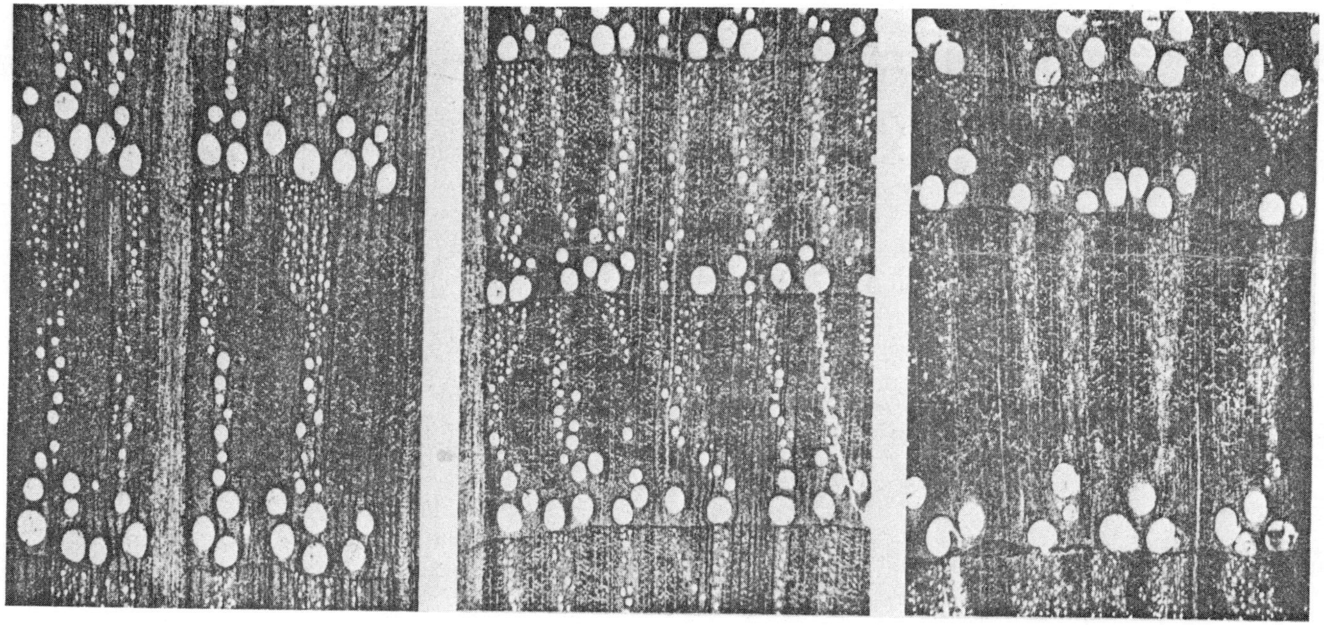

Fig. 1 Cross sections of (above) pedunculate and (below) sessile oak (after Huber, Holdheide and Raack, 1941, Figs. 7-12). They illustrate the decisive distinguishing features they found as well as the range of variation encountered.

STUDY WITH BAGLEY WOOD OAK

The method has been studied with 13 English oaks of recent growth and known species from adjacent stands that were planted in Bagley Wood, near Oxford, about 1916 by Prof. Schlich for demonstration purposes. The total area is about ½ acre, about two thirds on the south side occupied by the sessile plot. The site is relatively high (altitude c. 120 m) and near the Garden area of Compartment 16. The area had been Common Land until 1840 and the soil is plateau gravel. Advantage was taken of a silvicultural thinning in July 1976 to fell 5 pedunculate (Q. robur) and 8 sessile trees (Q. petraea) to obtain (i) wood specimens in the form of discs 6 to 8 inches in diameter taken at breast height; and (ii) botanical specimens. In this way wood samples of doubtful origin or with differences due to differing site factors were expected to be avoided.

Species identification

The botanical collection from the 13 trees was confined to leaf specimens in the absence of either fruiting or flowering material. This substantially reduces the taxonomic evidence as Cousens (1965) regards peduncle length to the first flower bract to be one of the two primary diagnostic features; the other, the ratio of the petiole length to the leaf length, was examined on several specimens from each tree.

Fig. 2 shows this ratio as a display diagram (cf. Cousens 1962) for 40 pedunculate and 50 sessile outer-crown leaves. There is a clear distinction between the two species: the range of mean values for each tree being 3.2 to 7.3% for the pedunculate and 12.2 to 15.6% for the sessile leaves.

Secondary features as follows (first fully described by Jones, 1959) also distinguished the species of these Bagley Wood trees by their leaves.

Leaf Feature	Pedunculate	Sessile
Auricles	Strongly marked	Absent, except very weak for trees SI and SVI
Abaxial pubescence	Absent	Present
Shape	Consistently obovate	Mostly ovate (slightly wider above the middle in 12% of the leaves)
Number of lobes	2-5-(6)	5-6-8
Type of lobe	Deep and irregular	Regular (a few exceptions), shallow
Veins to sinus	Always present	Present in a few only (only in one leaf was more than one such vein present)

The evidence does not conform completely to Wigston's criteria (1974) for reliable taxonomic assessment but any effect of introgression seemed likely to be sufficiently small for the material to be suitable for our purpose.

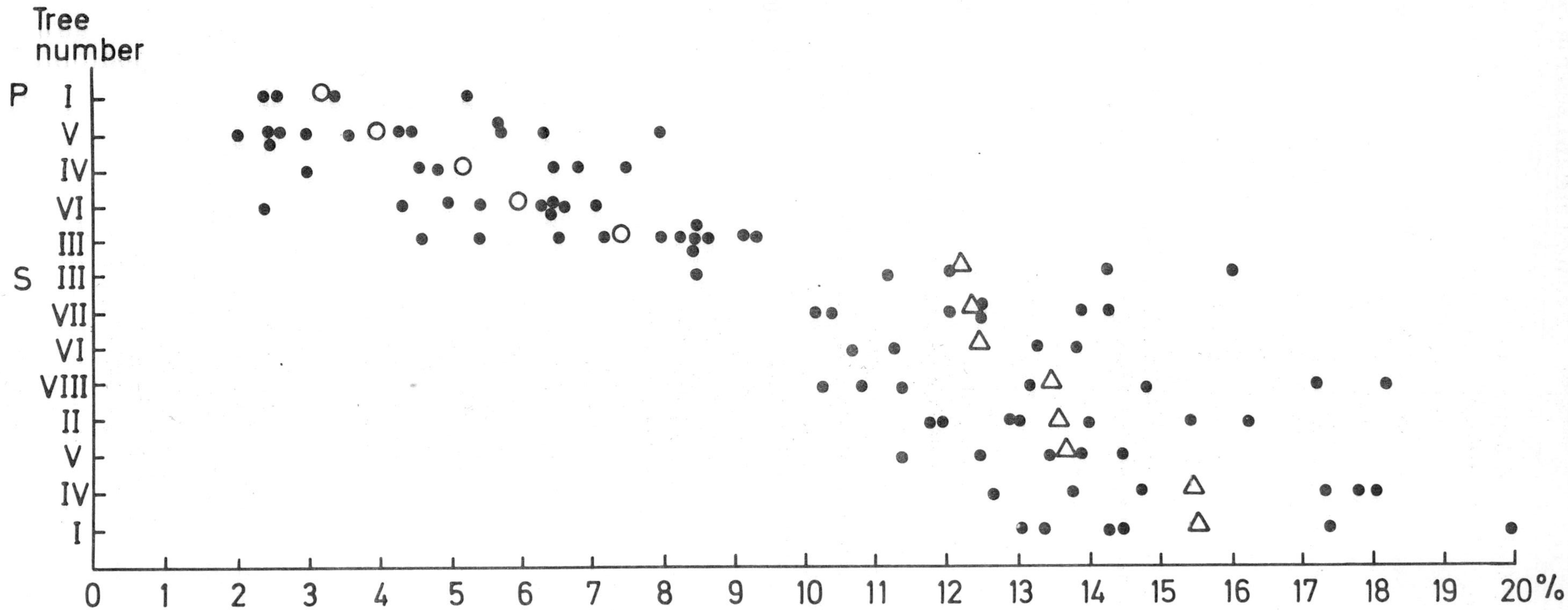

Fig.2. Ratio: Petiole per cent of leaf length of specimens from Bagley Wood Demonstration plantation in Compartment 16.
Mean Value for Tree: P, Pedunculate ○: S, Sessile △

Our findings on the decisive wood features of the Bagley specimens are set out below alongside those of Huber, Holdheide and Raach for their specimens.

A. PEDUNCULATE

Bagley: 5 specimens	Huber *et al*.: 40 specimens
1. The usual earlywood vessel shape, either clearly oval or distorted, is associated with crowding.	1. The earlywood vessel shape, clearly oval, predominated in 37 specimens: the ratio radial/tangential diameter ranged from 0.65 to 0.97, mean 0.75.
2. In rings over 2 mm wide (ignoring the first 5 years), there were two specimens with 4 rows of early vessels and three with not more than 3 rows.	2. In rings over 2 mm wide, many-rowed earlywood vessels (4 or more rows) predominated in 28 specimens.
3. In such rings the ratio earlywood vessel zone/total ring width exceeded 0.25 in three specimens.	3. In such rings the ratio earlywood vessel zone/total ring width exceeded 0.25 in 31 specimens.
4. The latewood vessel groups ("flames") were clearly marked and usually broadened towards the end of the years growth; or, if narrow and "straplike", frequently forked.	4. In 25 specimens the "flames" were wide and clearly marked, or faded, i.e., not easily differentiated from surrounding tissue. They were narrow and clear in 15 specimens.
5. The transition (in size) from earlywood to latewood vessels in all rings with 3 or more rows of earlywood vessels was usually gradual.	5. The gradualness of the transition in size was insufficient for use as a decisive feature.

B. SESSILE

Bagley: 8 specimens	Huber *et al*.: 36 specimens
1. Earlywood vessels were predominantly uncrowded and approximately circular in shape. Where oval the rarety of either crowding or distortion was distinctive.	1. Approximately circular in 14 specimens, oval in 9 and medial in 13: the ratio radial/tangential diameter ranged from 0.70 to 0.94, mean 0.83.
2. In rings over 2 mm wide (ignoring the first 5 years), there were no specimens with more than 3 rows of earlywood vessels and of the 57 rings concerned 50 had 1 or 2 rows.	2. In such rings, 27 specimens had an earlywood vessel zone with 1 or 2 rows: 3 (never more) were rare.

Our findings agree to a large extent with those of Huber, Holdheide and Raack, but we wish to include three additional characteristics:

(i) The earlywood vessels of sessile are relatively uncrowded compared to those of pedunculate oak; this feature can explain the greater proportion of oval or distorted vessels in pedunculate compared with sessile oak;

(ii) A greater degree of forking of the "flames" of pedunculate oak when they are narrow;

(iii) The more gradual transition (Item 5 above) in vessel size from earlywood to latewood in pedunculate compared with the normally abrupt transition in sessile.

These three characteristics are incorporated in Table 2, our slightly modified version of Table 1. The first two, as well as other differences, are illustrated in Fig. 3.

TABLE 2 Characteristics for Species Determination, Oxford 1978

Feature	Pedunculate	Sessile
1. Earlywood vessel shape	CLEARLY OVAL or CROWDED AND DISTORTED	Approximately circular, uncrowded
2. Number of rows of early vessels in rings over 2 mm wide and beyond year 5	3 or more	1 or 2, EXCEPTIONALLY AS MANY AS 3
3. In such rings, the ratio of the earlywood vessel zone to the total ring width	Over 0.25 predominantly	UNDER 0.25 PREDOMINANTLY
4. Appearance of latewood vessel groups, called "flames"	Wide and broadened towards the end of the year; if not, bifurcation frequent.	Narrow and straplike; bifurcation, if present, weak.
5. Transition from earlywood to latewood in rings over 2 mm wide	Gradual	Abrupt

Notes: Correct diagnosis is regarded as highly likely if there are at least 4 of the characteristics of the species and these include the essential ones (in CAPITALS)

The fading of flames in mature pedunculate oak observed by Huber (Table 1) has been omitted: the Bagley Wood oaks were not old enough to show this trait.

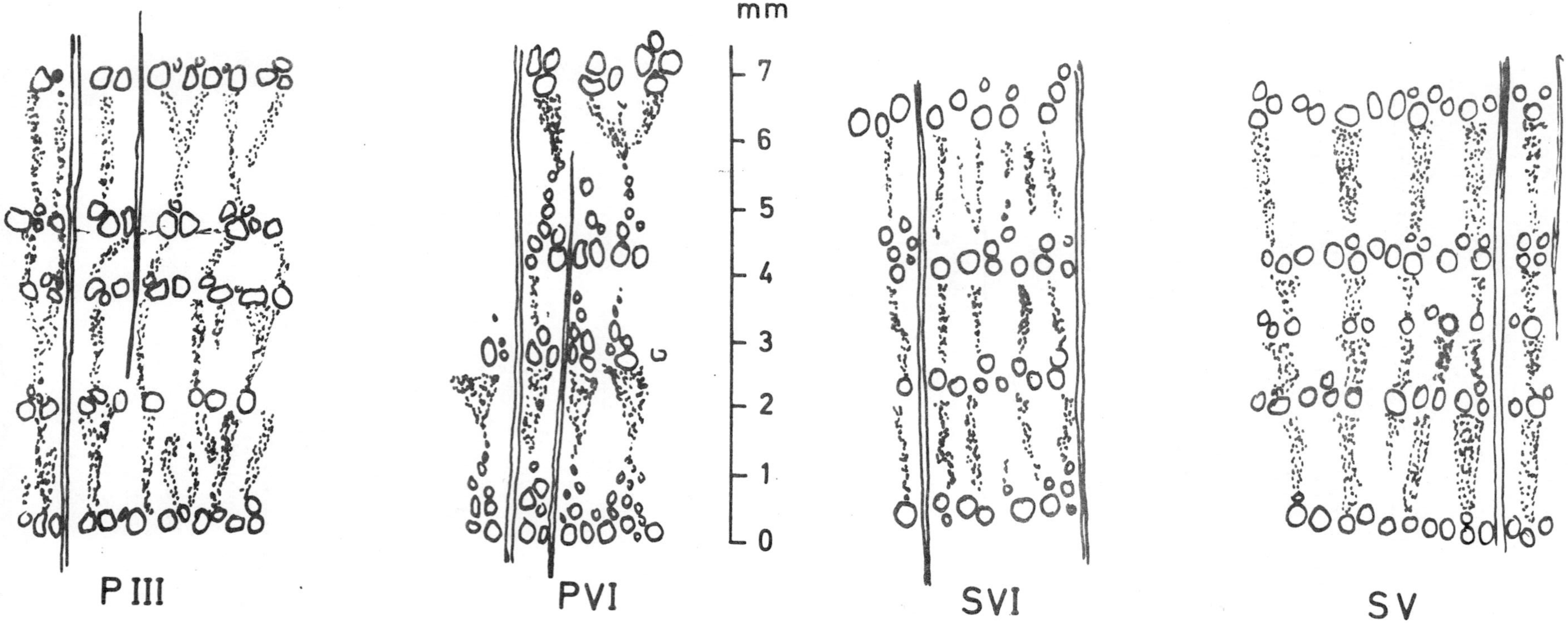

Fig. 3. Cross sections of specimens of Pedunculate (PIII and PVI) and Sessile (SVI and SV) Oak from Bagley Wood.

Pedunculate:

Sketches PIII and PVI show the essential vessel shape, namely (i) oval or (ii) crowded and distorted. Also notice latewood flames (i) broadened towards the end of the year's growth, or (ii) frequently forked.

Sessile:

Sketches SVI and SV show the essential limitation to a few rows of vessels in the earlywood (one or two rows, rarely three) and the low ratio of the earlywood zone to the total ring width.
Also notice uncrowded earlywood vessels, mostly nearly circular.

All 13 of the discs from the Bagley oaks showed the essential features required by Huber, Holdeheide and Raack and all but three showed a total of 4 of the 5 features listed in Table 2 as characteristic. The exceptions were pedunculate discs from three slow-grown trees with very few rings wide enough for diagnostic use by Features 2 and 3. When tested "blind", the Bagley specimens were diagnosed correctly except for one pedunculate disc (P V) in which vessel shape and frequency of forking of flames was misleading.

Microscope slides in the collection of the Dept. of Forestry, Oxford, from seven botanically identified trees (4 pedunculate and 3 sessile) conformed with the features described in Table 1, except for one specimen (pedunculate) in which there were relatively few rows of vessels.

Discussion and Applications

Studies of timber from known introgressed populations would be of interest to see if there is any relation with the characteristic wood features described here. They might give some indication of when introgression began and which were the original oaks of Britain.

Various sections cut from the oak timbers of buildings in southern England have been collected at Oxford by Fletcher to await dating by dendrochronology. They come from cathedrals, collegiate bodies, churches, houses and barns mainly not far from the Thames in its course from Oxford to Westminster. Their date of use spans the 13th to the 17th century.

I have attempted species identification, using X 20 magnification, on some of these and also on the cores collected in Wales in 1972 by the team from Munich and examined there by Siebenlist-Kerner (this volume, page 156); they were sent recently to Oxford as archive material. The results together with those from my examination on the edge of panel paintings during the course of recent ring-width measurements (Fletcher, this volume, page 303) are given in Table 3.

One problem encountered was the relatively few rings wider than 2 mm in some of the samples. To accommodate such circumstances and to allow for slightly conflicting evidence we have added classification as 'Affinity to Pedunculate' and 'Affinity to Sessile'.

Table 3. Attempted Species Identification on Building Timber and Panels.

Source	Number of Samples	Classified as: Pedunculate	aff. Ped.	aff. Sessile	Sessile
1. Buildings in southern England	13	4	3	3	3
2. Buildings in Wales	15	3	5	2	5
3. Panel Paintings 16th century					
English	6	5	1		
Flemish or French	3	1			2

Acknowledgements

I am grateful to Dr. J. M. Fletcher for suggesting this research; to Mr. T. Edmondson for information about the silvicultural thinning in Bagley Wood; to Mr. Patmoor and his staff for enabling us to take discs at the time of felling; and to Mr. P. Franklin for the slide sections.

References

Cousens, J. E. 1962. Notes on the status of the sessile and pedunculate oaks in Scotland and their identification, Scott. For., 16, 170-9.

Cousens, J. E. 1965. The Status of the Pedunculate and Sessile oaks in Britain, Watsonia, 6, 161-176.

Fletcher, J. 1978. Tree-Ring Bulletin, in course of issue.

Forest Products Research Bulletin No. 25. 1960. Identification of Hardwoods. A lens Key.

Godwin, H. and Deacon, J. 1974. The British Oak. Botanical Society of the British Isles, 51-61.

Huber, B., Holdheide, W. and Raack, K. 1941. Holz als Roh- und Werkstoff, 4, 373-380.

Jones, E. W. 1959. Biological Flora of the British Isles. Quercus L. J. Ecol., 47, 171.

Wigston, D. L. 1974. The British Oak. 47.

7.5 ABNORMAL SMALL EARLYWOOD VESSELS IN OAK. THEIR USE AS CHRONOLOGICAL INDICATORS AND THEIR RELATION TO ARRESTED HEARTWOOD FORMATION (INCLUDED SAPWOOD) AFTER CERTAIN COLD WINTERS

M. Tapper, J. Fletcher and F. Walker,
Oxford University

Summary

Abnormal small earlywood vessels in the rings of Quercus robur and Q. petraea growing in south-east England and Flanders in the period 1430 to 1450 form chronological indicators. Earlier research in France and Russia shows that this phenomenon is linked with 'included sapwood' and caused by very cold winters.

The outlook for tree-ring dating in England, where both climate and soil conditions are highly variable, was until recently regarded as unfavourable. It is surprising, therefore, that we have been able to match and date 225 (85%) ring-width sequences of the 265 boards used for panel paintings in the period 1350 to 1630 (Fletcher, 1977) that have been examined.

A factor contributing to this success was the recognition at an early stage in the research of an abnormality in the wood structure of a few specific calendar years. This feature has been present on 38% of the dated boards. It is comparable, as a reliable indicator, to the narrow rings on the skeleton plots for the South-West States of the U.S.A.

Occurrence

The need for careful preparation of the narrow edges of panel paintings, followed by detailed study with an eyepiece to identify each annual ring-width, led to the discovery of this abnormality (Fletcher, 1975). The occurrence in the mid-15th century of two such rings seventeen years apart has always dated the whole sequence correctly, the dating being confirmed by computer comparisons of the ring-width sequence with reference chronologies. When the abnormality has been present only once in rings likely to have been laid down in that period, it has provided a useful indication that the ring may have been formed in one or other of the two years, 1433 and 1450.

The feature has most frequently been found in years that lie in the second quarter of the 15th century. 185 of the 225 boards referred to above happen to cover that period so our information about those years is statistically very reliable. The occurrences in the period 1432-51 number 127, these being on 85 boards. In this period the percentages of boards showing the feature for the important years are as follows:

	Year (A.D.)			
	1433	1439	1441	1450
Boards with the feature/Total boards	30	9	4	23%

Although the number of samples is nearly as large in the decades on either side of 1432-51, occurrences are rare and mainly sporadic:

For the period 1350-1431, 8;
For the period 1452-1533, 15.

For the year 1344, however, 9 occurrences have been found. For Norman, Saxon and Roman times the feature has been noticed in the relatively few samples on which measurements have been made. No occurrences have been seen by us on samples from British oaks of recent growth.

The appearance of the abnormality is shown in Fig. 1. The vessels in the earlywood are half, or less than half, their normal average diameter. For occurrences in the year 1433 all the vessels in the earlywood are restricted in size. Those in the year 1450, however, usually show characteristics that are similar to those illustrated by Bolycevcev (1970) for pedunculate oak in his paper on 'included sapwood': namely, the vessels in the first row of earlywood are abnormally small, while those in the second row (which is distinctly separated from the first) are larger, i.e. the reverse of normal earlywood growth.

Included Sapwood

This related phenomenon was first described in 1737 by Buffon and Duhamel who noticed a band of lighter colour in sections of oak and linked it with the cold winter of 1709. They called the band a 'lunure' and attributed it to arrested heartwood formation. The effect was subsequently studied in detail in France by Henry (1896) and Mer (1897). They established that the colour and composition of the wood was intermediate between sapwood and heartwood. For the years concerned, even trees in which pale colouring was hardly noticeable proved, on examination under the microscope, to have imperfect heartwood in which starch was present. They linked its presence with the low temperatures during the month of December 1879. The rings of the immediately previous years (1878, 1877 and 1876) were not affected visibly, although chemical and microscopic examination showed abnormalities. Rings ten to fifteen years earlier than 1875 showed included sapwood and contained traces of starch. Mer also noticed that although the rings after 1879 formed normal heartwood, the first few that followed the climatic accident were distinctly narrower. Some of our samples have narrow rings for four to five years after the ring with abnormal small early vessels, while for others the widths of the subsequent rings are unaffected.

Bolycevcev's comprehensive study of trees from 50 to 250 years old growing in the Moscow region showed that included sapwood was present about once in every 50 years and occurred in trees that had experienced two successive dry summers followed by a very cold winter with prolonged frosts. It was accompanied, as illustrated in his paper, by abnormal small early vessels such as we have found in the year 1450, and thereby links the years in which we find the phenomenon with very severe winters. This we had already surmised (Fletcher, 1975) because the two examples most frequently found in our research occur in the period 1430 to 1450 when cold winters and hot summers were known to have occurred in Germany and N. W. Europe (Flohn, 1967).

Included Sapwood on Panel Paintings

We became aware of the French and Russian research only after measurements had been made on most of the 225 boards concerned. Examination of the backs of the boards for included sapwood has therefore been limited, but two examples, both associated with holes from escaping insects are of interest.

On the back of the large panel on which Holbein painted the version of Jane Seymour now in Vienna (see page 138), we noted a lighter coloured band of wood which had suffered insect attack. It was 1.5 to 2 cm in width and covered about ten to fourteen rings from c. 1425 to 1436. Abnormally small vessels are present in the earlywood of the year 1439, but not in 1433 and 1450.

The panel carrying the portrait of ? Lord Grey de Wilton by Flicke is made up of five moderately wide horizontal boards. The two boards measured have average ring-widths of 2.0 mm and 1.1 mm respectively. The feature is present in the ring for 1450 on the board (pedunculate oak) with wide rings and there is a light-coloured band of included sapwood along its entire length. This band covers the six wide rings for the years 1442 to 1447 and a few insect holes are present in it. The years of included sapwood bear the same relation to 1450 as do those in Mer's research to the cold winter of 1879. On other, narrow-ringed, board (aff. sessile oak), abnormal small earlywood vessels occur in the years 1433 and 1450, but no visible included sapwood was noticed.

Discussion

Clearly the presence on the trunks of oak of the features here described is likely to depend not only on the coldness of the winter but on other factors, such as the time of year of the low temperature, the exposure, the wind direction, the thickness of bark. When damage that leads to included sapwood occurs, the wood in the first half cm or so within the bark seems, from our two examples, to be relatively immune. This can be explained by the concentration of solutes in this band being sufficiently high at the time of cold spell to depress substantially the freezing point of the water not chemically bound.

Note: The presence of a 4-year periodicity in our oak chronology led to calendar dates that are 4 years later being attributed to each abnormality when they were reported in 1975.

Acknowledgment

We wish to thank Anne Watson for translating the paper by E. Mer.

References

Bolycevcev, V. G., 1970. Annual Rings in oak as an index of secular cycles of climatic fluctuation. Lesoved, Moskva, 15-23. (English translation in J. M. Fletcher and W. Linnard, Russian Papers on Dendrochronology and Dendroclimatology 1962, 1968, 1970, 1972, Research Laboratory for Archaeology, Oxford University, 1977.)

Fletcher, J. M., 1975. Nature, 254, 506-7.

Fletcher, J. M., 1977. J. Archaeological Science, 4, 335-352.

Flohn, H., 1967. In Die Schwankungen und Pendlungen des Klimas in Europa, 81-90.

Henry, E., 1896. Communication to Société des Sciences de Nancy, Feb. 1896 (quoted by E. Mer).

Mer, E., 1897. La Lunure du Chêne, Revue des Eaux et Forêts, 36, 417-430.

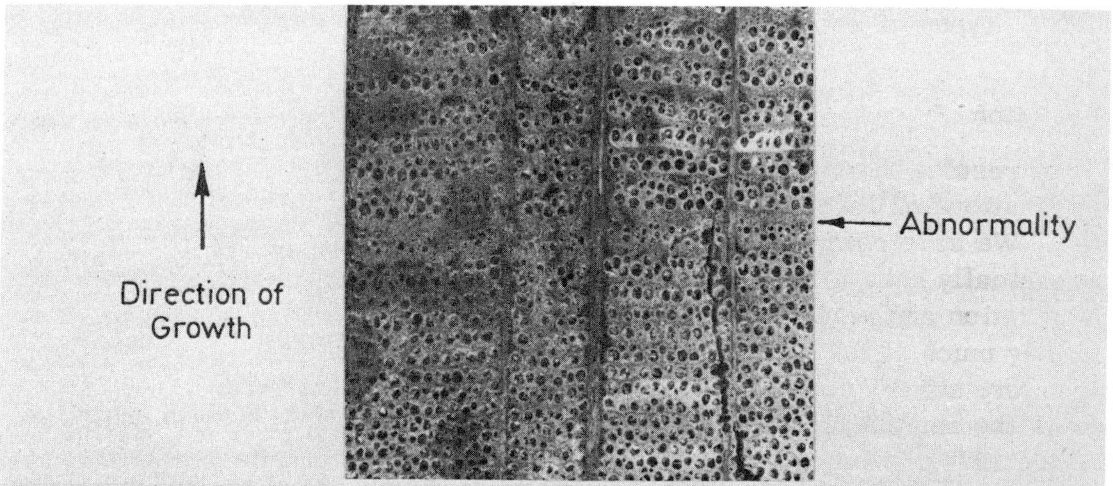

Fig. 1. Edge of oak cupboard door of c. 1518 showing earlywood with abnormally small vessels.

7.6 TRACE ELEMENT CONCENTRATIONS IN TREE RINGS

R. E. Tout and W. B. Gilboy,
University of Surrey

Summary

Research is in progress into trace element variations in Tree rings using the non-destructive technique of neutron activation. The uncertainties involved in interpreting the results such as those caused by biological changes within the tree or by disease are discussed for an elm tree from the Guildford area. The method offers the possibility of recognising particular regions such as the heartwood/sapwood boundary and also the outermost tree-ring.

The application of computer assisted X-ray imaging to produce almost instantaneous ring-width and density curves from various types of wooden object, without sample preparation, is also discussed.

Introduction

The results described in this paper are part of a project to determine the concentrations of several chemical elements that are present in individual tree rings. We have coined the term dendroanalysis to describe tree rings. It was originally hampered by the lack of published information on the content, concentration and variation of minor and trace elements in tree rings. Consequently much of the preliminary work involved in this study concerned an exploratory survey to establish typical values of such elemental concentrations. Some of the results already obtained have been published elsewhere (Tout et al., 1975 and 1977, Gilboy et al., 1976); this study deals with the results for the three dimensional distribution of particular chemical elements in an elm tree and their interpretation.

The second part considers the application of Computer Assisted X-ray Tomgraphy to both dendroanalysis and dendrochronology.

Experimental

An elm tree (Ulmus x hollandica 'Hollandica'), situated in the University grounds in Guildford, was felled and six transverse sections were cut from the trunk at one metre intervals. The lowest section was cut from the trunk at a distance of about 150 mm above the ground and contained 43 growth rings. The highest section was cut from the trunk at about 5 metres above the ground and contained 18 growth rings. The surfaces of the sections were prepared by a combination of sawing, sanding and planing. A steel boring tool, with flared saw teeth, was used to drill cores 2.5 mm in diameter. The ends of the cores were trimmed (to remove surface contamination) producing a length of 15 mm with an average mass of 70 mg. Samples were cut from every ring in all six sections, along a radial direction 60°E of N, and at intervals of 15° around growth rings formed in 1970 and 1964.

The felled tree, like most elms, had been infected by Dutch elm disease at various times. This had been most severe in the few years prior to felling, and it was noted that the most recently formed rings were often very narrow and in some setions partially missing, Some of the sections also emitted the characteristic pungent odour associated with the secondary infection known as 'Wet Wood' (Rishbeth, 1973). In elms the areas infected by 'Wet Wood' are stained, thus making it difficult to distinguish visually between heartwood and infected sapwood. It was noted that some of the sections from the tree displayed isolated dark coloured areas in the sapwood while other sections showed irregular dark wood extensions beyond the normal heartwood region.

After boring and trimming the samples were freeze-dried to remove residual moisture and sealed in polyethylene containers. Each was analysed by neutron activation (De Soete et al., 1972) using similar apparatus and conditions to those already described (Tout et al., 1977).

Neutron activation analysis essentially identified specific isotopes of the various elements, and gives very little chemical information. For this reason we use the term elemental concentration in this work.

Interpretation of Results

Results previously reported (Tout et al., 1975 and 1977, and Gilboy et al., 1976) have shown that systematic variations in the concentration of several elements exist between different areas of the xylem in a transverse section. In this work the radial and angular variations in concentration of sodium, magnesium, chlorine, argon, potassium, calcium, manganese, bromine, strontium, and iodine, were monitored in each section of the elm tree. Oxygen, sulphur, copper and rubidium were also detected but were only present in measurable concentrations in some of samples, thus prohibiting a meaningful interpretation of concentration gradients for these four elements.

The radial results obtained for chlorine in each section (Fig. 1), are similar to those previously obtained for the chlorine radial distribution observed in an earlier section of another elm from the Guildford area (Tout et al., 1977). The sharp drop in chlorine concentration at the sapwood/heartwood boundary is a phenomenon that has been noted in a number of samples (Tout, 1977). In this case the rapid drop in chlorine concentration occurs after the tenth (most recent) growth ring in the bottom four sections, after the fifth ring in the section second from the top, and after the second ring in the top section. The second, third and fourth sections contain a region of 10 visually distinguishable sapwood rings. The fifth section and the top section have five and two sapwood rings respectively, while due to staining the bottom section has no visible sapwood area along the radius which was sampled. These results suggest that the normal marked colour change (in elm) which is used to identify the sapwood/heartwood boundary coincides with a distinct change in elemental concentration for certain elements and that such concentration changes are not noticeably affected by the staining produced in 'wet wood'. The elemental distributions also confirm that the fifth and sixth sections do have smaller sapwood regions than thellower sections.

The results previously reported for potassium in a section of elm tree (Tout et al., 1977) showed that a typical radial distribution consisted of high levels of potassium in the outermost two or three rings of the sapwood, followed by lower concentrations in the rest of the sapwood rings. This is then followed

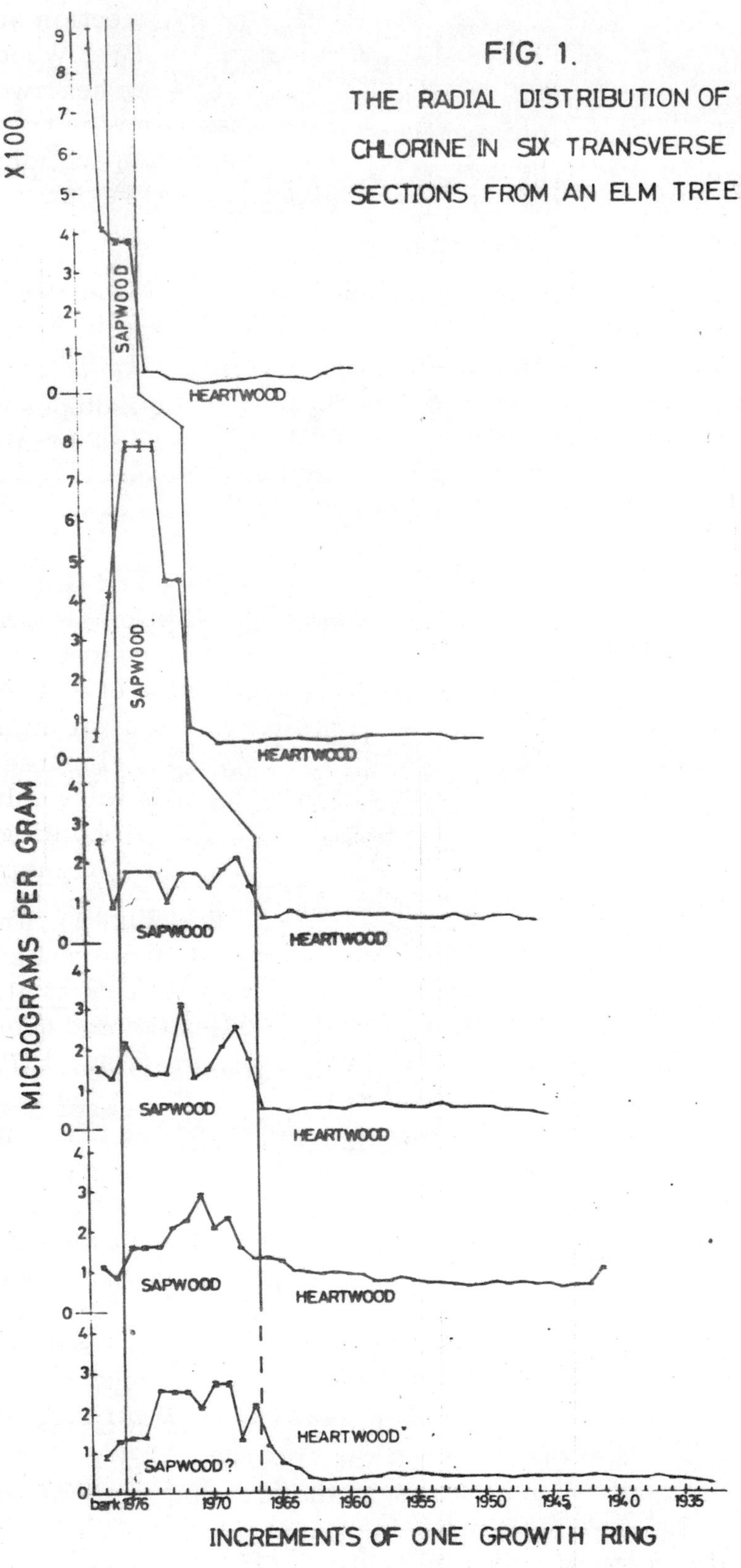

FIG. 1.

THE RADIAL DISTRIBUTION OF CHLORINE IN SIX TRANSVERSE SECTIONS FROM AN ELM TREE.

by an increase in concentration across the sapwood/heartwood boundary to relatively high levels in the heartwood. The same pattern appears to be present in four of the six transverse sections. In the radial distribution for the bottom section there is a step-like increase characteristic of the sapwood/heartwood boundary following the 10th growth ring from the bark. This supports the evidence of the chlorine distributions, i.e. that this is the actual position of the sapwood/heartwood boundary; and thus shows the marked changes in concentration between the sapwood and heartwood for certain elements are unaffected by the 'wet wood' infection.

In order to express some measure of the vertical variation in concentration of the various elements, mean concentrations over sapwood and heartwood regions were derived for each section. Due to the limited and variable number of sapwood rings, coupled with the strong variations in concentration in those regions, the sapwood average may be less meaningful than those calculated for the heartwood. Variations in the average heartwood concentrations where generally small or nonsystematic. However, the values of the average argon concentration increases with the height of the sections above the ground. The argon levels reflect the amount of trapped air in the cells (Tout, 1977) which therefore must increase in a similar manner with height. The concentration of iodine in the heartwood decreases with height.

Angular variations of the concentrations of certain elements have been previously observed around individual rings in ring porous trees (Tout et al., 1975 and 1977, Gilboy 1976) and similar results are observed for the present tree. The interpretation of these angular variations in this particular elm was greatly hindered by the irregular nature of the heartwood regions. The angular results from this tree (Tout, 1977) indicated that many of the small darker coloured areas within the sapwood had the characteristic elemental levels typical of normal heartwood. In rings which intersected these dark areas, the angular variation changes abruptly in a similar way to the elemental changes encountered at radial sapwood/heartwood boundaries.

The one element monitored which seemed to be unaffected by variations in the amount of sapwood and heartwood present, was iodine. It produced a consistent angular variation which shows a minimum concentration of iodine occuring at about the same position (approximately 60° W of N) on each section. Iodine shows a similar radial distribution for all six sections with a peak in concentration always occurring in ring 6 ± 1, (i.e. about the ring formed in 1970). The concentration of the peak value of iodine in each of the sections diminishes as the height of the section above the ground increases.

Discussion

Previous work has shown that well defined changes in concentration occur for certain elements across the sapwood/heartwood boundary and also between the outer few rings of the sapwood and the rest of the sapwood, In this work both these effects have been shown to persist in stained sections of wood where visual identification of the sapwood areas is rendered difficult.

We suggest that dendroanalysis could be a useful additional tool in helping to identify the presence and extent of sapwood. In bleached or discoloured samples the presence of a sapwood/heartwood boundary could be detected by looking for the characteristic sharp change in concentration of particular elements, such as chlorine and potassium. Also manganese shows a particularly high concentration in the outermost ring and this may be used to identify that region.

The various tree sections were generally allowed to dry out at ambient temperature before sampling, and it is conceivable that processes such as osmosis and ion exchange could give rise to spurious concentration gradients. We have investigated this possibility by extracting a core from a live oak tree and sectioning it immediately to avoid moisture loss. Such experiments yielded similar elemental distributions (Tout 1977) to those described above which indicates that drying-out effects are not large.

APPLICATION OF X-RAY TOMOGRAPHY TO DENDROCHRONOLOGY

A large number of the above type of samples must be evaluated to characterise the three-dimensional distribution of elements in wood. Further, although the method of neutron activation analysis is specific and sensitive, in this application is can be time consuming and expensive. Therefore, we have considered alternative methods of measuring elemental distribution which promise to be suitably sensitive, cheap and rapid. These considerations led us to use an X-Ray tomographic scanner for some trial runs.

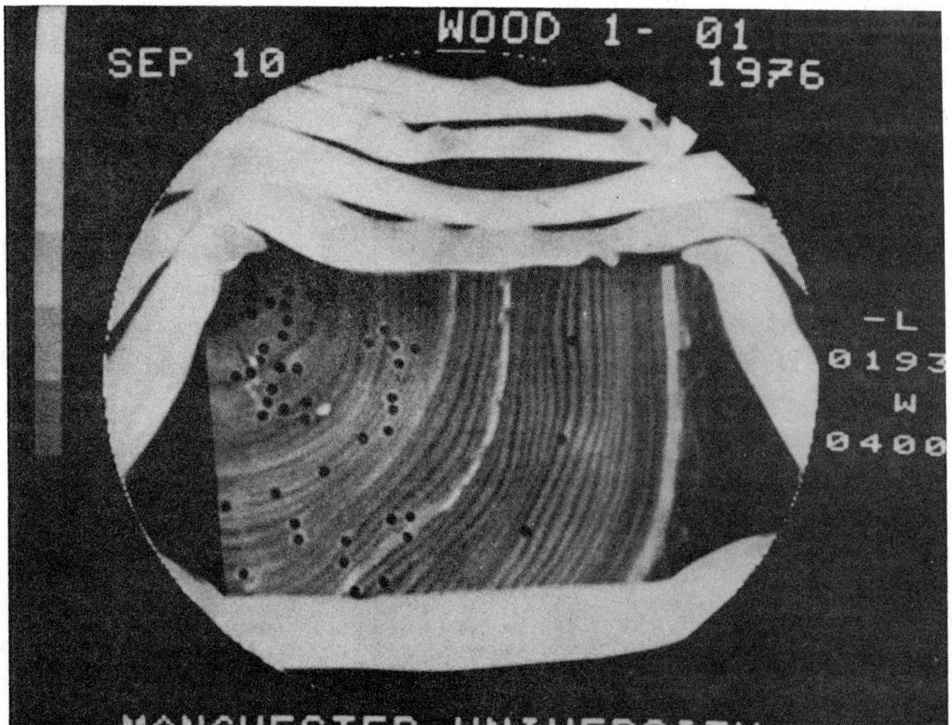

Figure 2. A computer assisted X-ray tomograph of a cross-section through the first elm analysed from the Guildford area.

With the permission of Professor B. R. Pullan, we were able to use the EMI computer assisted tomographic body scanner at Manchester University. This instrument utilises narrow beams of X-rays which are usually used to scan across sections of the human body for diagnostic radiology purposes (Brooks, et al., 1976). The absorption of the X-rays is measured for many beam orientations, and the X-ray attenuation coefficients for the chosen cross-section are computed and displayed by the computer either digitally, or as a grey-scale image. Figure 2 shows the scanner image though a rough cut section of our earliest elm tree sample. The ring structure shows up particularly clearly since the X-ray 'density' is related to the density of the wood; it is also affected by the atomic number of the material, and at the X-ray energy used the photo-electric interaction (Z^5) dominates at high atomic numbers. This shows up strongly in the 1950 ring of this sample, which the neutron activation results had previously shown (Tout et al., 1977) to contain anomalously high levels of magnesium, calcium, strontium and barium. By this method the extent of the 2-dimensional distribution of heavy elements in the wood matrix, can be revealed over the whole section in a single run which is completely non-destructive.

These X-ray scans require 20 seconds irradiation, followed by 3 to 4 minutes of computation. The results can be expressed digitally, and the values of the data along any chosen chord of the scan can be displayed to that ring-width and density curves can be derived very rapidly, with no sample preparation. Since the method is non-destructive, it can be applied to valuable museum pieces and objects d'art. Present scanners are large and expensive machines and are in great demand for medical purposes. The costs and likely availability for non-medical purposes are not yet clear, but we are currently studying these aspects. We are also planning a portable tomograph scanner for field use, and among other applications we hope to measure ring-widths and densities for living trees in situ.

Acknowledgements

We would like to thank our colleague Mr. N. M. Spyrou who participated in the earlier stages of the work. We are also indebted to the Director, Mr. M. Kerridge of the University of London Reactor Centre, and his staff for the use of the activation analysis facilities.

References

Brooks, R. A., di Chiro, G., Phys. in Med. and Biol. 21 (5) (1976) 689.

De Soete, D., Gijbels, R., Hoste, J., Neutron Activation Analysis, Wiley (1972).

Gilboy, W. B., Tout, R. E., Spyrou, N. M., Proc. ERDA X and Gamma Ray Symp. Ann Arbor, MI, (1976).

Rishbeth, J., Newsletter of the British Tree Surgeons Assoc. (1973) 6.

Tout, R. E., Gilboy, W. B., Spyrou, N. M., 4th Symp. Recent Developments in Activation Analysis, Cambridge (1975).

Tout, R. E., Gilboy, W. B., Spyrou, N. M., Radioanal. Chem. 37 (1977) 705.

Tout, R. E., Ph.D. Thesis, University of Surrey, (1977).

DISCUSSION—Section 7
New Techniques and Tree-Ring Features

Mr. Walker asked how X-ray densitometry coped with the existence of slippage of the equivalent annual ring along a medullary ray. Mr. MILSON and Mr. M. HUGHES described how the densitometer was channelled between two rays and then, after the sample had been turned manually, the scan repeated with an overlap of 10 to 15 rings. The ring-widths from the X-ray negatives were used to form a chronological sequence.

Fleming asked the speakers their opinion of alternative techniques such as (i) neutron activation (to overcome problems presented by heavy elements in the wood); and (ii) X-ray back scattering (to avoid the depth problem in ray displacements). Milsom and Hughes explained that they were not trying to develop a new technology but to adapt the X-ray technique pioneered elsewhere. The creation of the X-ray negative as a permanent visual record formed an important part of the method. The development of a new technique for densitometry of special samples, e.g. fossil wood, would be a lengthy process.

In reply to Schove, who asked how suitable the "moving beam" method was for use with oak it was said to be suitable but time consuming.

Chaloner and Gilboy asked about the width of the X-ray trace and the precision obtained on scanning the X-ray negative. Polge drew attention to the loss of accuracy, particularly in assessing maximum and minimum density. Milsom and Hughes explained that the collimator on the X-rays and the slit on the densitometer both had their long axes at right angles to the radius of the tree, and therefore parallel to the ring boundaries. The speed was slow, $16\frac{1}{2}$ cm per minute.

Finally the question of the resolution required to gain further knowledge by this technique was raised. Was it necessary to achieve a high degree of accuracy if the natural inhomogeneity of the material being examined did not warrant it? Might there be an overall gain if the method was made quicker?

Bauch expressed the view that differentiation between extremes in the wood anatomy of sessile and pedunculate oak was one thing. Hybridisation, however, had been considerable in the last 300–400 years and their experience was that, for panels, identification of the species was likely to be limited to 10 to 20%. Mr. WALKER pointed out the method had been tested in the first place on an experimental plot in which Prof. Schlick (then Head of the Dept. of Forestry at Oxford), had planted sessile and pedunculate oaks side by side. The panels being examined were more than 300–400 years old so the effect of recent introgression did not arise.

He stressed that as rings wider than 2 mm were needed for two of the characteristics, a number of the slow grown oaks used in dendrochronology could not be tested adequately.

Polge considered that identification within a particular forest where the two species both existed would not be simple. Prof. Dimbleby said he had recently read a thesis (N. Flower, 1977) that dealt with the New Forest, a large area and one of our few remaining native woodlands. Both species grow there in abundance and both plantations and semi-naturally regenerated woodlands exist in the forest. On only one site of the 24 sampled had it been possible to identify hybrids. This raised the question whether hybridisation in oak varies in different parts of Europe.

M. Hughes suggested that having looked first at well-defined samples, it would be advantageous to look at trees with different degrees of introgression. Certain oaks in Cumberland form a classic example of hybridisation and would be of interest. Becker mentioned that they had found that the ring-width series of pedunculate and sessile oaks in the same geological region had more or less the same ring-width pattern.

Bauch referred to the abnormal early vessels that he had seen when working at Windsor Castle and said that subsequently they had found some examples on Netherlandish panels for years in the 16th and early 17th century. Madame Trenard asked if the abnormality occurred all round the trunk of a tree. Dr. FLETCHER said that as they worked mainly with thin panels of radial section, they did not know over what fraction of the cross-section of a tree the phenomenon occurred. In one instance, with three panel paintings almost certainly from the same part of the same tree, the effect occurred on all three.

Chaloner enquired whether there was an explanation for the iodine peaks. Mr. TOUT and Dr. GILBOY replied that iodine behaved differently from the other elements analysed. It was not smoothed out within the tree and offers the possibility of identifying specific rings. Hibbert enquired about the chemical form and location in the cell of an element such as iodine and its relative mobility in a tree. Did iodine still exist in fossil wood? Dimbleby considered the method gave us no information about the state of the elements, nor in what tissues the elements measured were present. It would make a difference to their mobility if they were in living tissue, such as the ray cells, rather than in dead tissue. Asked by Polge about the advantage of the tomograph scanner, Tout replied that the data were in computer form and capable of re-analysis and presentation in many ways.

CONTRIBUTORS

<u>France</u>

Dr. H. Polge
Station de Recherches sur la Qualité des Bois,
Centre National de Recherches Forestières,
Champenoux 54280 Seichamps

Mme F. Serre
Laboratoire de Botanique historique et Palynologie,
Rue Henri Poincaré,
13397 Marseille Cedex 4

<u>Germany, D.R.</u>

Dr. M. Jährig
Zentralinstitut für alte Geschichte und Archäologie,
Akademie der Wissenschaften der DDR,
DDR 108 Berlin, Leipziger Strasse 3/4

<u>Germany, West</u>

Prof. Dr. J. Bauch, Prof. Dr. D. Eckstein and Prof. Dr. W. Liese
Ordinariat für Holzbiologie,
Universität Hamburg,
Leuschnerstr. 91 d
205 Hamburg 80

Dr. B. Becker
Institut für Botanik
Universität Hohenheim, 7000 Stuttgart 70

Dr. A. Delorme
Institut für Forstbenutzung,
Büsgenweg 4, D-3400 Göttingen

O.St.R. E. Hollstein,
Rheinisches Landesmuseum Trier,
55 Trier, Ostallee 44

Frau V. Siebenlist-Kerner
Steina 170, 8341 Asenham.

<u>Great Britain and N. Ireland</u>

Dr. M. G. L. Baillie
Paleoecology Laboratory, Queen's University of Belfast,
Belfast, N. Ireland BT7 1NN

Mr. D. W. Brett
Bedford College, Regents Park,
London NW1

Mr. N. B. M. Carlton
School of Geography, Mansfield Road,
Oxford

Dr. J. M. Coles
Department of Archaeology, Downing Street,
Cambridge

Dr. J. M. Fletcher, Mrs. M. C. Tapper and Mr. F. S. Walker
Research Laboratory for Archaeology and the History of Art,
6 Keble Road,
Oxford

Dr. W. B. Gilboy and Mr. R. E. Tout
Department of Physics, University of Surrey,
Guildford, Surrey

Mr. A. Heyworth
Department of Geography, University College of Wales,
Aberystwyth.

Dr. F. A. Hibbert
Department of Biological Sciences, Portsmouth Polytechnic
King Henry I Street,
Portsmouth

Mr. J. F. Hughes
Department of Forestry, South Parks Road,
Oxford

Dr. M. K. Hughes, Mrs. P. Leggett and Mr. S. J. Milsom
Department of Biology, Liverpool Polytechnic,
Byrom Street,
Liverpool BL3 3AF

Dr. S. McGrail
National Maritime Museum,
Greenwich, London SE10 9NF

Dr. R. A. Morgan
Department of Prehistory, Sheffield University,
Sheffield S10 2TN

Miss B. J. Orme
Department of History,
The University, Exeter

Mr. J. Schofield and Mr. G. Milne
Museum of London, Department of Urban Archaeology,
71 Basinghall Street,
London EC2

Dr. D. J. Schove
St. David's College, Justin Hall,
Beckenham Rd,
West Wickham, Kent BR4 OQ3

Prof. G. C. Varley
Hope Dept. of Entomology,
University Museum,
Oxford

Poland

Dr. M. J. Dabrowski
Institute for Material and Cultural History,
Polish Academy, 00-140 Warszawa, ul. Swierczewskiego 105

Sweden

Dr. T. S. Bartholin
Dept. of Quaternary Geology,
Tornavägen 13, S-223 63 Lund

Mr. A. Bråthen
Lunnevägen 2,
S-46100, Trollhättan

Switzerland

Dr. O. U. Bräker, Dr. F. Röthlisberger, Mr. E. Schär and Dr. F. H. Schweingruber
Swiss Federal Institute of Forestry Research,
Ch-8903 Birmensdorf ZH

Lic. Phil. G. Lambert and Mr. C. Orcel
Musée d'Archéologie,
Peyrou 7, 2000 Neuchâtel

U.S.A.

Dr. A. C. Barefoot and Dr. W. L. Hafley
School of Forest Resources, North Carolina State University,
Raleigh, N.C., 27607

U.S.S.R.

T. T. Bitvinskas
Laboratory of Dendroclimatochronology,
Academy of Sciences of the Lithuanian SSR,
233000 Kaunas, Laisves al. 53

L. A. Kairiukstis and A. Juodvalkis
Research Institute of Forestry,
Academy of Sciences of the Lithuanian SSR,
232600 Vilnius - MTP, Lenino 3

SUBJECT INDEX